Water Management at Abandoned Flooded Underground Mines

Christian Wolkersdorfer

Water Management at Abandoned Flooded Underground Mines

Fundamentals, Tracer Tests, Modelling, Water Treatment

Habilitationsschrift zur Erlangung des akademischen Grades

Doctor rerum naturalium habilitatus

Dr. rer. nat. habil.

vorgelegt der Fakultät für Geowissenschaften, Geotechnik und Bergbau
an der TU Bergakademie Freiberg

von Dr. rer. nat. Christian Wolkersdorfer

geboren am 17. Februar 1964 in Schwabach/Mittelfranken

Freiberg/Sachsen, am 9.3.2006

Springer

International Mine Water Association
Bob Kleinmann (Editor-in-Chief)

Dr. rer. nat. Christian Wolkersdorfer
Ludwig-Maximilians-Universität München
AG Hydrogeologie und Umweltgeologie
Luisenstr. 37
80333 München
Germany
christian@wolkersdorfer.info

ISBN 978-3-540-77330-6 e-ISBN 978-3-540-77331-3

DOI 10.1007/978-3-540-77331-3

Library of Congress Control Number: 2007943146

Production: LE-TEX Jelonek, Schmidt & Vöckler GbR, Leipzig, Germany
Cover design: deblik, Berlin

Printed on acid-free paper

9 8 7 6 5 4 3 2 1

springer.com

We all seek an ideal in life.

Charlotte Brontë

1 Acknowledgements

Many colleagues have helped and assisted me in my research. Some accompanied my way for just a short time; others have been with me since I started to work in the mine water field, and again others were (and are) my teachers and friends. Yet, the list must start with four colleagues and friends who, mainly but not exclusively from a scientific point of view, made me what I am today:

Miran Veselič from Ljubljana/Slovenia was the first person to help me understand what happens when a mine is flooded. I owe him many ideas in many areas of science. Bob Kleinmann from Pittsburgh/USA (United States of America), then, was my mentor in what has to be known when one works in the mine water field. His knowledge helped to solve many of the problems I had. Next comes Paul Younger from Newcastle upon Tyne/United Kingdom. It is hard to copy him, nearly impossible, but it is worth trying because the way he thinks and works stimulates me to try harder all the time. Finally, Broder Merkel from Freiberg/Germany gave me the freedom to conduct research in the areas I deemed to be necessary for solving the difficulties related to mine water tracers and stratification. His "Why-do-we-need-that?"-questions were more than once the key to developing better procedures.

Well, this first part was easy while the second part is virtually impossible. So many people have helped me and there is no way of ranking them in an appropriate way. It would bother me very much if you looked for your name here and found it missing when you expected it to be here. Therefore, I thank all of you who contributed to my research and my growth as a scientist: thanks for whatever you did for me. Your ideas were important to me! Without your help – and even if it was only a reference list or a copy of a hard to find article – this work would not have been possible! Furthermore, I appreciate very much the assistance of all my colleagues and students at the Technische Universität Clausthal, the Ingenieurbüro für Geotechnik/Bautzen, the Technische Universität Bergakademie Freiberg, and the Ludwig-Maximilians-Universität München – which is a chronological ranking! Finally I thank my friends from IMWA – International Mine Water Association and the Springer Verlag Heidelberg who made this publication possible.

Yet, there are two person to whom I owe much of the fun, strength, knowledge and freedom that I had during the years since I started this work: my wife Ulrike and my friend Jana. They never lost their love and belief in me and my work!

2 Preface

Many articles and excellent books have been written on mine water issues. This publication focuses on some of the gaps in those publications, specifically:

- why and how to close a mine
- the hydrodynamics of a flooded mine
- tracer techniques for flooded mines, and
- stratification patterns in the mine water body

Since my first contact with mine water issues in 1991, I realized that scientists and consultants from many different backgrounds are working in this field of science: e.g. mining engineers, geologists, biologists, chemists, civil engineers, environmentalists, lawyers, and regulators. For some of them, mine water issues are their daily job; others work on a project for a short time period and than never come back to mining in their career or work. Most of the books I read about mine water issues – and when I first started to think about this publication, there were not many books available – were intended for either one audience or the other.

I have attempted to bridge the gaps that exist without repeating a lot of material already published elsewhere. Therefore, there are very detailed parts of this book, which address issues that are not well covered in the mine water literature, and other sections that simply provide an overview for the novice in this field. So, some facts are covered in detail (e.g. tracer tests or mine water stratification), others are discussed briefly (e.g. treatment techniques, surface mining) and others are just missing because covering them in detail would have been beyond the scope of this publication (e.g. salt mining, solution mining). Most of the book assumes that the reader is a practitioner with a basic knowledge of natural science.

But why should you care about mine water stratification and mine water tracer tests?

Since the 1960s, it was known, that water in a flooded underground mine might stratify (Stuart and Simpson 1961; Fig. 98). Many researchers since then have mistakenly thought that stratification (which is the layering of water with different physico-chemical properties) could be used to extract unpolluted mine water from a flooded mine. Although numerous cases of flooded and stratified mines exist and have been made public, no comprehensive synopsis of the density stratification problem has been published so far. This publication will fill this gap by describing the main physical laws involved, categorizing the types of stratification, and describing several case studies.

When looking through the references, you might notice that many of them have been written in a language that you may not be very familiar with: German. This might be unusual for an English publication, but it arose from two simple but decisive factors:

The Technische Universität Bergakademie Freiberg has organized several mine water related symposia and conferences. During the organization of such conferences, it became clear that most German-speaking colleagues from industry and the authorities depend on publications written in their own language. This is probably why many aspects of passive treatment were reinvented in Germany, without detailed studies of all the English publications available. Therefore, the list of references contains German as well as English publications, giving German-speaking colleagues several references to publications in their own language.

On the other hand, if you are a reader with English as your mother tongue, you will have the opportunity to learn about research done in Germany that you might find useful.

Contents

Abbreviations

ABA	acid-base accounting
AD	*anno domini*
ADP	Adenosindiphosphate
ALD	Anoxic Limestone Drain
AMD	acid mine drainage
AMLIS	Abandoned Mine Lands Inventory System
ANOVA	analyses of variance
ARD	acid rock drainage
ARD/ML	acid rock drainage/metalliferous leachate
ASCII	American Standard Code for Information Interchange
a.s.l.	above sea level
ATP	Adenosintriphosphate
BATNEEC	best available technology not entailing excessive costs
BRGM	Bureau de Recherches Géologiques et Miničres
BS	British Standard
CAVE	Carbonate Aquifer Voids Evolution (computer software)
CFD	computational fluid dynamics
DESD	UN Department of Economic and Social Development
DIN	*Das ist Norm* / Deutsches Institut für Normung
DMT	Deutsche Montan Technlogie GmbH
DNA	deoxyribonucleic acid
DO	dissolved oxygen
DQO	data quality objective
DVWK	Deutscher Verband für Wasserwirtschaft und Kulturbau e.V.
EAI	environmental impact assessment
EHTD	Efficient Hydrologic Tracer-Test Design
EN	European Norm
ENRA	elevated natural radiation areas
EPA	Environmental Protection Agency
eq. / eqs.	Equation / equations
ERMITE	Environmental Regulation of Mine Waters in the European Union
EU	European Union
EURATOM	European Atomic Energy Community
FDEM	frequency domain electromagnetic
FEFLOW	Finite Element Subsurface FLOW System
FEM	finite element method
FIZ-GEO	Fachinformationszentrum Geowissenschaften (geoscientific information centre)
GALILEO	European Global Satellite Navigation System
GaRD	Gas Redox and Displacement

GARD	Global Acid Rock Drainage Guide
GeoRef	Geoscientific References
GIS	geographical information system
GLONASS	Global'naya Navigatsionnaya Sputnikovaya Sistema (global navigation satellite system)
GmbH	Gesellschaft mit beschränkter Haftung (Inc.)
GPM	U.S. gallons per minute
GPS	global positioning system
GRAM	Groundwater Rebound in Abandoned Mineworkings
GSF	Gesellschaft für Strahlen und Umweltforschung
GVV	Gesellschaft zur Verwahrung und Verwertung von stillgelegten Bergwerksbetrieben mbH
HDPE	high density polyethylene
HDS	high density sludge
HERO	Hydrogeochemical Engineering Research and Outreach
HLNRA	high level natural radiation area
IC	ion chromatography
ICMM	International Council on Mining & Metals
ICP-AES	Inductively-Coupled-Plasma Atom Emission Spectroscopy
ICP-MS	Inductively-Coupled-Plasma Mass-Spectrometry
ICRP	International Commission on Radiological Protection
IMP	Inorganic Media Passive Systems
IMWA	International Mine Water Association
INAP	International Network of Acid Prevention
IR	infra red
ISI	Institute of Science Information
ISL	in-situ leaching
ISO	International Organization for Standardization
JPEG	Joint Photographic Experts Group
LAWA	Länderarbeitsgemeinschaft Wasser
LDO	luminescent dissolved oxygen
LDS	low density sludge
LED	light emitting diode
LLNRA	low or normal level natural radiation area
LydiA	*Lycopodium* Apparatus
m a.s.l.	meters above sea level
MAC	Maximum Allowable Concentration
MDM	multi-dispersion-model
MeFiSTo	Multiple Filter Storage Tool
MEND	Mine Environment Neutral Drainage
MIFIM	Mine Water Filling Model
MIRO	Mineral Industry Research Organisation

MLNRA	medium level natural radiation area
MMSD	Mining, Minerals and Sustainable Development
MOTIF	model of transport in
MPC	Mineral Policy Center
MS	Microsoft™
Mt	mount
NGO	non governmental organisation
NNP	net neutralization potential
NORM	naturally occurring radioactive materials
NR	Nebenrechnung (side calculation)
NURBS	Non uniform rationale B-splines
NW	northwest
OPAC	Online Public Access Catalogue
PADRE	Partnership for Acid Drainage Remediation in Europe
PC	Personal Computer
PCB	poly chlorinated bi-phenyls
PDA	Personal Digital Assistant
pers. comm.	personal communication
PE	polyethylene
PIR	Passive in-situ remediation
PIRAMID	Passive In Situ Remediation of acid Mine/Industrial Waters
PTFE	polytetrafluorethylen
PUF	polyurethane foam
PVC	polyvinyl chloride
QA	quality assurance
QC	quality control
RAPS	Reducing and Alkalinity Producing Systems
REE	rare earth elements
SAPS	successive alkalinity producing system
SC	side calculation
SCI	Science Citation Index
SDAG	Sowjetisch Deutsche Aktiengesellschaft (Soviet-German-Corporation)
SFB	Subsurface-Flow Bacterial Sulphate Reducing Systems
SFDM	Single Fissure Dispersion Model
SL	sampling location
SME	small and medium enterprises
TDS	total dissolved solids
TIC	total inorganic carbon
TinA	Tracer injection Apparatus
TOC	total organic carbon
TU	Technical University

UK	United Kingdom
UNCED	United Nations Conference on Environment and Development
UNDESA	United Nations Department of Economic and Social Affairs
UNEP	United Nations Environment Programme Industry and Environment
UNSCERA	United Nations Scientific Committee on the effects of Atomic Radiation
US/U.S./USA	United States of America
UTM	Universal Transversal Mercator System
VEB	Volkseigener Betrieb (nationally owned enterprise)
VHLNRA	very high level natural radiation area
WGS84	World Geodetic System 1984
WHG	Wasserhaushaltsgesetz (water management legislation)
WTP	Wetland-Type Passive Systems
Zl.	Zahl/Zähler (number)
ZPC	zero point of charge

Abbreviations not listed here are either the names of companies, of computer software, or standard abbreviations. The authors of computer software are usually not referred to in the text.

3 Introduction

3.1 Mining Issues

"Everything we use is either grown or mined", as miners say, and therefore, abandoned, closed, or orphaned mine sites cover approximately 240,000 km^2 of the Earth's surface and can be a hazard to both humans and the environment (Fig. 1). Countless papers about such sites and their remediation have appeared, most of them focused on a single mine site, and on acid mine/rock drainage/metalliferous leachates. This book attempts to provide a more comprehensive presentation of the water management that is necessary at such sites. It focuses on general aspects of flooded underground mines and aims to improve the understanding of the processes involved in underground mine flooding and mine water treatment. Furthermore, it explains some of the hydrodynamic aspects of flooded underground mines, which can be used to plan new in-situ and on-site treatment options. To reach that goal, the following chapters will:

- summarise the necessary measures for the management of abandoned and flooded underground mines
- explain the fundamental hydrodynamics within flooded underground mines, and
- describe the reactive transport mechanisms within the flooded voids

While the first part is mainly a compilation of previous investigations of mine water research, the second part consists of observations in flooded underground mines and my own, innovative tracer techniques for flooded underground mines.

Since the 1990s, the number of mine water related articles has burgeoned and a complete outline of all techniques and investigations is hardly possible. A database retrieval in the four major databases relevant for mine water (GeoRef, Geobase, FIZ-GEO, SCI Expanded) listed more than 10,000 entries with the key words '(mine or mining) and water' (Wolkersdorfer 2004). If you think your work or somebody else's work is missing in this book, please just let me know and it will be considered for inclusion in future editions.

To illustrate the techniques and the behaviour of water in flooded mines, several case studies will be presented and all the details that are necessary to transfer the outcomes of those investigations to other mines sites will be described. Although several thousand mines have been closed in recent decades, few were described in the literature; therefore I have included results of both published and unpublished mine closures that are relevant to water management.

Fig. 1. Abandoned Tyrconel/Australia gold mine site. The headframe stands about the flooded main shaft.

Hámar et al. (2002) concluded that "illegal and orphan mines pose the highest potential risk to the environment over the long term", whereas "orphaned" according to their definition means a closed, not remediated mine which has no owner. Similar, if not more pessimistic – but challenging – views have been given by Balkau (1999), who stated that "among the outstanding environmental problems confronting the mining industry, that of abandoned mine sites has been particularly slow to be tackled." Such publications categorise "unused" mines as either 'closed', 'abandoned', or 'orphaned'. Though such a nomenclature might be helpful on a legal basis, all such sites have one thing in common: mining is no longer taking place and somebody has to take care of the legacy. The geochemical and hydrodynamic processes do not care about such definitions and therefore no special focus will be given in this book to those terms. Wherever they appear in this book they can be considered as synonyms, as there is no generally accepted worldwide definition of these words, as was observed during a European Commission Workshop about "Mine and Quarry Waste – The Burden from the Past" (Puura et al. 2003).

In the USA, several programmes for reclaiming abandoned mine land and polluted mining-related superfund sites have been undertaken at the national and regional levels. The U.S. Department of the Interior's "Office of Surface Mining" hosts the "Abandoned Mine Land Reclamation" programme, which is financed through an "Abandoned Mine Land Fund" ($1.7 Billion U.S. in 2006). Another programme is hosted by the "Bureau of Land Management" within the same department. Most of those activities are linked through the "National Association of Abandoned Mine Land Programs" and the "National Coalition for Abandoned Mine Reclamation". Although these programmes are collecting data about abandoned mine lands and potential hazards from such sites, they do not carry out research on mine flooding or acid mine drainage (Office of Surface Mining 2002). A different programme is the U.S. Geological Survey's "Abandoned Mine Lands Initiative", which does conduct research (U.S. Geological Survey 2000).

Canada runs the "Orphaned abandoned Mines Initiative" with the "National Orphaned/Abandoned Mines Advisory Committee" coordinating various initiatives (Castrilli 2002); in Europe, PADRE (Partnership for Acid Drainage Remediation in Europe) focuses on acid mine remediation works in Europe (Wolkersdorfer et al. 2004).

To date, no reliable statistics about the total number of abandoned mine sites exists, though their total worldwide number might go into the millions (e.g. International Institute for Environment and Development 2002). In the USA, the internet database of the Abandoned Mine Land Inventory System (AMLIS) lists about 20,000 abandoned mine sites that have been investigated, of which only 5% may pose a significant threat to watersheds or the environment (Office of Surface Mining Reclamation and Enforcement 2004; Anonymous 2006). Copa et al. (1995) estimated that only 30,000 or 6% of the estimated 500,000 mining sites in the USA might cause potential risks to the public or the environment. Nevertheless, the States of Pennsylvania and West Virginia identified acid mine drainage as their № 1 threat to water quality (Diz 1998). Wireman et al. (2001) reported on tens of thousands of abandoned mines in the Rocky Mountains, and Benner et al. (1997) summarise several investigations in their introduction and come to similar conclusions. Furthermore, ten U.S. States reported that between 8,000 and 16,000 km of streams were polluted by mine drainage (U.S. Environmental Protection Agency et al. 2001). During the European Union PECOMINES project, the mine inventory of the candidate countries at that time (Bulgaria, Czech Republic, Estonia, Hungary, Latvia, Lithuania, Poland, Romania, Slovakia, Slovenia) identified several thousand potentially dangerous mine sites – in just ten European countries (Jordan et al. 2004)! Also the "Abandoned Mines Working Paper" (International Institute for Environment and Development 2002) refers to only 5 countries (Australia, Canada, United Kingdom, United States, South Africa) though the numbers of four "other countries" (Ireland, Sweden, Japan, Chile) are listed. In addition, what is missing in many of these statistics is the scope of pollution emanating from small-scale mining, though some reports try to evaluate the problem (Wrobel et al. 2000; Hentschel et al. 2002). Zamora (1999) summarised some of the problems of small-scale mining and concluded that

> it cannot be assumed that small-scale miners are knowledgeable about environmental technologies or that they can adopt international standards or codes of conduct. They do not generally have access to financing, so cannot invest in new technologies and systems. They cannot be treated like any other miners.

Though every mine site is special in itself and nearly every mine is – to borrow a mathematical term – a singularity, the question is whether, to some degree, the hydrogeochemical behaviour of an abandoned mine can be predicted and if that classification can be used to plan remediation strategies. As in weather forecasting, many non-linear equations determine the hydrogeochemical and the hydrodynamic development of a flooded mine so that (as has often been shown) small changes in conditions can result in big differences at the end. There is a similar unpredictability on the atomic scale: the decay of a radioactive element. Though the lifetime of a single atom cannot be predicted, a huge quantity of them can be described with a characteristic constant: their half-life. The question, therefore, is whether flooded underground mines have similarities that can be used to estimate their behaviour after they have been flooded, and that can then be used to set up an environmentally sound mine water management plan.

When reading the following chapters, you must bear in mind that elevated metal concentrations may occur in both acid and neutral drainage. Although commonly referred to as "acid mine/rock drainage" or AMD/ARD, it is more accurate to refer to the phenomenon as "acid mine/rock drainage/metal leaching: AMD/ML or ARD/ML". Therefore, whenever the term acid mine drainage is used in the following sections, it must be kept in mind that this includes elevated concentrations of metals as well. In this work, the term 'metals' is commonly used for both semi-metals and metals, whereas the term 'heavy metals' is avoided whenever possible, because several of the so-called heavy metals are neither heavy nor are they metals in our common understanding (Duffus 2002). Yet, if the term 'metal' is used for all elements that show cationic behaviour in aqueous solution or that have an acid soluble oxide (as defined by Parish 1977), then the term becomes clearer, since this also includes Al and Sb as metals. Only arsenic, being a metalloid, does not fall in this category, so if one wants to be exact, one should write "metals and arsenic" instead of "heavy metals".

Some readers might be bothered by the fact that standard hydrogeological methods are not described. A thorough hydrogeological and hydrological understanding is essential for all mine water research. However, how can (and why should) the already existing books about geological practises be superseded by a book focused on mine water hydrogeology for flooded underground mines? It is therefore beyond the scope of this book to repeat what has already been written about good hydrogeological practices. The interested reader – and obviously, you belong to this group – is referred to the already existing literature (e.g. Freeze and Cherry 1979; Fetter 1988; Jordan and Weder 1995; Brassington 1999; Domenico and Schwartz 1998; Deming 2002; Langguth and Voigt 2004; Hiscock 2005; Hölting and Coldewey 2005).

Not all the literature cited in this publication is readily available in all libraries or can be bought at a bookshop. This publication covers many issues connected with mine water management and mine closure to give the interested researcher or consultant an insight into special cases as well as general practices. Publications concerning the latter can usually be found in most well equipped libraries. All the others should be accessible through the internet or through national or international book exchange systems. Also, persistent digital object identifiers have been avoided since at least two competing systems are available; which of these will survive the next decade is still unclear. As web sites are changing quickly, the reader will be referred to the excellent Search Engines or library OPACs available on the internet if he or she wants to find more information about a single issue.

3.2 How to read this book

Depending on your personal background, you might find some parts of this book more interesting than others. First, I will briefly introduce you to each of the chapters and then I will let you know which chapters might be of interest to you.

The overall idea was to describe the processes related to mine abandonment from a hydrogeologic perspective, following the path that a single drop of water might take through an abandoned mine. First, it flows through the fractures, fissures and bedding planes into the mine where it comes in contact with distinctive minerals. This process is described in chapter 4, "Hydrogeochemistry of Mine Water", which also describes some attenuation and precipitation processes that might happen. Then we will look into the necessary measures that lead to the "enclosure" of that water drop in the abandoned mine; chapter 5: "Insights into Mine Closure" describes why a mine is closed and what is necessary for an environmentally safe mine abandonment. A short chapter, which attempts to sketch some important "Do's and

Don't's of Mine Closure", follows. This chapter mainly resulted from hearing statements like "the next time we close a mine, we should …". It is a summary of recommendations that did not fit into the other chapters but that I thought would be useful for people not directly involved in the closure process. In the following chapter, the water drop meets all the other water drops that are in the mine. They start to move around and therefore, chapter 7, "Hydrodynamics", describes the governing laws that force those water drops to flow through our abandoned mine workings. To understand how the water drops flow through the mine, how they interact with the rocks and each other, and how their behaviour could be predicted is the subject of chapter 8 "Mine Closure Flow Models and Geochemical Modelling". Because all our models are simplifications of reality, we now have to prove if our water drop behaves as we predicted that it would. Chapter 9, "Monitoring and Sampling", summarises the procedures that are necessary for sampling mine water. Because mine water is such a unique liquid, the pitfalls of mine water sampling and monitoring are given special consideration in this chapter. Because sampling and monitoring does not explain how, in which direction, and at which speed our water drop flows through the mine, we have to trace its way by conducting tracer tests. A comprehensive description of the relevant procedures for conducting mine water tracer tests is given in chapter 10, "Tracer Techniques for Mines". Then, after minutes or decades in the dark and in contact with the rocks and minerals, our water drop leaves the mine. Perhaps it has been lucky and has a 'clean record'; otherwise, chapter 11, "Mine Water Treatment and Ground Water Protection", comes into play. Finally, in chapter 12, thirteen well-investigated case studies are described, and the interaction of the material discussed in the previous chapters is integrated. Because stratification could possibly be seen as a key instrument for in-situ treatment of mine water, case studies relevant to that topic take up significant space in this chapter.

After the references, you will find three appendices. I am not fond of such annotations, but as in an interesting letter, the post-script adds that extra something. Appendix I describes a simple procedure for checking the accuracy of water analyses. By comparing the calculated electrical conductivity against the measured one (hopefully, you calibrated your conductivity meter!), you will get a better reliability than by the usual cation/anion balance (eqs. 62 and 63). Appendix II addresses flow measurements. While putting together equations for flow measurements, I realised that the books I read used different equations, even for a simple mechanical structure like a V-notch weir. Flow measurement is far more complicated than most people think, and because flow measurement is so essential to our work as mine water researchers, this highly physical and mathematical appendix had to be added so that you would have simple and reliable equations that you can readily use. Finally, the last appendix summarises the Berlin I and II Guidelines that have been drawn up to guarantee environmentally sound mining and mine closure as well as sustainable mining and post-mining sites.

Well, after that short outline, where should you start to read or what should you read? Each chapter assumes that you are coming from a different background. I have balanced the chapters as I did based on the questions that I received during 13 years of lectures and conference presentations. However, all of the chapters assume that you have at least a basic knowledge of chemistry, hydrogeology, mining issues, mathematics, and physics.

If you are new in the field of mine abandonment and would like to read about some general ideas, start with chapter 4 and proceed to chapters 5, 9 and 11. From the case studies, choose sections 12.3, 12.4, 12.6, and 12.10.

Should you wish to understand the hydrodynamics of an abandoned and flooded mine, or if you are planning to conduct a tracer test, start with chapter 7 and then read chapters 9 and 10. Out of the case studies you might find 12.4, 12.6, 12. 7, and 12.10 helpful.

If you are going to close a mine, you will want to read chapters 5, 6, 8, 9, and 11. Choose sections 12.4, 12.7, 12.11, and 12.12 to get an idea of the processes that can occur during mine flooding.

As a regulator or a lawyer, you might find parts of chapter 4 useful, as well as sections 5.4, 5.9, 6, 9.1, 9.3, 10.1, and 11. The case studies are probably more technical than you require, but you might like to read appendix III.

4 Hydrogeochemistry of Mine Water

4.1 Introduction

Problems with mine drainage generally develop when ground and surface waters enter mine sites and come in contact with primary and secondary minerals under oxic conditions. During contact with those minerals, the water may dissolve components of the ore and rock-forming minerals. Most of the processes involved are controlled by the E_H-pH conditions at the water-mineral contact. If even relatively small amounts of disulphides, as pyrite or marcasite, oxidise and dissolve, the acidity formed can dissolve other water contaminants. Because metal or coal deposits are so diverse, the metals and the composition of mine water are unique at every mine site. Commonly, carbonate minerals are present and neutralize (buffer) the acid, leading to neutral or even alkaline mine water, which might still be harmful to surface and ground waters in a catchment area. Barnes and Clarke (1964) nicely summarised the problem associated with disulphide oxidation: "if mine workings could have been isolated hydrologically, operators, who must pump as much as 23 t of water per ton of coal from the anthracite mines of Pennsylvania, could have saved a great amount of money". That was in 1964 and even today, after nearly all of the questions discussed by Barnes and Clark have been resolved, we are still far away from being able to definitively stop acid mine drainage and metal leaching. These problems were, in fact, described by Agricola (1556), and clearly existed long before that.

Though this chapter focuses on metal and semi-metal pollution of the mining environment, there are obviously other pollutants that might negatively affect watersheds. One of the most threatening is high salinity, which is mainly connected with salt mines but is also associated with deep coal mines as in Germany and Poland and with mines cutting across mineral water aquifers. In addition, ammonium, nitrate, organics, hydrocarbons (from machinery used underground), PCBs, radium, and barium are contaminants that can sometimes be found at mine sites (e.g. Bench 2000).

Besides the above-mentioned processes, microbial activity also has to be considered. As the outstanding work of Baas Becking et al. (1960) shows, there is nearly no area in our environment which is not inhabited by some kind of bacteria (Fig. 4). Interestingly, when Baas Becking et al. (1960) wrote their work about the E_H-pH conditions of waters and the environment in which microorganisms live, they found "only two published readings on mine waters" but "fortunately, unpublished information increases the total number of readings to fifty-six". Furthermore, by the time they wrote their paper, they realised that "oxidation of pyrite is a chemical system which has been neglected by geochemists" though "investigations by us have shown

Fig. 2. Iron- and manganese oxihydroxide as well as other metal precipitates as stalactites in the Reiche Zeche Freiberg/Germany mine. Width of picture: 30 cm.

quite conclusively that the first stages of oxidation are entirely inorganic". And even more interestingly, this work is hardly ever cited by mine water researchers, though it contains interesting data about the environments in which bacteria prefer to do their job! Luckily, the last 50 years saw a considerable change in that field of research, though the present fashion to describe the reactions of pyrite oxidation detailed in *every* small article about mine water issues should be resisted – a reference to Sigg and Stumm (1994), Stumm and Morgan (1996), or Younger et al. (2002) provides an adequate overview. Most people – in the meantime – know how the process is working. The following section has been written for those who are interested in the governing reactions. Others can directly skip ahead to section 4.3. If you do not like chemical formulae at all, you can go directly to chapter 5 – yet you will miss understanding all the processes that make up our interesting and often colourful work (Fig. 2)! A potential classification of mine waters is given in section 9.3.3.

4.2 The Process Starts: Abiotic Disulphide Weathering

It is commonly accepted that disulphide weathering and especially the oxidation of pyrite and marcasite are the initial reactions in the formation of acid mine drainage/metalliferous leachate (Fig. 3):

$$FeS_2 + 7/2\ O_2 + H_2O \leftrightarrow Fe^{2+} + 2\ SO_4^{2-} + 2\ H^+ \qquad (1)$$

$$Fe^{2+} + 1/4\ O_2 + H^+ \leftrightarrow Fe^{3+} + 1/2\ H_2O \qquad (2)$$

$$Fe^{3+} + 3\ H_2O \leftrightarrow Fe(OH)_3 + 3\ H^+ \qquad (3)$$

$$FeS_2 + 14\ Fe^{3+} + 8\ H_2O \leftrightarrow 15\ Fe^{2+} + 2\ SO_4^{2-} + 16\ H^+ \qquad (4)$$

As can easily be seen, the overall reaction produces protons, and thus releases acid into the mine water (Singer and Stumm 1970). For that reason, pyrite weathering is the strongest acid-producing process of all oxidation processes known to occur in the natural environment (Stumm and Morgan 1996). If no buffering minerals are present, the pH of evaporating mine water can reach extremely low values, sometimes as low as –3.6 (Nordstrom and Alpers 1995; Stumm and Morgan 1996; Nordstrom et al. 2000). As the reaction is exothermal, the mine air and mine water can reach unusually high temperatures (air temperatures up to 40 °C were recorded in the Iron Mountain mine, D. Kirk Nordstrom, pers. comm.; Nordstrom 2004c). Consequently, owing to buoyancy effects, the warm oxygen-depleted air will move into the upper parts of a mine, creating a 'chimney effect': fresh, cold and oxygen-rich air is drawn into the mine through the portals, fostering additional disulphide oxidation.

Due to its prominent role in many reactions that are important in the formation and attenuation of mine water, the pH has been called the "master variable" (Stumm and Morgan 1996):

$$pH = -\log_{10}\{H^+\} \qquad (5)$$

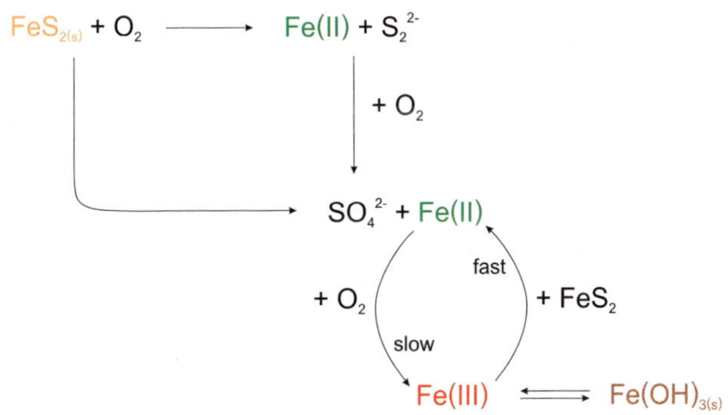

Fig. 3. Schematic representation of the pathways during abiotic pyrite or marcasite oxidation (modified after Kester et al. 1975 in Stumm and Morgan 1996; Singer and Stumm 1970).

For those not familiar with the conventions in chemistry, numbers in [ ] are concentrations, *c*, and numbers in { } are activities, *a*. They are related by the activity coefficient, *f*, according to the following relation:

$$\{a\} = f \times [c] \tag{6}$$

All chemical-thermodynamic codes use this relation for calculating the species distribution in analysed water (Stumm and Morgan 1996; Nordstrom 2004c).

Reactions 1 and 2 are catalysed by bacterial activity (see chapter 4.4). Reaction 3 is abiotic and often does not occur until the water has left the mine. The above reactions have a net proton production and therefore can cause the pH to decrease rapidly. As the pH reaches about 4, ferric iron becomes the predominant oxidant and reaction 4 governs the rate of pyrite oxidation (Singer and Stumm 1970). Though this might imply that pyrite oxidation can occur without oxygen involvement, oxygen is the initial oxidant either directly for pyrite or indirectly to generate ferric iron by reaction 2. Pyrite does not oxidise in anoxic environments unless previously generated ferric ion is present in the solution. The four reactions also prove that pyrite oxidation occurs as long as enough of the reactants are available, as has nicely been shown by Nordstrom (2004b) for the Iron Mountain Superfund site, where there is enough pyrite for another 3,200 years of acid generation!

The above disulphide weathering reactions require pyrite or marcasite, oxygen, and water. With the exception of some limestone quarries, phosphate mines, or sand and gravel operations, those three reactants can be found in nearly every surface and underground mine and because all it takes to keep the process going is 1–2% pyrite in the rocks exposed to the mine air, this process occurs worldwide. A large mineral surface area is preferable for the decomposition of the minerals: the smaller the crystals, the faster the process and the quicker the water gets acidified. Generally, if the pyrite crystals can be seen with the bare eye, the problem will be less intense than in cases where the same amount of pyrite is dispersed as sub-micron sized grains throughout the rock matrix. Of course, higher concentrations of pyrite make the problem worse. However, as long as the ground water is percolating through undisturbed rock, through open pore space, or natural fractures without human interference, and as long as the ground water is in a reducing state, no significant pyrite oxidation occurs. Therefore, acid rock drainage is rarely a problem in undisturbed natural environments, though there are exceptions, such as the Austrian Rötlbach, where the first natural occurrence of schwertmannite was observed (section 4.6, Schwertmann et al. 1995; Rüde 2004), or the Colorado/USA South Fork/Lake Creek near Red Mountain in the Grizzly Peak Caldera (Neubert 2000). The opposite is also true: if only one of the reaction components is missing, the reactions will stop and no acidity will be produced. That is the basis for most in-situ or on-site remediation procedures (section 11.6).

Even "dry" mines can contain a large amount of water that can oxidise pyrite. Such "dry" mines are dry because all the water that enters the mine through the rock matrix evaporates as it contacts the mine air. Brewitz et al. (1985) measured the amount of this evaporating water in the Konrad/Germany mine and compared it with results from the Stripa/Sweden experiments. While the Konrad mine gained a volumetric evaporation rate of 0.016 m^3 min^{-1} m^{-2}, it was 0.076 m^3 min^{-1} m^{-2} at Stripa (cited after Brewitz et al. 1985).

Mining significantly changes the natural ground water conditions and as the ground water table falls and large underground openings are created, oxygen enters a formerly reducing environment. While the pre-mining conditions may have supported the formation of pyrite from iron-enriched ground water, the process now is reversed and pyrite will be oxidised. This process occurs throughout the dewatered area but especially in the proximity of the mine workings, where the rocks are fragmented because of blasting and excavating. Disaggregation zones ('excavation disturbed zones') generally extend about 2 m around the mine workings, but can extend up to 8 m (Müller-Salzburg 1978; Jacobi and Everling 1981; Brewitz et al. 1985; Militzer et al. 1986; Stoll and Bauer 1991; Roschlau 1994; Wolkersdorfer 1996). Blasting, up to now, has generally not been seen as a factor that affects disulphide weathering. However, as destressing blasting protects the rock and thus minimises blasting induced fractures in the rock, it can reduce the formation of acid mine drainage/metalliferous leachate production in underground workings.

Kinetics and consequently mineral weathering rates are important in understanding the evolution of different mine waters. Different authors report different mineral weathering rates but a discussion of those important investigations is beyond the scope of this publication. A thorough discussion and summary of those investigations, which were conducted under various atmospheric conditions, has been given in Salmon and Malmström (2002), Younger et al. (2002), and Höglund et al. (2004).

4.3 The Process goes on: Metal Dissolution

Because pyrite is always found with other minerals (paragenesis), predominantly other sulphides, but also oxides, vanadates, arsenates, phosphates, and carbonates, mine water is commonly enriched in many elements. Their concentrations mainly depend on the mine water pH, with maximum concentrations typically found at lower pH values (Table 2). These minerals may weather in a congruent, incongruent, or redox process (Stumm and Morgan

1996). Congruent weathering processes are reversible processes, such as the dissolution of gypsum ($CaSO_4 \cdot 2H_2O$). Incongruent weathering processes, on the other hand, are irreversible because one of the products either vaporises, precipitates, or forms other, stable species in the mine water. Redox weathering reactions, similar to the incongruent weathering reactions, are irreversible under normal atmospheric conditions. When mine workings are eventually inundated with water (see section 5.4), several of these processes, including the dissolution of secondary mineral precipitates, again play an important role and eventually create the 'first flush' (section 5.4.1).

Like pyrite, most of those minerals are stable at the E_H-pH conditions and the oxygen content of the ground water before mining. When oxygen enters the ore veins or the rock matrix, the stable conditions change and those mineral phases start to oxidise to more mobile species, which are then transported with the water to the point of the lower hydraulic pressure, which is usually the open space of the underground mine or the mine sump (Me in equation 7 stands for any metal):

$$Me^{n+}S_s + n\ O_{2aq} \leftrightarrow Me^{n+} + \tfrac{2}{n}\ SO_4^{2-} \quad (7)$$

Because this looks rather theoretical, here are some examples of the oxidation of primary minerals commonly found in underground mines:

Galena: $$PbS_s + 2\ O_{2aq} \leftrightarrow Pb^{2+} + SO_4^{2-} \quad (8)$$

Sphalerite: $$ZnS_s + 2\ O_{2aq} \leftrightarrow Zn^{2+} + SO_4^{2-} \quad (9)$$

Greenockite: $$CdS_s + 2\ O_{2aq} \leftrightarrow Cd^{2+} + SO_4^{2-} \quad (10)$$

Table 1. Selection of weathering processes that occur in mines (modified after Stumm and Morgan 1996). FeS_{2s}: pyrite; $CaSO_4 \cdot 2H_2O_s$: gypsum; $K[AlSi_3O_8]$: orthoclase; $KAl_2[(OH)_2|AlSi_3O_{10}]$: muscovite.

Type	Reaction
Congruent	$CaSO_4 \cdot 2H_2O_s + H_2O \leftrightarrow Ca^{2+} + SO_4^{2-} + 3\ H_2O$
Incongruent	$3\ K[AlSi_3O_8]_s + 2\ H_2CO_3 + 12\ H_2O \leftrightarrow$ $2\ K^+ + 2\ HCO_3^- + 6\ H_4[SiO_4] + KAl_2[(OH)_2|AlSi_3O_{10}]_{(s)}$
Redox	$FeS_{2s} + \tfrac{7}{2}\ O_2 + H_2O \leftrightarrow Fe^{2+} + 2\ SO_4^{2-} + 2\ H^+$

Covellite: $CuS_s + 2\ O_{2aq} \leftrightarrow Cu^{2+} + SO_4^{2-}$ (11)

Chalcopyrite: $CuFeS_{2s} + 2\ O_{2aq} \leftrightarrow Cu^{2+} + Fe^{2+} + 2\ SO_4^{2-}$ (12)

Millerite: $NiS_s + 2\ O_{2aq} \leftrightarrow Ni^{2+} + SO_4^{2-}$ (13)

The second process is induced by disulphide oxidation and the lowering of the pH. Furthermore, the release of Fe^{3+} is a process induced by disulphide oxidation (Nordstrom 1977; Tuovinen and Bhatti 1999):

$$CuFeS_{2s} + 16\ Fe^{3+} + 8\ H_2O \leftrightarrow Cu^{2+} + 17\ Fe^{2+} + 2\ SO_4^{2-} + 16\ H^+ \quad (14)$$

$$ZnS_s + 2\ O_{2aq} \leftrightarrow Zn^{2+} + SO_4^{2-} \quad (15)$$

$$UO_{2s} + 2\ Fe^{3+} \leftrightarrow UO_2^{2+} + 2\ Fe^{2+} \quad (16)$$

Gottschalk and Buehler (1912) were amongst the first authors to systematically investigate the interaction of sulphide minerals and pyrite or marcasite. They realised that non-iron metal concentrations in the solution increase in the presence of pyrite or marcasite though they were unable to satisfactorily explain their findings since they were not aware that microorganisms were partially responsible for their findings, and that those microorganisms more easily 'nibble' on pyrite or marcasite than on other sulphides. Nevertheless, their work clearly shows that the release of metals and semi-metals from other sulphides increases if pyrite or marcasite are abundant in the system.

Viewed the other way: if mine water has elevated concentrations of Fe, Zn, Cu, Pb, and Cd, it can be assumed that pyrite, sphalerite (Cd always substitutes for Zn in sphalerite), galena, and chalcopyrite are relevant minerals in the ore deposit from which it emanates. It is thus likely that it is flowing out of a lead-zinc-mine. Chemical-thermodynamic codes have been developed that include the relevant stability constants for the formation and dissolution of mineral species and calculate, using iterative computation, which mineral phases is, or was, in equilibrium with the water.

Arsenic, for example, is a very mobile metalloid commonly found in polluted mine water that has provoked considerable interest in recent years (e.g. Merkel and Sperling 1998;Williams 2001; Zhu et al. 2003). It is mainly released from arsenopyrite (FeAsS), in a reaction similar to equation 1 but also from other common As-minerals such as enargite or tennantite. Many uranium mines and nearly all hydrothermal gold deposits show elevated concentrations of As in their mine waters (e.g. Straskraba and Moran 1990; Wolkersdorfer 1996; Schneider et al. 2001; Bowell 2001).

Table 2. Composition of different mine waters with the most predominant mine water constituents. Sources: Alpers et al. (1992); Yelpatyevsky (1995) [according to a pers. comm. by Vera Pavlovna Elpatyevskaya, the values in Yelpatyevsky's publication are wrongly given in mg L^{-1} instead of µg L^{-1}]; Wolkersdorfer (1996); Banks et al. (1997); Nordstrom et al. (2000); Younger (2000b); Wolkersdorfer and Hasche (2001a); Banks (2004); unpublished data from author; concentrations in mg L^{-1}.

Locality	pH	SO_4^{2-}	Fe_{tot}	Al	Mn	Zn	Cu
Iron Mountain California (copper)	0.4	108,000	18,600	2,320	–	2,060	290
Iron Mountain California (copper)	1.1	41,000	7,820	1,410	11	1,860	360
Pyrite mine (anonymous)	2.5	5,110	1,460	84	3	1	0.2
Smirnovsky tin mine (Sichote Alin)	3.2	–	132	–	14	25	5
abandoned coal mine (anonymous)	3.6	1,044	101	17	4	0.2	0.007
Ynysarwed Wales (coal)	4.2	1,554	180	< 0.5	6	0.06	–
waste rock dump (coal)	5.5	146	287	1	5	0.05	< 0.01
Gernrode Harz Mountains (fluorite)	5.9	78	22	–	4	0.14	0.016
Straßberg Harz Mountains (fluorite)	6.3	359	31	–	6	0.9	0.08
Dunston Chesterfield (coal)	6.3	210	11	< 0.05	1.3	<0.01	–
abandoned coal mine (anonymous)	6.3	83	5	0.08	0.4	0.05	0.005
metal mine (anonymous)	6.5	124	15	0.1	2	0.003	–
Niederschlema Erzgebirge (uranium)	7.1	1,138	3	0.4	3	0.1	0.03
Frazer's Grove Durham (fluorite)	7.7	260	11	–	–	0.4	0.04
Mine № 3 Svalbard (coal)	8.2	7	< 0.01	< 0.02	0.004	0.055	< 0.01
Schwaz Austria (dolomite, fahlore)	8.4	13	< 0.01	–	0.002	0.022	0.04

4.4 The Process is Speeded up: The Role of Microorganisms

4.4.1 Introduction and Metabolism

Microorganisms play an important role in the processes of natural and man-made environments (Nordstrom 1977; Singh and Bhatnagar 1988; Smith et al. 1988; Benedix et al. 1991; Ledin and Pedersen 1996; Salmon and Malmström 2002; Ebena 2003). Most microorganisms in mine water are bacteria, but fungi, algae, archaebacteria and diatoms can also be found (Perrin et al. 1992; Douglas et al. 1998; Ohimain 2003; Rohwerder et al. 2003). A good synopsis of microbial activities in mine water is given by Hallberg and Johnson (2005) and Mills (1999). For those who are interested in a more detailed approach, Edwards et al. (2000), Johnson (2003), or Nordstrom (2000) are recommended. Herbert Jr. (1999) and Rawlings (2005) give an excellent review of microbial activities in mine wastes. The main interactions necessary to understand microbial activity in mine water is given in the following paragraphs, underlain by some basic references to published work.

Because of their diversity, microorganisms cover nearly all eco-systems on our planet (Fig. 4), and if found in really extreme conditions, reports about such environments find their way into popular journals (e.g. Siefer 2000). Some microorganisms prefer extremely acidic conditions (Robbins et al. 2000) whereas others like to live under extremely alkaline environments (Takai et al. 2001), and are called extremophiles because they live in conditions which are considered to be extraordinarily unusual for normal organisms (Madigan et al. 2003; Hallberg and Johnson 2005). Due to their ability to catalyze the dissolution of metal sulphides, they are commonly used in bioleaching processes to extract metals from the host or mother rocks of low grade deposits or waste rock dumps (Ebner and Schwarz 1973; Bosecker 1980; Lundgren and Silver 1980; Beyer 1986; Duarte et al. 1990; Brunner et al. 1993; Ahonen and Tuovinen 1995; Gourdon and Funtowicz 1995; Modak et al. 1996; Cassity and Pesic 1999; Vasan et al. 2001; Rawlings 2002; Rohwerder et al. 2003; Sand and Gehrke 2006). The role of microorganisms in mine water is twofold: they catalyse pyrite oxidation and help to remediate polluted mine water.

Until the year 2000, every mine water researcher was able to name the common bacteria most responsible for catalysing and accelerating the reactions that cause acid mine drainage: *Thiobacillus ferrooxidans* and *Thiobacillus thiooxidans*. Then Kelly and Wood (2000) reclassified *Thiobacillus* into three new genera: *Acidithiobacillus*, *Halothiobacillus*, and *Thermithiobacillus*, based on the 16S rRNA gene sequence. To distinguish the non-acidic "true" *Thiobacillus* from the acidophilic species, *Thiobacillus ferrooxidans* and *Thiobacillus thiooxidans*, which prefer pH-ranges from 0 to 5, were re-

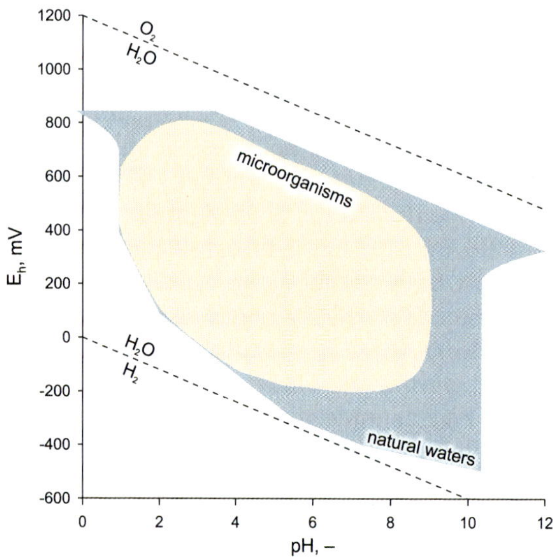

Fig. 4. Areas of natural E_H- and pH-conditions including mine waters and the preferred fields for commonly occurring microorganisms in mine waters (modified after Baas Becking et al. 1960).

named *Acidithiobacillus ferrooxidans* and *Acidithiobacillus thiooxidans* (Hallberg and Johnson 2005, Table 3). The new names underline the acidic environment in which these bacteria prefer to live (all references to these bacteria in this publication were renamed using this convention, independent of the name in the original literature). In circum-neutral to alkaline conditions, *Acidithiobacilli* become less common and *Thiobacilli* take on their role, as can be seen in the Niederschlema/Alberoda uranium mine, where circum-neutral mine water dominates (Table 4).

Most of the bacteria responsible for the development of acid mine drainage are chemotrophic bacteria, which use organic or inorganic electron-donating sources as their energy supply. Chemo*litho*trophs use only inorganic chemicals (Fig. 5), and because CO_2 is the source of their cellular carbon, they can further be classified as autotrophic – as opposed to microorganisms that use organic carbon, which are called heterotrophs. Furthermore, lithoautotrophs use inorganic electron donors whereas phototrophs obtain energy from sunlight.

Whichever primary energy source the microorganisms use, they must phosphorylate ADP (adenosine diphosphate) to ATP (adenosine triphosphate), because all cell reactions depend on ATP as an energy source. The reaction process to produce ATP requires +32 kJ of energy and involves the enzymatic transfer of protons through the inner cell membrane:

$$\mathrm{ADP} + \mathrm{P}^- \leftrightarrow \mathrm{ATP} \qquad \Delta G^{0'} = +32\ \mathrm{kJ} \tag{17}$$

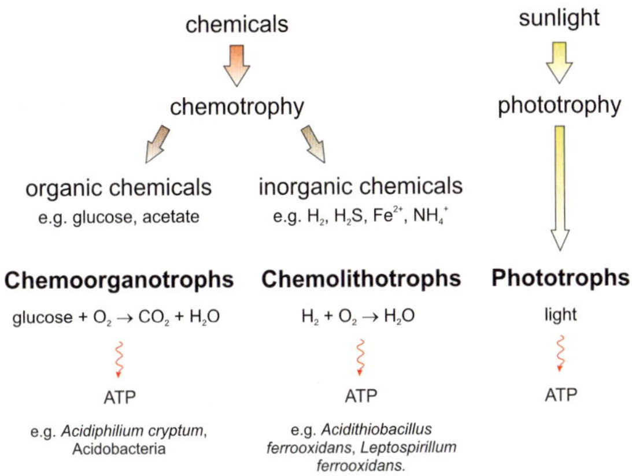

Fig. 5. Classification of some common bacteria occurring in mine water according to their metabolism. Chemical reactions just show one of the principle processes involved. Acidobacteria are a group of microbes that have yet to be named, but are closely related to *Acidobacterium capsulatum* (modified after Madigan et al. [2003] and adopted after Kevin Hallberg, pers. comm.).

For chemolithotrophs, this energy is obtained from the oxidation of sulphides, primarily pyrite and marcasite, but also from other sulphides (Table 5). Inside the bacteria cell, the following exothermal reactions (exemplified by *At. thiooxidans*, *Beggiatoa* and *Thiotrix*) take place:

$$2\,S^{2-} + 4\,H_3O^+ + O_2 \leftrightarrow 2\,S + 6\,H_2O \qquad (18)$$

$$2\,S + 6\,H_2O + 3\,O_2 \leftrightarrow 2\,SO_4^{2-} + 4\,H_3O^+ \qquad (19)$$

The bacteria gain $\Delta G^{0'} = -498$ kJ of energy, which is used for producing ATP, in which the energy is stored until it is needed in endergonic metabolic processes, such as kinetic, chemical, osmotic, electrical, or thermal processes.

The biotic iron oxidation occurs at a rate which is 100 to $1 \cdot 10^6$ times faster than abiotic iron oxidation (Singer and Stumm 1970; Rackley 1976; Lundgren and Silver 1980). Though the bacteria accelerate the reactions, they cannot change the end points of reactions or their thermodynamics, but by changing the rate of reactions, they change the natural endpoint.

Four microbiologically catalysed processes are important for mine water: two (sulphur and iron oxidation) are responsible for the formation of acid mine drainage/metalliferous leachate and two (sulphate and iron reduction) for the attenuation of the water (Mills 1999). Furthermore, the biotic activities of bacteria are responsible for many mechanisms that lead to metalliferous leachate, which is used for bioleaching (Lundgren and Silver 1980).

Table 3. Acidophilic prokaryotic microorganisms associated with mine water (from Hallberg and Johnson (2005), Kevin Hallberg, pers. comm.) Meso: mesophiles ($T_{optimum}$ < 40 °C); Mod Thermo: moderate thermophiles ($T_{optimum}$ 40–60 °C); Ext Thermo: extreme thermophiles ($T_{optimum}$ >60 °C); [1]: inferred ability to oxidise minerals (via production of sulphuric acid).

Mineral-degrading Acidophiles	Thermal classification	Phylogenetic Affiliation
Iron-oxidisers		
Leptospirillum ferrooxidans	Meso	Nitrospira
L. ferriphilum	Meso	Nitrospira
L. thermoferrooxidans	Mod Thermo	Nitrospira
"Thiobacillus ferrooxidans" m-1	Meso	β-*Proteobacteria*
"Ferrimicrobium acidiphilum"	Meso	*Actinobacteria*
Ferroplasma acidiphilum	Meso	*Thermoplasmales*
"Fp. acidarmanus"	Meso	*Thermoplasmales*
Sulphur-oxidisers		
Acidithiobacillus thiooxidans	Meso	β/γ-*Proteobacteria*
At. caldus	Mod Thermo	β/γ-*Proteobacteria*
Thiomonas cuprina	Meso	β-*Proteobacteria*
Hydrogenobaculum acidophilum	Mod Thermo	*Aquifacales*[1]
Iron- and sulphur-oxidisers		
Acidithiobacillus ferrooxidans	Meso	β/γ-*Proteobacteria*
Iron-reducers		
Acidiphilium spp.	Meso	α-*Proteobacteria*
Iron-oxidisers/reducers		
Acidimicrobium ferrooxidans	Meso	*Actinobacteria*
Iron-oxidisers/reducers and sulphur-oxidisers		
Sulfobacillus spp.	Meso | Mod Thermo	*Firmicutes*
Heterotrophic acidophiles (non mineral-degrading)		
Acidocella spp.	Meso	α-*Proteobacteria*
Acidisphaera rubrifaciens	Meso	α-*Proteobacteria*
Acidobacterium capsulatum	Meso	Acidobacterium
Acidomonas methanolica	Meso	α-*Proteobacteria*
Alicyclobacillus spp.	Meso	*Firmicutes*
Picrophilus spp.	Mod Thermo	*Thermoplasmales*
Thermoplasma spp.	Mod Thermo	*Thermoplasmales*

Table 4. Results of a microbiological investigation in the Niederschlema/Alberoda/Germany uranium mine (Wolkersdorfer 1996); –: not measured, n.d.: not detectable, w: data measured by Wismut GmbH (microbiological analysis: Dr. M. Graff, Institut für Mikrobiologie der Technischen Universität Braunschweig 1996).

	Shaft 366 b, level -540,	Shaft 366 b, onsetting station level -720	Cross drift Q918, level -540	Surge tank, inlet, level -540
Air temperature, °C	–	13.6	22.8	16.0
Water temperature, °C	31.3^{w}	13.0	24.4	17.5
pH, –	9.45^{w}	8.10	7.33	7.73
Oxygen, mg L^{-1}	2.5	10.0	7.2	9.5
Oxygen, % saturation	33	94	85	99
Redox potential, mV	315^{w}	345	335	365
neutr. thiobacteria (S_0-Ox.), L^{-1}	$4.5 \cdot 10^3$	$1.5 \cdot 10^3$	$4.5 \cdot 10^4$	$2.5 \cdot 10^4$
acid. thiobacteria (S_0-Ox.), L^{-1}	n.d.	n.d.	n.d.	n.d.
acid. iron oxidisers (Fe^{2+}-Ox.), L^{-1}	n.d.	$0.4 \cdot 10^3$	n.d.	$0.4 \cdot 10^3$

Table 5. List of metal sulphides that were described to be oxidised by *At. ferrooxidans* or *At. thiooxidans* (Gottschalk and Buehler 1912; Nordstrom 1977; Lundgren and Silver 1980; Bosecker 1980; all chemical formulas according to Strunz and Nickel 2001).

Mineral	Chemical formula	Mineral	Chemical formula
antimonite	Sb_2S_3	galena	PbS
arsenopyrite	FeAsS	greenockite	CdS
orpiment	As_2S_3	marcasite	FeS_2
bismuthinite	Bi_2S_3	millerite	β-NiS
bornite	Cu_5FeS_4	molybdenite	MoS_2
chalcopyrite	$CuFeS_2$	pentlandite	$(Ni,Fe)_9S_8$
chalcocite	Cu_2S	pyrite	FeS_2
cinnabar	HgS	pyrrhotite	FeS
covellite	CuS	sphalerite	α-ZnS
enargite	Cu_3AsS_4	tetrahedrite	$Cu_3SbS_{3.25}$

4.4.2 Mechanisms of Organism–Mineral Interaction

As discussed earlier, the chemolithotrophic bacteria require sulphur (*At. thiooxidans*, *At. ferrooxidans*) or iron (*At. ferrooxidans*, *Leptospirillium ferrooxidans*) as well as oxygen to produce energy and to maintain their metabolism. They oxidise the disulphides pyrite and marcasite (FeS_2) as well as other sulphide minerals (e.g. PbS, ZnS, CdS) under acidic conditions. Though the bacteria catalyse the chemical reactions, they are not able to initiate chemical reactions that would be thermodynamically impossible (Stumm and Morgan 1996). Other mechanisms catalyzed by bacteria are sulphate and iron reduction, especially in passive treatment systems. So far, not much is known about chemolithotrophic microorganisms in alkaline mine water environments (Hallberg and Johnson 2005). Takai et al. (2001) describe the bacteria *Alkaliphilus transvaalensis* from a borehole in an ultra-deep gold mine near Carletonville/South Africa, which grows over a temperature range of 20–50 °C and a pH-range of 8.5–12.5. Perry et al. (2001) described *Gallionella ferruginea,* which thrives in neutral mine water discharges.

There has been an ongoing discussion how the microorganisms catalyze metal-sulphide oxidation: by direct or indirect mechanisms (e.g. Bennett and Tributsch 1978; Nordstrom and Alpers 1999). It was found that some bacteria are attached to the mineral surface and others are not and therefore, it was concluded that there must be a biotic and an abiotic process involved. In the meantime it seems to be clear that abiotic processes start the reaction chain (Sand et al. 1995; Rawlings 2005). Depending on the type of mineral, two mechanisms are relevant: iron disulphides release ferrous iron because of the oxidation through ferric iron or oxygen whereas other metal sulphides are attacked by the acid in the mine water and release metal and sulphates into solution.

Where direct contact between the mineral and the bacteria has been observed, it seems as if the cell surface is important in this reaction process. This is indicated by the fact that bacteria with mineral contact have other cell surfaces than those without bacteria-solid contacts (Sand et al. 1995 Herbert Jr. 1999) *At. thiooxidans* or *At. ferrooxidans* oxidise the sulphur to sulphate and *At. ferrooxidans* oxidises ferrous iron (Fe^{2+}) to ferric iron (Fe^{3+}). Those two processes also occur without a bacteria-solid contact and inhibit a concentration gradient from the solid into the liquid which even releases more iron and sulphur from the pyrite or marcasite. Finally, the bacteria use the ions in solution to maintain their metabolic processes, according to equations 17 to 19 and through the oxidation of ferrous iron to ferric iron (equation 2).

4.5 The Process Slows Down: Buffering of the Acidity

Besides the minerals and micro-organisms responsible for acid production, there are also minerals that have the ability to consume the protons produced during the above-described processes (eqs. 1–4). These minerals are usually present in the gangue or in the host rock of the raw material deposit. From a chemical point of view, these minerals are proton acceptors; they buffer the acidity and prevent the mine water from becoming too acidic or acidic at all. Buffering reactions control the pH as well as the level of acidity in the mine water. Since the pH, in turn, controls the bioavailability and mobility of most metals, buffering reactions are fundamental for keeping the pH of mine water in an acceptable range.

Carbonate minerals, such as calcite, dolomite, or siderite, are well known to buffer mine water pH, but oxi-hydrate and silicate minerals also buffer pH when carbonates are not present or have been consumed (Table 6; Blowes et al. 1994; Jurjovec et al. 1995). The consumption of acidity by carbonate minerals can be seen in equations 20 and 21:

$$CaCO_3 + H^+ \leftrightarrow Ca^{2+} + HCO_3^- \tag{20}$$

$$CaMg[CO_3]_2 + 2\ H^+ \leftrightarrow Ca^{2+}\ Mg^{2+} + 2\ HCO_3^- \tag{21}$$

Table 6. Minerals that buffer protons produced during disulphide weathering (Blowes et al. 1994, Geller et al. 2000b, Goethite, K-Jarosite, and aluminosilicates from Jurjovec et al. 1995). Buffering ranges of K-Jarosite, and alumosilicates deduced from column experiments. The values are possibly too low as reaction kinetics have not been considered (Paul Younger, pers. comm.).

Mineral	Formula	pH-Buffer Range
calcite	$CaCO_3$	6.5 – 7.5
dolomite	$CaMg[CO_3]_2$	6.5 – 7.5
siderite	$FeCO_3$	4.8 – 6.3
mixed carbonates	$(Ca, Mg, Fe, Mn)CO_3$	4.8 – 6.3
gibbsite	$Al(OH)_3$	4.0 – 4.3
ferrihydrite	$Fe(OH)_3$	< 3.5
goethite	α-FeOOH	1.3 – 1.8
K-jarosite	$KFe_3[(OH)_6\|(SO_4)_2]$	1 – 2 (experimental)
alumosilicates		1 – 2 (experimental)

Other mechanisms that consume acidity are ion-exchange and surface protonation, but they are comparably slow and have limited capacity compared to the mineral dissolution reactions. It must be clear that the buffering minerals do not prevent pyrite oxidation (reactions 1–4). However, if the buffering occurs in the same environment as the pyrite oxidation, virtually all of the Fe^{3+} is precipitated as a hydroxide before it can dissolve more pyrite, which greatly decreases the rate of pyrite oxidation. Alternatively, the buffering can occur down-gradient of the pyrite; this can produce circum-neutral mine water that is enriched in sulphate in the same way as acid mine water (Table 2). Usually, the buffering ranges can be attributed to specific minerals and the mechanism of that process is referred to as sequential pH buffering. These processes were first observed in mine tailings investigations (Blowes et al. 1994), but the pH-ranges of most acid mine waters prove that the same processes are occurring within most mines and in the receiving streams.

Calcite can also react with the Fe^{2+} produced during disulphide oxidation, forming secondary siderite (Blowes et al. 1994):

$$CaCO_3 + Fe^{2+} \leftrightarrow FeCO_3 + Ca^{2+} \tag{22}$$

Once calcite and dolomite are consumed, the pH rapidly drops to be buffered by the next occurring mineral phase in the system (Fig. 6). If all the buffering minerals are consumed, the pH drops below 1 and can reach the extreme pH values reported in section 4.2. Another factor in buffering the acidity are the different kinetic processes involved. Carbonate weathering rates, compared to those of pyrite, biotite, feldspar, or chlorite, are comparatively high. Therefore, as long as carbonates are present, the acidity is quickly buffered. After the carbonates are consumed, the silicates buffer at a lower pH range (Salmon and Malmström 2002).

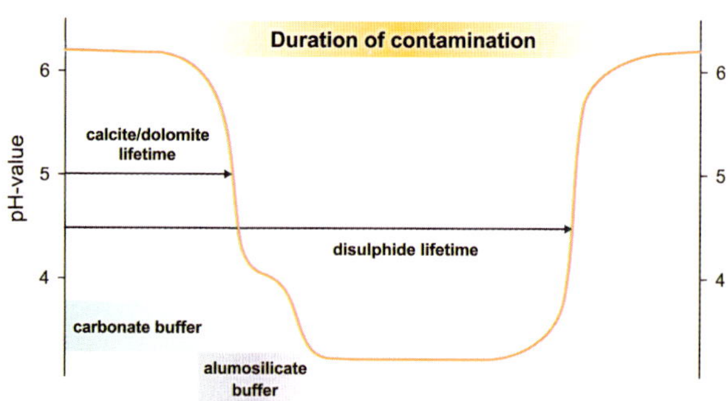

Fig. 6. Potential development of pH of a mine water in a mine with more pyrite than carbonates. The second rise in pH might be the result of different buffering processes and the fact that acidity is not formed any longer (modified after Younger et al. 2002).

In acidic mining lakes, the buffering mechanisms might be slightly different, as the ions in the lake water act as reaction partners for the protons. A low pH buffering range lies between pH 2.6 and 2.9, controlled by hydrogen ions and hydrogen sulphate. A second one ranges from pH 2.9 to 4.3 and is controlled by iron sulphate. Aluminium sulphate buffers the next step, from pH 4.3 to 5.5. Finally, surface exchange of sulphate and basic cations are buffers above pH 5.5 (Geller et al. 2000b; Uhlmann et al. 2004).

4.6 The Process Ends: Mineral Precipitates

One of the predominant features of mine water and one of the reasons why dealing with mine water problems also has an aesthetic aspect are the precipitates associated with acid mine water/metalliferous leachate (Fig. 2, Fig. 7). Precipitation of mineral species or amorphous substances occurs when the relevant solubility products of the species in solution are exceeded (Alpers et al. 1994). Such a situation can arise under several conditions:

- the water evaporates
- reduced species are oxidised
- the pH changes

Fig. 7. Epsomite (Mg $[SO_4]\cdot 7H_2O$) stalactites and stalagmites in the Idrija/Slovenia mercury mine (the stalagmites are about 1 m high). Secondary minerals, such as epsomite, are dissolved during the flooding of the mine and – especially during the first flush – contribute to the acidity and metal content of the mine water.

- the redox potential changes
- flocculation or coagulation occurs
- two waters of different chemical compositions are mixed
- CO_2 is stripped off the water

Most of the secondary minerals in underground mines are iron minerals, as iron is usually the predominant cation in the water around a mine. This can be seen in the list of common secondary minerals in the mining environment (Table 7).

Two different groups of precipitates can be distinguished from each other: the minerals that are usually referred to as "ochre" and the water soluble minerals that form in the open space of a mine, called secondary minerals (Fig. 7, Fig. 8). Most authors discuss them in separate chapters but chemically and physically, the processes forming them are similar and in both cases they are formed by mine water. Therefore, they are described together here. Also, when I write: "The process ends," this refers to a single ion's perspective, as it is dissolved, transported, and then precipitated. Because many of the secondary minerals can easily be dissolved when the water table in an abandoned mine rises, they could then restart their travel through geological space and time.

Secondary minerals can also develop on mine heaps and tailings where they sometimes form low-permeable encrustations (e.g. Rammlmair and Grissemann 2000; Jung 2002; Regenspurg et al. 2005). As has nicely been shown by the Quaking Houses/County Durham incident, destroying that protective cover at mine heaps can lead to local mine water problems that can be difficult to resolve (Kemp and Griffiths 1999). Therefore, old encrustations at mine heaps should be left undisturbed during remediation, if possible, to avoid initiating mine water problems that did not exist before (Stefanie Hurst, pers. comm.).

Fig. 8. Ochre precipitates in a mine water discharge pipe of the abandoned Straßberg/Harz fluorspar mine.

Table 7. Some of the secondary minerals commonly found in the mine water environment and responsible for the first flush and their mineralogical classification; a more comprehensive review of such minerals is given in Alpers et al. (1994) and Nordstrom and Alpers (1999).

Mineral	Chemical formula	Colour
class V Carbonates		
malachite	$Cu_2[(OH)_2\|CO_3]$	green
cerussite	$PbCO_3$	white
hydrozincite	$Zn_5[(OH)_3\|CO_3]_2$	white
class VI Sulphates		
jurbanite	$Al[OH\|SO_4]\cdot 5H_2O$	white
basalumnite	$Al_4[(OH)_{10}\|SO_4]\cdot 5H_2O$	white
melanterite	$Fe[SO_4]\cdot 7H_2O$	blue-green
szomolnokite (schmöllnitzite)	$Fe[SO_4]\cdot H_2O$	white, green
coquimbite	$Fe_2[SO_4]_3\cdot 9H_2O$	purple, white
alunite	$KAl_3[(OH)_6\|(SO_4)_2]$	white, orange
jarosite	$KFe_3[(OH)_6\|(SO_4)_2]$	orange, red
epsomite	$Mg[SO_4]\cdot 7H_2O$	white, orange
anglesite	$Pb[SO_4]$	white
goslarite	$Zn[SO_4]\cdot 7H_2O$	white

The most common precipitates associated with mine water are called "ochre", "yellow boy" or iron oxihydroxides, because they are a mixture of crystallised and amorphous substances with a yellowish to red colour. They can exist as soft gelatinous flocculants, as suspended colloids (as long as the water velocity is high enough), as a medium soft mud, as foam on the water's surface, as coatings on rocks, and as hard, thick encrustations. Though, ironically, those minerals sometimes cause interesting and attractive features in abandoned underground mines or remote areas, the same minerals cause an ugly staining of surface streams or wetlands. Furthermore, the staining covers microorganisms, cause their die back and consequently macroorganisms leave those parts of the streams. Thus, such streams or rivers are often biologically dead for several hundreds of meters and kilometres. It has to be kept in mind that "biologically dead" is only true for the usual macroorganisms that can be found in surface streams with a good ecological status. As discussed earlier, though the biodiversity in mine water affected environments significantly decreases, other micro- and macroorganisms take over the niche and can be very efficient in such extreme environments, as has been verified in heavily polluted river catchments (Ariza 1998; Batty 2005).

"Ochre" mainly consists of the following mineral species: goethite (yellowish brown), lepidocrocite (orange), ferrihydrite (reddish brown), schwertmannite (yellow), jarosite (straw yellow) and, sometimes, also Mn minerals. All of those species are commonly found in acidic environments where iron is abundant as well. They form when Fe^{2+} in the solution is oxidised to Fe^{3+} and finally precipitates as a colloid or a mineral:

$$4\,Fe^{2+} + O_2 + 4\,H^+ \leftrightarrow 4\,Fe^{3+} + 2\,H_2O \tag{23}$$

$$4\,Fe^{3+} + 12\,H_2O \leftrightarrow 12\,H^+ + 4\,Fe(OH)_3\downarrow \tag{24}$$

the sum of those two reactions is:

$$4\,Fe^{2+} + O_2 + 10\,H_2O \leftrightarrow 8\,H^+ + 4\,Fe(OH)_3\downarrow \tag{25}$$

As can be seen from equation 25, this process releases a total of 2 protons per mole of iron (which is the reason, why acidic waters being "treated" in aerobic wetlands often decrease in pH). Those protons, if not buffered by proton-accepting processes, lower the pH until ferric iron no longer precipitates as a hydroxide, which occurs at a pH of around 2.2 (Stumm and Morgan 1996). Predominance diagrams (also called speciation diagrams, E_H-pH-diagrams, and Pourbaix-Diagrams), as shown in Stumm and Morgan (1996) or Appelo and Postma (1994) nicely show the relationship between the iron species and the E_H-pH conditions in natural environments. In Merkel and Planer-Friedrich (2002), the calculation of such diagrams with PHREEQC is explained and a small software tool for simple diagrammatic representations is provided. More sophisticated software tools for constructing predominance diagrams are FactSage (Thermfact/CRCT, Montreal/ Canada; GTT-Technologies, Aachen/Germany) and Thermo-Calc (Thermo-Calc Software, Stockholm/Sweden).

Rüde (2004) observed that during the course of the Rötlbach/Austria, while pH and sulphate concentrations of the water decreased, first jarosite, then schwertmannite, and finally goethite formed encrustations in the stream. Similar conditions can be found in many mining-impacted streams. It appears that the formation of these minerals is controlled by the pH and sulphate concentrations. Williams et al. (2002) showed that the pH of the water can, in fact, be estimated using the colour and spectral reflectance of the minerals that form. As the ochre-minerals precipitate, other metals and semimetals co-precipitate. Again, Rüde (2004) showed that in the Rötlbach, the concentrations of As, Ni, and Pb in the precipitates increased while their concentrations in the water decreased to less than detection limits after approximately 1,200 m. Co-precipitation of Al, Cr, Co, Cu, Pb, Mn, Ni, REE, Sc, U, Y, and Zn similarly occurs in mine waters (Bigham 1994; Fuge et al. 1994; Lottermoser 2003). Stumm and Morgan (1996) and Appelo and Postma (1994) give the following degree of affinities for co-precipitation with Fe- and Mn-oxides:

Fe-oxides: As(III) – Pb ≥ Cu > Zn> Ni ≥ Cd > Co > As(V)
Mn-oxides: As(III) – Pb – Co > Cu > Zn > Ni > Cd > As(V)

Co-precipitation is an important mechanism in the natural attenuation of post-mining landscapes and in mine water treatment. It results from pH and metal dependant surface complexation reactions (Dzombak and Morel 1990) and the zero point of charge (pH_{ZPC}; Stumm and Morgan 1996). Many metals (e.g. Pb, Cu, Mo, U, Sb, Mn, Se) and As co-precipitate with colloids formed by iron or manganese oxihydroxides as well as minerals of the alunite-jarosite-group. As, Sb, and Se tend to precipitate at acidic pH while other elements precipitate at alkaline pH (Alpers et al. 1994; SENES Consultants Limited 1994; Skousen et al. 1998; Webster et al. 1998; Rothenhöfer et al. 2000; Ball and Brix 2000; Noubactep et al. 2002; Tutu et al. 2003; Merkel 2004).

Yelpatyevsky (1995) quantified this effect in an acidic "red stream leaking from [an] adit at [a] cassiterite-sulphide deposit" with a pH of 3.2 in south-east Russia (the Smirnovsky mine in the Rudnaya river basin, Vera Pavlovna Elpatyevskaya, pers. comm.). He observed the following removal factors after a river length of 300 m: As 78%, Pb 45%, Fe 38%, Cd 27%, Zn 10%, Cu 10%, Mn 1.2% (initial element concentrations are given in table 2). In the Italian Libiola Fe-Cu mining area, Dinelli and Tateo (2002) observed that after the precipitation of the Fe-mineral-associations, Al-, and finally Cu-phases formed in the stream.

Besides the minerals that form ochre, many other mineral phases are formed when water enters a mine and oxidises primary minerals within the ore veins or the host rock. Most, but not all, of those minerals are soluble in water: cerussite, malachite, jarosite, goslarite, melanterite, szomolnokite, epsomite, alunite, jurbanite, coquimbite, basalumnite, and anglesite. Other minerals that precipitate from mine water upon oxidation are gibbsite, and Zn-hydroxides. Secondary minerals such as these can commonly be found on the adit and shaft surfaces (because of the oxygen and the humidity within the mine workings) or where the mine water drains into the receiving stream.

The general process forming those minerals results from lowering the ground water table during mining, allowing oxygen and water to contact the primary or ore minerals. Those minerals are oxidised or dissolved. Then the water transports the ions in the direction of the lowest hydraulic pressure, usually towards the mine pumps or the open mine workings. There, the water collects in the sump, spillways, drainage gullies, or puddles within the adits. If the water is flowing slowly through the rock matrix or small fractures, the mine air is able to evaporate the mineralised water and carbonate, hydroxycarbonate, sulphate, hydroxisulphate, or oxihydroxides minerals form on the surfaces of the mine workings. Because the metal ions forming those minerals are proton acceptors, they store the protons and consequently the acidity. As long as those minerals are in contact with the mine air and the overall situation does not change significantly, this acidity will be stored in the minerals *ad infinitum*. Yet, after mining ceases and the mine is abandoned, if it is inundated, the water-soluble carbonates, hydroxycarbonates, sulphates, and hydoxisulphates will be dissolved by the mine water. Thus, the mine water, which commonly is already acidic, gets enriched in the ions constituting the secondary minerals (section 4.3); this causes what is known as the first flush (section 5.4).

4.7 Predicting the Post-Mining Chemical Composition

Predicting the post-mining chemical composition of the mine water is an integral part of nearly all mine closure procedures. The procedures commonly used are geochemical modelling, empirical conclusions (see chapter 8), acid-base-accounting (ABA), calculation of net neutralization potential (NNP), or static and kinetic leach tests. Good descriptions of these procedures and references to other methods of how to predict post-mining water chemistry can be found in Henton (1976), Sobek et al. (1978), MEND [Mine Environment Neutral Drainage Program] (1991), Werner et al. (1999), Perry (2000), Prediction Workgroup of the Acid Drainage Technology Initiative (2000), Younger (2000d), Younger and Robins (2002), and Kruse and Younger (2005). Younger and Sapsford (2006) discuss several of the shortfalls of these techniques and conclude that a good conceptual model and site specific investigations are more important than the before-mentioned tests (see section 4.5).

The differences in mineral weathering rates are a potential pitfall for researchers and consultants and can lead to incorrect predictions of future pH values. Many mine water investigations are based on column or batch experiments and last several days, weeks, or months (the duration of a master's or diploma thesis). Seldom do such investigations last more than a year. In the initial phase of those experiments, the carbonates quickly dissolve and buffer the acidity produced by disulphide weathering in the column, so that the leachate has a pH between 6.5 and 7.5. Some researchers therefore tend to conclude that the final mine water discharge will be circum-neutral and negative environmental impacts should not be expected. If only those researchers had employed a PhD-student who would have had *two years* to conduct his or her experiments! They would then see the total depletion of the carbonates and watch the pH drop to acidic conditions. They would conclude that the final pH of the mine water would be acidic – which is true for the significant time period, though eventually the disulphides will be consumed and the pH will rise to circum-neutral conditions again. This example illustrates that all mine water investigations must include site investigations, geochemical modelling, acid-base accounting and, if possible, any other method mentioned in Prediction Workgroup of the Acid Drainage Technology Initiative (2000) and The Pennsylvanian Department of Environmental Protection (1998). Younger and Sapsford (2006), after critically reviewing the relevant literature, conclude "that there has been an excessive concentration on simple pollution potential assay tests (acid-base accounting, humidity cell tests) at the expense of rational assessments of contaminant transport pathways". Furthermore, it must be kept in mind that most column tests are not standardised (U.S. Environmental Protection Agency 2003) – applying the results from such tests to the real world is not an easy task.

A key feature of many abandoned mines is that the concentration of contaminants increase shortly after the mine water discharges to the receiving water courses and then decreases thereafter. This typical scenario is called the first flush (see section 5.4). Knowledge about the first flush, its peak concentrations and the pattern of how it decreases, are essential for the planning of remediation technologies, especially mine water treatment plants. Many active treatment plants work economically only with high concentrations of metals, while passive treatment plants require relatively low concentrations and loads. If, therefore, the duration of the first flush can be predicted to a reasonable degree, the time when an operator can change from active to passive treatment can be estimated (ERMITE Consortium et al. 2004).

This was done at the German Straßberg/Harz mine (unpublished data). All available data, such as weekly flow, iron measurements, and time series analyses of up to 10 years were used (Fig. 9). The left graph shows the measured and predicted decrease in the iron concentration and the right one shows the expected loads until the year 2015. The following commercially available software was used: statistical analyses (SPSS), correlation analyses (TableCurve 2D), spreadsheet software, and scientific graphic software (Sigma Plot 2D). Similarly, Burke and Younger (2000) predicted the first flush scenario for the South Yorkshire Coalfield by applying expert knowledge and the numerical code GRAM (Groundwater Rebound in Abandoned Mine Workings). Another method was described by Younger (2000d), who analysed more than 80 British mine water discharges and came up with a flow chart to predict the medium and long term mine water qualities with respect to

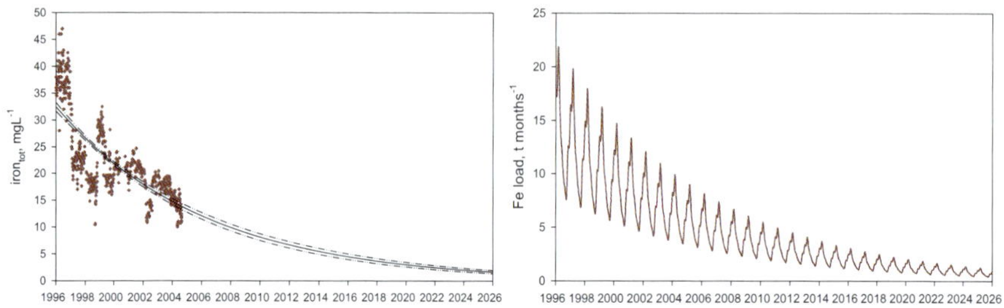

Fig. 9. Prediction of future Fe_{tot} concentrations and loads in the Straßberg/Germany Fluorspar mine until the year 2026 by using past data and a black-box model assumption (raw data kindly provided by BST Mansfeld GmbH and GVV).

iron concentrations. However, the latter flow chart is coal mine drainage biased and focuses on the United Kingdom, and has not been tested for many other areas of the world, except in French coalfields (Paul Younger, pers. comm.). Depending on the total sulphur content of the mined strata, which ranges between below 1 and 5 weight percent, Younger (2002d) estimates total peak iron concentrations of 0.01–1,500 mg L^{-1} (Table 10). However, as only the disulphide content of a stratum is important for the release of acidity and consequently iron, it is more appropriate to use the total disulphide content instead of the total sulphur content of a stratum.

4.8 Radionuclides in non-Uranium Mines

Enriched radionuclide concentrations in mine waters from uranium mines are a frequent occurrence (Merkel et al. [1997] and many others), and the radionuclides commonly found in mine waters are thorium, radon, radium, and uranium. In non-uranium mines, such radionuclides are usually not considered by researchers or authorities, though the phenomenon of naturally occurring radioactive materials (NORM) has been known for a long time. Water from hot springs and effluent from water separators from gas and oil wells commonly contain such contaminants (e.g. Sohrabi 1998; Chałupnik and Wysocka 2000; Gillmore et al. 2002; Dowdall et al. 2004; Skowronek et al. 2004). This does not imply that non-uranium mine waters are of no danger to human beings, it just means that there is very often a lack of awareness concerning the radioactivity of such mine waters, which Paschoa (1998) can also see in other industrial areas. Yet, the problem should not be overemphasised as the concentrations of the radionuclides – from a radiotoxic and very often also from a chemotoxic point of view – are usually very low, though several sites are known with elevated radionuclide concentrations (Sohrabi 1998; Gillmore et al. 2002; Schmid and Wiegand 2003). Besides the

Table 8. Summary of classification of natural radiation areas according to Sohrabi (1998). LLNRA: low or normal level natural radiation area, MLNRA: medium level natural radiation area, HLNRA: high level natural radiation area, VHLNRA: very high level natural radiation area, UNSCERA: United Nations Scientific Committee on the effects of Atomic Radiation, ICRP: International Commission on Radiological Protection.

Classification	Criteria: potential public exposure	Remedial action
LLNRA	≈ 5 mSv y^{-1} (twice natural average global effective dose, UNSCEAR)	None
MLNRA	≈ 5 – 20 mSv y^{-1} (present ICRP limits)	Within 5 years
HLNRA	≈ 20 – 50 mSv y^{-1} (former ICRP limits)	Within 5 years
VHLNRA	≥ 50 mSv y^{-1}	Urgent, possible evacuation

radionuclides in the water phase, there is the issue of radon in the mine air, in the vicinity of ventilation shafts, and mine adits ('mine gases'), sometimes referred to as elevated natural radiation areas (ENRA; Sohrabi 1998). Radon is known to cause lung cancer, originally called the *Schneeberger Lungenkrankheit* (Schneeberg lung disease; Enderle and Friedrich 1995; Boice et al. 2003), and therefore the public must be protected against elevated concentrations of radon in their environment (Gillmore et al. 2002). However, radon is also reputed to have a positive effect, and there are numerous radon spas throughout the world (section 11.7); at which concentration radon becomes toxic is still a matter of discussion, even if many countries, for obvious reasons, have set up legal standards for radon concentrations in houses or at the working place (e.g. U.S. Environmental Protection Agency 2000; 2001/928/EURATOM; Austrian Dinking Water Guidelines). Finally, elevated radon concentrations might also be expected in zones with fissures in volcanic rocks (Davis 1983; King et al. 1996).

According to Sohrabi (1998), who gives an excellent summary of areas with elevated levels of natural radioactivity, NORM should be classified into 4 categories and depending on the mean annual potential public exposure, no remedial action, remedial action within 5 years or urgent remedial action should be taken (Table 8). This is just one of many such classifications known, but it provides clear guidelines of potential actions that should be taken in areas with elevated natural radioactivity.

Recent publications have addressed radionuclides in the Polish and German Ruhr coal mining districts (Lebecka et al. 1997; Chałupnik and Wysocka

2000; Chałupnik et al. 2001; Schmid and Wiegand 2003) and elevated radon concentrations in British tin-copper-uranium mines (Gillmore et al. 2002). In the Silesian coal mining area, two types of mine waters exist: water with a higher mineralization and water with a comparatively lower mineralization (Lebecka et al. 1994, 1997). Both types contain elevated concentrations of ^{226}Ra and ^{228}Ra, with typical ^{226}Ra activities of $62.8 \cdot 10^3$ Bq L^{-1} for the first and $3.4 \cdot 10^3$ Bq L^{-1} for the second. Concurrently, the more highly mineralized water contains high Ba concentrations with typical values around 1,500 mg L^{-1}. To prevent the precipitation of (Ba, Ra)[SO_4] in pipelines or river banks (Pluta 2000), the mine water is chemically treated with different methods (Lebecka et al. 1994; Chałupnik and Wysocka 2000). Elevated radioactivity is also known to occur in the flooded Stockheim coal mine in Franconia/Germany (Ebner and Schwarz 1973; Ziehr 1983).

Gillmore et al. (2002) present an interesting study of extreme radon concentrations in abandoned mines in Devon and Cornwall/UK, though these mines were also mined for uranium, and so, strictly speaking, do not belong in this section. Yet, their conclusions concerning radon emanations from those abandoned mines are also relevant for non-uranium mines. For the South Terras mine in Cornwall, which was abandoned and flooded in 1928, they reported a radon concentration of $3.9 \cdot 10^6$ Bq m^{-3} at a distance of 70 m from the mine entrance. Within the mine, they measured doses for a duration of one hour which equalled the 18 year personal dose for a member of the public. Everybody working in even long abandoned mines where uranium was a by-product must be aware that extreme radon concentrations might be expected. High radon concentrations were used by the Soviet occupiers of Saxony to locate potential uranium sources in the silver/lead mines of the Erzgebirge/Germany. They knew of the radiological investigations in the early 20th century by Schiffner and Genser (Schiffner 1908; Schiffner and Weidig 1909; Schiffner et al. 1911; Weidig 1912) and used those data to concentrate their exploration at localities with the highest radioactivity and temperature anomalies in the working and abandoned polymetallic mines (Table 9). It should be noted that the water of the Hindenburg and Bismarck springs was used in the Bad Schlema Spa, which was the world's most radioactive spa water at the time (Genser 1933).

Potential dangers of radionuclides can be seen for people living very close to places where radon emanates from flooded underground mines, for researchers working in abandoned mines, for ecologists taking care of bats, for private archaeologists, and for mineral collectors. Schmid and Wiegand (2003) calculated the doses for the public along the German river Lippe between the Auguste Victoria mine and its confluent with the river Rhine for external exposure and ingestion of contaminated, Ra-barite rich soil. External exposure, depending on their scenario, ranged between 0.03 mSv a^{-1} and

Table 9. Radioactivities above 100 Bq L^{-1} of mine waters in the Markus-Semmler adit near Schlema, Erzgebirge/Germany before large scale uranium mining started. Localities ordered by their activities (Wolkersdorfer 1996 after Schiffner 1908; Schiffner and Weidig 1909; Schiffner et al. 1911; Weidig 1912). [IF]: Institute Fresenius (H. Zerbe, pers. comm.) analysed an activity of 246,500 Bq L^{-1} in the 1930s. The German names are typical old fashioned miner's names. Ort: face; Flügel: wing (of a main gallery); Spat: ore vein in easterly direction; Morgengang: ore vein in north-easterly direction; Flachen: ore vein in south-easterly direction; Stehend: ore vein in northerly direction.

Date	Location	Radioactivity, Bq L^{-1}
1932	Radiumort, Hindenburg Spring[IF]	181,818
1932	Radiumort, Bismarck Spring	40,404
1932	Friedrichflügel	10,774
1932	Heinrichflügel, Drill Hole III	10,101
7.2.1911	Radiumort, 25 m inside Markus-Semmler-Adit	9,400 – 11,000
1932	Radiumort, Drill Hole I	8,350
19.10.1911	Radiumort, 25 m inside Markus-Semmler-Adit	7,886 (n = 7)
19.9.1911	Radiumort, drainage into Markus-Semmler-Adit	5,375 (n = 16)
23.11.1910	Radiumort, 27 m inside Markus-Semmler-Adit	4,539
1932	Radiumort, Radium Shaft	4,444
1932	Radiumort, Drill Hole II	4,108
1932	Radiumort, B-Flügel	4,040
1932	Heinrichflügel, Heinrich Shaft	3,771
19.10.1911	Radiumort, 35 m inside Markus-Semmler-Adit	2,979 (n = 5)
1932	Wolfgangflügel	1,886
1932	Gleesbergflügel	1,886
1932	Johannesflügel	1,347
23.3.1909	Maximilian Spat	1,048
7.01.1909	Eva Spat	640
1932	Jung-König-David-Flügel	337
27.3.1909	Fleischer Morgengang	226
30.3.1909	Neuglück Flachen	218
30.3.1909	Quergestein close to Gottes Schickung	191
30.3.1909	Türck Flachen	121
30.3.1909	unnamed Spat at the crossing with Glück Flachen	110

2.4 mSv a^{-1}, while ingestion of the sediments could result in an additional 0.01–15.5 mSv a^{-1} (this number largely depends on several parameters, including the isotopic composition of the sediment). They conclude that underground precipitation of the radionuclides would effectively prevent the drainage of radium into the anthroposphere.

Another very special case are the South African gold mines in the Witwatersrand Basin, where elevated concentrations of uranium are known to occur in the ore body as a by-product within the carbon leaders (Pohl 2005). Pollution comes from the drained mine water as well as from the seepage water of the tailings dams and can locally reach high concentrations of uranium (Coetzee 1995; Tutu et al. 2003), which are sometimes treated by bioremediation (Weiersbye and Witkowski 2003). Wendel (1998) describes the different sources of radioactive mine waters in South Africa and recommends developing cost effective prevention and remediation methods based on further investigations. Because more and more mines in the Rand will be closed during the next decades and their mine water will drain into the surrounding areas, there is a real need for the South African mining industry and the regulators to invest more research on NORM for those cases.

None of the authors gives a general recommendation of how to prevent elevated concentrations of radionuclides in the anthroposphere. Yet, the Polish experiences show that high radionuclide concentrations can be prevented in non-uranium mines using underground treatment facilities. Furthermore, it is possible to diminish the mine water's radium concentration by active treatment with barium chloride (Skeaff and Campbell 1980; SENES Consultants Limited 1994). In most of the other cases, a thorough monitoring of the radon exhalation in or at abandoned mines in which uranium or thorium minerals were found is possibly the only precautionary measure that can be taken. Interestingly, the U.S. Environmental Protection Agency (2000) clarifies that only incremental radioactivity associated with the abandoned mine must be considered when speaking about regulated sites. Elevated background radioactivity due to geological reasons, as occurs in many granitic settings, must be subtracted from the total radioactivity. However, it is known from many localities that even if the radon concentrations outside the abandoned mine workings are less than the national limits, closed rooms, especially cellars, can be contaminated with radon and therefore have to be treated in a special way. From a water managerial point of view, if uranium or thorium minerals are known to occur in the mine or its vicinity, uranium and radium levels should be measured in the mine water and radon measured in the mine air.

5 Insights into Mine Closure

5.1 Reasons for Mine Closure

Mining is an economic activity that provides a growing world population with raw materials such as metals, fuels, and building materials. Depending on the available technology and the mineral mix to be extracted, the lifetime of a mine might last for several years or millennia, as in the case of the Tyrolean Schwaz silver-dolomite mines, the Salzburg Mitterberg silver-uranium deposit, or the Spanish Rio Tinto mining area (Zschocke et al. 1932; Salkield 1987; Ammann 1990; Günther et al. ca. 1993; Leblanc et al. 2000). However, eventually, all mines must close, and nowadays closure procedures have to meet particular standards that have been established by the various governments and regulatory agencies. These environmental requirements are usually drawn up by the authorities based on expert reports or on environmental impact assessments, in contrast to already abandoned mines that usually did not undergo such procedures. Nevertheless, in both cases, the available remediation options and the time scales or financial efforts necessary to remediate these sites, which cover about 240,000 km^2 of our planet (Furrer et al. 2002), are very similar.

The reasons for mine closure that Lässl listed in 1556 in his *Schwazer Bergbuch* (Lässl 1556; Winkelmann 1956) have not changed considerably. He stated that "four things ruin a mine": war (and accidents), diseases, inflation (economy), and listlessness. In addition, two more reasons now exist, of which the first was known by the miners of that time (Agricola 1556), but never was a matter of real concern: environmental factors and deposit depletion.

5.2 Post-Mining Sustainability of Mine Sites

The various economic and financial, social and community, legal, political and psychosocial, environmental, technological aspects of mine closure should be addressed on a site or catchment scale and resolved by involving all the relevant stakeholders, and keeping in close communication with all stakeholders. Responsible mine closure, as outlined by Sheldon et al. (2002), should guarantee that plant and equipment (if not an industrial heritage) as well as hazardous substances are removed, the used land is reclaimed, and all reasonable measures are taken to exclude potential ground- or surface water pollution. This does not only apply to the mine site but to all installations that were used by the mine operator. Too many sites are known worldwide where the mine operator just switched off the lights and left the mine site without any after care. In this age of sustainability, mine operators have to ensure that they leave their mine sites as environmentally sound as possible.

Sustainability as a technical term was introduced by the Saxonian Oberberghauptmann (senior chief-inspector of mines) Hannß Carl von Carlowitz. Like Georgius Agricola, he came from the German Erzgebirge and in 1713, he published *Sylvicultura Oeconomica,* where he first described sustainability in forestry (Carlowitz 1713). Since then, forests in Germany have generally been maintained in a sustainable manner, which means that no more trees are cut than are growing, or the land is reforested at the same time. In the 19th Century, the principles of sustainable forestry became the method of choice in nearly all developed countries of the world. Because sustainability is so important in today's mining and mine closure processes, von Carlowitz' first mentioning of the word shall be cited here. Interestingly, it stands in a direct context to mining (the words are written exactly as they appeared in the original publication):

> Aber da der unterste Theil der Erden sich an Ertzen durch so viel Mühe und Unkosten hat offenbahr machen lassen, da will nun Mangel vorfallen an Holtz und Kohlen dieselbe gut zu machen; Wird derhalben die größte Kunst, Wissenschaft, Fleiß, und Einrichtung hiesiger Lande darinnen beruhen, wie eine sothane *Conservation* und Anbau des Holtzes anzustellen, daß es eine *continuier*liche beständige und nachhaltende Nutzung gebe, weiln es eine unentberliche Sache ist, ohne welche das Land in seinem Esse nicht bleiben mag. Denn gleich wie andre Länder und Königreiche, mit Getreide, Viehe, Fischereyen, Schiffarthen, und andern von GOtt gesegnet seyn, und dadurch erhalten werden; also ist es allhier das Holtz, mit welchem das edle Kleinod dieser Lande der Berg-Bau nehmlich erhalten und Ertze zu gut gemacht, und auch zu anderer Nothdurft gebraucht wird. (Hannß Carl von Carlowitz, *Sylvicultura Oeconomica*, Capter VII. Von Nothwendigkeit und Nutzen des Holtz-Anbaues, p. 105–106; Leipzig 1713).

Translated, this means:

> Now that the ores in the lowermost part of the earth, through effort and expenses, were revealed, a shortage of wood and coal for their beneficiation occurs. Thus, the greatest skills of art, science, diligence, and institutions of the local states will examine how a conservation and cultivation of wood shall be done so that a continuous, lasting, and sustainable use is given, because it is an indispensable thing, without which this state will not continue to exist. Since other countries and kingdoms are blessed by God with crops, cattle, fishing, ships, and others, and survive because of that; so, here it is wood that is the precious jewel of these states: it allows mining to stay alive and ore to be beneficiated, and also used for other affairs.

Since the Berlin round table on Mining and the Environment (United Nations Department of Technical Co-operation for Development and Development Policy Forum of the German Foundation for International Development 1992, see appendix III) and the UNCED conference in Rio de Janeiro (1992) which resulted in the Agenda 21 (United Nations Conference on Environment and Development [UNCED] 1992), sustainability and environmental sound mining also became crucial for mine operators. Furthermore, these principles are a fundamental part of the UNDESA and UNEP guidelines for environmentally safe mining and post-mining procedures (United Nations Environment Programme [UNEP] 1994). A two year project called "Mining, Minerals and Sustainable Development (MMSD) Project" focused on many aspects of sustainable mining, with a critical view on the mining companies, but although the project produced a lot of material on many different aspects of mining, mine water issues were not covered in great detail (Mining Minerals and Sustainable Development [MMSD] Project 2002).

Post-mining sustainability regarding mine water means that mine water quality after mining will reach a good ecological status and that the water quality can be sustained without further input of energy or financial subsidy. This does not only apply to the mine water *sensu strictu* but also to the surface and ground waters that could be affected by the closed mine. This is true for both the chemical constituents and suspended solids, which in some areas of the world account for a large part of the mining-impacted pollution, as in the Indonesian Grasberg-Ertsberg (Lottermoser 2003), the Papua New Guinea Ok Tedi (Miller et al. 1990; Hettler 1995; Lottermoser 2003), or Misima (Jones and Ellis 1995), the Tasmanian Mount Lyell mines (Harries 1990; Sustainable Development Advisory Council and Copper Mines of Tasmania 1995; Featherstone and O'Grady 1997), or the sludge from mine water treatment operations (Fig. 10). It is difficult to know which is worse: precipitated solids or acid mine drainage.

Fig. 10. Ore processing sludge from the Alegria mine, Minas Gerais/Brazil.

MPC, the Mineral Policy Center, an NGO based in the USA, called for a complete ban on submarine tailing disposals in 1999. Indeed, submarine tailing disposal are usually not permitted in the USA. One exception are the Alaskan A-J Kensington, and Quartz Hill mines where exceptional circumstances had to be considered (U.S. Environmental Protection Agency 2003). In Indonesia, submarine tailing disposal is commonly used due to limited land space and the potential for earthquakes that might cause disastrous tailing dam failures (Natural Resources Management Unit Division "Protection of the Environment and Natural Resources"1999). Those cases show that submarine tailing disposal, compared to alternative options, might be comparatively harmless, and from a water managerial point of view is usually of no great concern (Ellis et al. 1995). Yet, submarine tailings disposal needs all the environmental precautions that are necessary in a up-to-date mining operation.

During the last couple of decades, the need to protect the environment and the need for standards regarding mine closure became obvious. Environmental protection groups, international organizations, regulators, NGOs, consultants, and the mining industry itself were all considering appropriate procedures to close a mine. Furthermore, while conducting several mine closures, there was a general consensus that since post-mining remediation can be extremely expensive, measures should be taken to prepare for sustainable mine closure during the active mining process. As a result, nowadays, most mining operations have to provide mine closure plans at a very early stage of the environmental assessment plan, before being given permission to mine. Concerning water quality, those closure plans must consider very long time scales as the longevity and potential changes in mine water quality must be properly investigated. This leads to the demand for "Post-mining Sustainable Land Use Planning" instead of "Closure Plans", which assumes that regulators will accept a sustainable land use other than the pre-mining conditions (Robertson and Shaw 1999). Within such plans, it could well be that some parts of an abandoned mine have to be managed (for example slopes of tailings dams or open pits), while others would be sustainable. A sound understanding of both the physico-chemical aspects that could occur after mine closure and the mass transport within the flooded underground mine is necessary for post-mining sustainable land use planning. Therefore, post-mining sustainability requires many of the methods needed for acid mine drainage prediction. It is not always possible to predict whether a self sustainable state will be reached after mine closure. Also, it is sometimes unclear what self sustainability means: many long abandoned mining areas, such as the Rio Tinto area, are arguably self sustainable, but if a mine operator proposed to leave a mine site in this condition, he would never get a permit for mine closure by the authorities. Similar conditions, from a legal point of view, occur at the German Freiberg/Saxony Rothschönberg or the Burgfey/Eifel (Fig. 11) drainage adits, which are draining polluted water (Zoumis et al. 1999; Mair 2002; Weyer 2003). In all three cases, natural attenuation creates sustainable conditions (see section 11.5 for a description of "natural attenuation").

Using mine water in an environmentally safe manner after mining ceases is considered an excellent example of post-mining sustainability (see section 11.7). Sustainable uses of mine water include drinking water (Burbey et al. 2000; Teaf et al. 2006), spa water (e.g. Bad Suderode, Feengrotten Saalfeld, Alexisbad/Harz Mountains, Bad Gastein/Austria), turbine water (e.g. Max-Braun-Adit, Biberwier/Tyrol, Fig. 12), and recreational purposes (e.g. Passagem de Mariana Gold mine, Minas Gerais, Brazil; Banyan Tree Phuket, Thailand). Many mine sites are examples of excellent post-mining sustainability (Fernández Rubio 2004). Such sustainable sites are not always as spectacular as the Cornish Eden Project (Macadam and Shail 1999; Harris 2004), the Swedish Dalhalla Stone Quarry (Mattsson and Stenbacka 2004), the Yorkshire Rother Valley Country Park (Brook and Cresswell 1989), the German EuroSpeedway Lausitz (Lusatia), the German Rammelsberg World Heritage lead-zinc-mine (Roseneck 2001; World Heritage site 623), or the Japanese Iwami Ginzan World Heritage silver mines (World Heritage site 1246).

One of the world's nicest health resorts is the Banyan Tree Phuket resort at Laguna Phuket in Thailand, which was not affected by the December 2004 Indian Ocean tsunami because it was situated on the lee side of the wave. Tin dredging in the Ohang-Nga Bay and Bang-Tao Bay started in 1907, and led to the mangrove swamp being dredged to a depth of 30 m, resulting in the destruction of the mangrove environment. Even six years after an area was dredged, the natural mangrove habit did not recover at all, and sediment plumes moving into the open sea subjected off-shore coral reefs to siltation

Fig. 11. The portal of the 7.5 km long German Burgfey/Eifel mine water adit discharging 15–19 m^3 min^{-1} of polluted mine water of the abandoned Mechernich mining area into the Veybach stream.

Fig. 12. Turbine house of the Austrian Biberwier/Tyrol hydropower plant using mine water of the Max-Braun-Adit located in the Silberleithe hills that can be seen in the background of the photograph. The first turbine to produce energy was installed in 1904; the currently operating turbine was installed in 1936.

and devastation (Aranyakanon and Yongsakul 1981; Chansang 1988). Due to the large tin reserves in the Ko Phuket area, a tin smelter, accounting for 10% of the world's tin production, was erected. Its proximity to a reef, called the Tin Smelter Reef, was the reason for several biological studies, which revealed a good adaptation of the corals to the high metal contents in the water (Brown and Holley 1982; Scoffin and Bradshaw 2000). Because of the increasing conflict between the interests of the tourist industry and the shallow and coastal tin dredging activities, as well as low tin prices, mining ceased in 1985 (Carlin 1999). Chansang (1988) concluded that even after the dredging operation had ceased, the "problem of effects, caused by increases in suspended solids and turbidity will remain". For a long time during and after the active dredging operations, an intense, well-documented monitoring of the area was conducted by the Phuket Marine Biological Research Centre. In 1977, the United Nations Development Plan stated that the area is a toxic wasteland too severely damaged to support vegetation and consequently "too environmentally damaged to have any development potential". Nevertheless, two years later, the Tourism Authority of Thailand wrote: "although the damage was probably beyond repair, effort should not be spared to try to restore the environment". A $200 million (U.S.) restoration programme was initiated, starting with importing fertilised topsoil, which was used for planting casuarinas, palm, fruit and flowering trees as well as other vegetation. This measure successfully stopped the erosion of the abandoned mine and tin smelter land. At the same time, large quantities of contaminated soil were removed, pit lakes were flooded with fresh and salt water, and environmental impact studies were undertaken for new hotel resorts. Still, previous offshore tin mining in Bangtao Bay continues to affect the region as millions of Baht have to be used to replace the sand swept away into the holes on the ocean floor left by the dredging operations (adopted from Rung-Aroon Wichitwatee, Public Relations & Advertising Manager, Laguna Phuket Media Contacts).

Miller et al. (2005) investigated mine and mine treatment discharges for their potential use in aquaculture in West Virginia, where, especially in the southern parts, mine water is also used as drinking water with little or no treatment. They concluded that mine waters that meet the standards for fishing waters might be used for aquaculture as long as the CO_2 levels are not too high. Furthermore, abandoned treatment facilities might be converted to aquaculture, as shown by the Guyses Run acid mine drainage plant, which started operation in 2001. In 2004, six aquaculture facilities produced salmonids in Pennsylvania, Maryland, and West Virginia/USA. Minnesota had more than 20 abandoned mine sites being used for aquaculture, but several of them had to be closed as the increase of nutrients and turbidity affected the ecological status of the watersheds (Axler et al. 1998).

The Felsendome Rabenstein near Chemnitz/Saxony is a visitor's mine in which the two lowermost levels have been flooded since the beginning of the 20th century (see section 12.10). It has been a local attraction since 1936. During the last couple of years, the site was converted into an attractive recreation area, and the underwater part is now being used by divers. The "Alte Kalk-Bergwerk" limestone mine in Miltitz/Saxony has been used for various purposes between 1571 and 1967, and diving in that abandoned and flooded, up to 65 m deep underground mine, started in 1997 (Udo Krause, Dresden, pers. comm.). Yet, compared to the flooded Bonne Terre/USA lead mine, the world's largest fresh water diving resort, 100 km south of St Louis, Missouri, both of those mines are relatively small. A very economically working visitor's mine is the Schwaz/Austria visitor mine with more than 100,000 visitors annually (Palme and Ingenhaeff-Berenkamp 1993; Palme et al. 2002).

Inert material as well as radioactive waste is commonly disposed of underground as a post-mining use, at sites where the ground water around the mine will not be negatively affected. Usually, the substances stored in abandoned underground mines are inert materials, such as fly ash or residues from different industrial activities. Radioactive waste disposal, though it is a post-mining use, cannot be listed under the definition given at the beginning of this section, because it requires maintenance.

Underground mines are also used for geothermal energy production. The potential of this was first investigated back in the 1970s, though that investigation in Montana/USA yielded only seven anomalies with discharge temperatures between 5 and 12 °C (Lawson and Sonderegger 1978). In Canada's Nova Scotia, water in the Springhill coal mines with a temperature of 18 °C is used for geothermal energy production (Jessop et al. 1995; Jessop 1998). Furthermore, Saxony/Germany is currently investigating the potential for geothermal energy from abandoned underground mines (Workshop Geothermie Bad Schlema 2004), and a EU project called "minewater project" investigates potential energy sources from mine water in the Netherlands (Heerlen) and Scotland (Midlothian). Another geothermal project using warm mine water has been initiated in the Aachen/Germany coal mining area (Clauser et al. 2005). Other mines where mine water is used for heating purposes include Park Hills/USA, Follda/Norway, Shettleston/Scotland, and Ochil View/Scotland (Banks et al. 2003). Future geothermal projects at mine sites with elevated water temperatures could reduce CO_2 emissions into the atmosphere, though their potential for that purpose has not yet been fully developed. Finally, modern heat pump technology has eliminated the need for the water to be warm; Watzlaf and Ackman (2006) indicate that annual heating and cooling costs could be reduced by 67 and 50% respectively in the Pittsburgh/Pennsylvania/USA area.

A residential area for prosperous inhabitants of Belo Horizonte/Minas Gerais was constructed directly on the shores of the flooded Brazilian Águas Claras mine pit lake. Most pit lakes within the Quadrilátero Ferrífero (Iron Quadrangle) in Minas Gerais can be seen as environmentally friendly from a water quality perspective, because the abandoned iron mines are usually situated within itabiritic rocks that do not generate acid mine drainage. Several kilometres away, the abandoned Mangabeiras iron mine was transformed into a public park, which is now the biggest green area within the borders of Belo Horizonte (Grandchamp et al. 2001; Amorim et al. 2001).

One of the most spectacular European open-cast mine regeneration projects is the Cornish Eden Project. It is an excellent example of co-operation and an ecologically sound after use of a mine site. Cornwall contains 80 km^2 of active and abandoned china clay mine workings and numerous abandoned tin mines (Younger and LaPierre 2000). Many of the abandoned mine sites are of public concern because the metal loads pollute surface streams. Once, the Bodelva Mine was a prosperous china clay mine in the central part of Cornwall, and extracted the clay for about 150 years. A total of $7 \cdot 10^6$ m^3 of material was removed from the open pit leaving behind a 30–70 m deep hole. In 2001, the Eden Project began and the remediation and dewatering work of three years could be presented to the public. Today, Rio Tinto, Anglo America, MIRO, Eden Project and English Nature are working together in a collaboration to establish an educational centre for post-mine regeneration at this site (Macadam and Shail 1999; Harris 2004).

If mine water is polluted, a sustainable treatment scheme has to be undertaken to ensure that the environment is protected. The European Union PIRAMID R&D project investigated the sustainable installation of passive mine water treatment schemes (PIRAMID Consortium 2003), and similar studies have been conducted elsewhere. The use of polluted mine water for irrigation, e.g. for sugar beans and wheat, has been investigated by several authors (e.g. Du Plessis 1983; Voznjuk and Gorshkov 1983; Annandale 1998; Annandale et al. 1999, 2001; Jovanovic et al. 2001; Annandale et al. 2002; Jiries et al. 2004). Yet, the long term effects of such irrigation on ground or surface water quality have not been studied in detail.

In the case of the German Rammelsberg lead zinc mine, the mine operator presented a closure plan to the authorities that also included the treatment of the acidic mine water. It was unquestionable that the mine water had to be treated for a given time period but the requisite time period could not be agreed to by the authorities and the mine operator. Therefore, the case went to court and the court ruled that the mine water has to be treated by the mine operator until the quality reaches the limits given by the authorities – if necessary, *ad infinitum* (Anonymous 1994, 1995). This case clearly proves how important post-mining water quality is regarded by the authorities.

5.3 Previous Mine Flooding Studies

While predicting flooding is highly uncertain, the prediction of mine water quality and mine water stratification is even more challenging. Currently, no case is known where the stratification of a mine was predicted precisely and where remediation methods based on stratification predictions were successful. The most comprehensive study of mine water stratification in a single mine known so far was conducted at the Niederschlema/Alberoda/ Germany uranium mine (Wolkersdorfer 1996). Between June 1992 and December 1994, some 115 depth profiles in seven underground shafts were measured and the evolution of stratification in shafts was observed. A thorough investigation of flooded mine shafts on a regional scale was conducted by Coldewey et al. (1999), who studied all accessible flooded hard coal shafts in the German Ruhr Area, summing up to 19 shafts. Another comprehensive study of ten flooded underground shafts was conducted by Ladwig et al. (1984) in Eastern Pennsylvania/USA. Playton et al. (1980) investigated seven flooded mine shafts in the Picher/USA Mining District.

In chapter 12, several case studies of flooded and stratified underground mines will be described using examples found in the literature and my own investigations. Only mines that are not backfilled will be discussed. Physicochemical parameters (temperature, salinity, pH, redox potential, oxygen content), in addition to chemical parameters, will be presented, a preliminary interpretation of the stratification patterns will be given, and an attempt to classify stratified mine shafts will be discussed.

The situation in metal, non-metal, and coal mines differs significantly from flooded salt mines. Nevertheless, one should bear in mind that the controlled flooding and monitoring of the German Hope Salt mine near Hannover was one of the first investigations studying the full range of mechanical and chemical processes during the flooding of an underground mine (GSF – Gesellschaft für Strahlen- und Umweltforschung 1985, 1986; Flach and Heick 1986; Herbert 1989). Therefore, some case studies of flooded salt mines will be presented in chapter 12, as well.

Some investigators have proposed that mine water stratification can be used as a remediation measure for polluted mine water (e.g. Heidenreich et al. 1991; Kindermann 1998; Gatzweiler et al. 2002). In fact, mine water stratification can be used to aid remediation methods, but only where thermal or forced convection will be reliably eliminated. Incorrect assumptions in the Straßberg case (12.11) and the sudden breakdown of stratification in the Frazer's Grove mines (12.6) clearly prove the importance of a thorough understanding of mine water stratification.

5.4 Effects Contributing to Mine Water Constituents

5.4.1 The First Flush

In most situations in surface streams, concentrations of water constituents inversely correlate with the stream flow rate (Stumm and Morgan 1996). Such a scenario can often be seen after runoff events, where the electrical conductivity, as a measure of total dissolved solids, decreases while the flow increases. However, at most flooded surface and underground mines, there is a steep increase in the concentration of mine water contaminants after the mine water reaches the outflow level (Fig. 13, Fig. 14). At partially flooded mine sites, this commonly occurs after every major storm event (see section 5.4.2), while at mines that are completely flooded, the contaminant concentrations usually decrease until they stabilise at a relatively constant value –

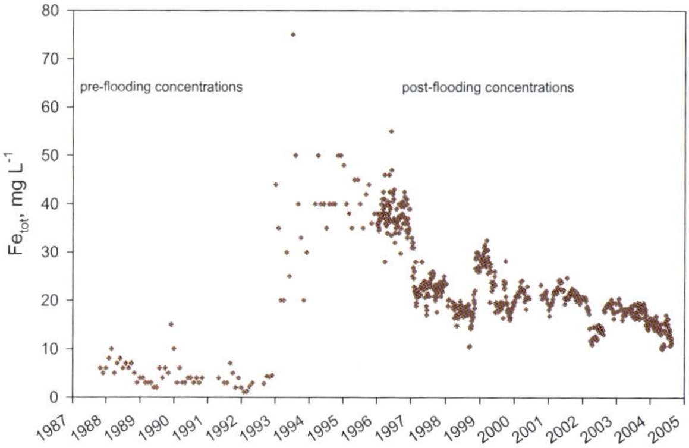

Fig. 13. Time series of changing Fe_{tot} in the Straßberg/Germany fluorspar mine before and after the flooding of the mine in 1993 (data kindly provided by BST Mansfeld GmbH and GVV).

which hopefully is the background concentration of that constituent. Though this phenomenon of mine water quality change has long been known, it was not before 1997 that it got a name: the "first flush" (Younger 1997). Commonly, the compounds leached during the first flush are more easily soluble geogenic or anthropogenic compounds (U.S. Environmental Protection Agency 2003).

Depending on the mine geometry and volume, the type of mineralisation, and the amount of precipitation, the mine water within a flooded mine will recover to background values after several years or several decades, sometimes centuries (Younger 1997; Demchak et al. 2004). The reasons for the initially elevated concentrations of the mine water constituents can be assigned to one of the following three mechanisms:

- production of vestigial acidity
- production of juvenile acidity
- buffering

Vestigial acidity (Younger 1997) comes from the dissolution of the secondary minerals formed during and after disulphide oxidation (section 3.2). Most of those minerals are easy soluble sulphate salts and store a large amount of acidity, sulphate, and metals. They are therefore commonly referred to as 'acid generating salts' (Bayless and Olyphant 1993). Because of their high solubility, they are quickly dissolved by the rising mine water and thus contribute to the peak of the first flush scenario.

Juvenile acidity (Younger 1997) results from three sources: proton acidity from the disulphide oxidation within the fluctuation range of the mine water table, proton acidity from the disulphide oxidation in the unflooded mine workings above the water table, and mineral acidity from the dissolved

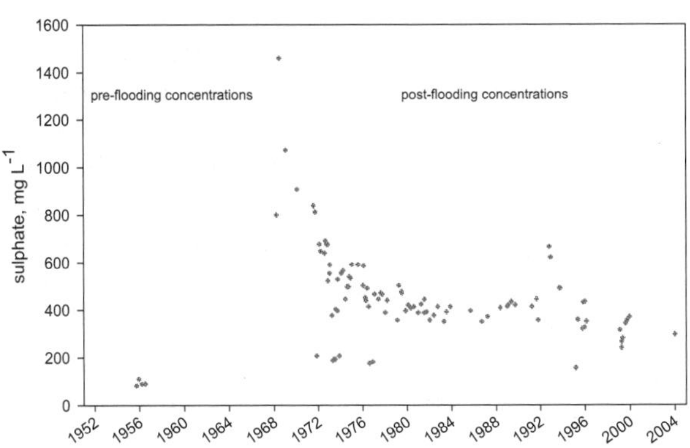

Fig. 14. Time series of changing sulphate concentration at the *Rothschönberger Stollen*/ Germany adit (modified after Merkel et al. 1997; Baacke 2000; Beuge et al. 2001; Yefimochkina 2004).

metals, such as Fe, Mn, Al, and Zn, which are usually present as hydroxide or sulphate salts. This phase is characterised by fresh water flowing into the system and reacting with the remaining disulphide and the before-mentioned minerals.

Besides the acid-producing component, there is also an acid-consuming component that occurs during the first flush scenario. As described in section 4.5, a range of minerals react as proton acceptors and neutralize acidity. Such buffering can lead to neutral or circum-neutral conditions in the mine water, after all the acidity is flushed out or consumed by the buffering minerals (Demchak et al. 2004).

Most first flush scenarios show a characteristic time lag after which the pollution load reaches its maximum. This time lag can range from several hours to several days, and has not been thoroughly investigated. Yet, physico-chemical measurements in shafts during the flooding process often show that the electrical conductivity in the upper meters of the flooded shaft is less than it is several meters beneath the water surface. Wolkersdorfer (1996) shows that this low conductivity layer (zone № 1) is commonly between 1 and 18 m thick, though in one case it was 190 m. It is followed by zone № 2, where the electrical conductivity decreases within a thickness of 1 to 30 m; finally comes zone № 3, where the electrical conductivity is relatively constant or slightly decreasing (see also Fig. 113). The reason for this behaviour is that the polluted mine water is commonly overlain by less polluted infiltration water. When the mine water finally discharges into the anthroposphere, first the less polluted mine water drains out of the mine followed by the more polluted water, which leads to the characteristic first flush.

The duration of the first flush, t_f, mainly depends on the acidity removal, aci_{rem}, by buffering or dissolution, by the rate, r_w, at which acid-containing minerals weather, the volume, V, of the inter-connected workings, the hydraulic connection, conductivities, K, of the mine workings, and the ground water recharge, R_{GW}. According to the findings of the HERO research team (University of Newcastle upon Tyne), the duration of the first flush after the overspill of the mine water can be calculated (in minutes) using the following empirical equation (Younger 2000d):

$$t_f = f(aci_{rem}, r_w, V, K, R_{GW}) \approx (3.95 \pm 1.2) \times t_r \tag{26}$$

with t_r the time it takes for the voids to be flooded ("rebound time"), min

Wood et al. (1999) commented that t_f is a function of time dependent and time independent factors, whereas the lithological setting and the extend of the workings belong to the first, and the removal of vestigial acidity, car-

Table 10. Disulphide contents of mined coal seams and their relation to observed peak iron concentrations in the United Kingdom. The numbers can be used to predict peak iron concentrations for the first flush period. Modified after Younger (2002d).

Total disulphide content, wt. %	Observed ranges of peak Fe_{tot}-concentrations, mg L^{-1}
< 1	0.01–0.5
1–2	0.5–100
2–3	100–350
3–4	350–1,200
4–5	1,200–1,500

bonate dissolution, iron hydroxide armouring, bacterial reduction of sulphate, and change in water flow rate belong to the latter. For the 32 coal mine discharges they analyzed, they were able to show that over the long term, the pH ranged between 6 and 7 and total iron decreases to less than 20 mg L^{-1}. A similar approach was used by Younger (2002d), who related the peak concentration of Fe_{tot} to the total disulphide content of the mined coal seams (Table 10).

Alkaline mine water can revert to acidic conditions when all the buffering phases are consumed and the production of juvenile acidity cannot be buffered any longer. Such a scenario has been observed at the Yorkshire Bullhouse mine, which became acidic after 60 years of being alkaline (Laine and Dudeney 2000; Younger 2000a). For every mine flooding scenario, it is therefore essential to calculate the potential net neutralization capacity of the rocks in contact with the flooded mine.

5.4.2 The Effects of Storm Events

Closely related to the first flush is the increase or dilution of contaminant concentrations at a mine water discharge. At partially flooded mines, a storm event can raise the water level in the mine enough to dissolve the secondary minerals that have formed since the last storm event, significantly (in some cases) affecting the discharge quality. Mines that have relatively constant water levels are not subject to this effect. Interestingly, there are many publications touching the matter of increased pollution and sediment loads in mine water affected streams, but no general work about that interesting subject, though Bayless and Olyphant (1993) describe many of the effects that cause stream pollution after surface runoff at mine sites. Most pollution comes from the erosion of tailings, waste rock piles, or from adit mines and the subsequent transport of the sediments and metals into the receiving streams, the flooded mine workings, or pit lakes. Several measures to prevent such runoff by covering waste rock piles were described in Integrated Geotechnical Engineering Services Specialists in Unsaturated Zone Hydrology (2003) and U.S. Environmental Protection Agency (2000). Field studies, monitoring of surface runoff at abandoned coal mine sites, and the modelling of the runoff were described by Hodgkinson and Armstrong (1996).

In general, the impacts of storm events at abandoned mine sites can be minimised by using a thorough remediation strategy. Many regulations include precautions against surface runoff from active mines, and the recommendations should be kept in mind for mine closure activities as well (U.S. Environmental Protection Agency 2003). The necessary measures to be taken before, during, and after a storm event should be compiled in a stormwater management plan, similar to all the other management plans for an active or abandoned mine site (Jacobs and Denham 2001). Such procedures include covering of tailings or waste rock piles, covering of potential infiltration zones above the mine workings, water diversion, regrading of slopes, or revegetation. Some of those measures are described in section 11.4.7. Retention ponds hold back suspended solids. If diversion ditches are constructed to guide the runoff around waste rock piles, they must be large enough to collect all the water, even during high storm or snowmelt events. At the Aitik/Sweden mine, diversion ditches were too small and erosion caused damage to the cover (Salmon and Destouni 2001). Especially in tropical regions, stormwater management is critical, as shown at the Batu Hijau mine, located on Sumbawa Island, Nusa Tenggara Barat, Indonesia. Up to 50 mm in a 15 min interval and 200 mm in a 24 hour interval were measured at the mine site and if no measures were taken to drain the water – even during the post-mining period – severe damage to the surrounding environment could occur. Sediment traps were installed to combat the problem (Jacobs and Denham 2001; Jacobs and White 2001).

Besides the erosion of solids and its transport in the runoff water, rainfall also raises the water table in tailings or waste rock piles and can release the easily soluble secondary minerals in tailings (Al and Blowes 1996). Loadings during and after rainfall events can be up to 20% higher than during normal flow situations. Especially after long rainfalls events, when the tailings become flooded, significantly high discharges might be expected.

Possibly one of the most interesting sites where such phenomena occur is the Tasmanian Mount Lyell mining area, which affects the King River and, at its mouth, trout farming in Macquarie Harbour (Lake et al. 1977; Locher and Keller 1995). Several hundreds of kilometres downstream, the aquatic life was severely affected by the effects of surface runoff, sediments from the remnants of copper mining, and a high daily load of copper (Parliament of the Commonwealth of Australia 1970; Taylor et al. 1996; Twining and Cameron 1997). The sediments transported by the Kings and Queen Rivers filled up the river mouth with nearly 1.5 m of sediments, so that boats were hindered from entering the river's mouth (Parliament of the Commonwealth of Australia 1970).

Proper regrading and revegetation is an effective means to minimise acid mine drainage at surface operations (The Pennsylvanian Department of Environmental Protection 1998). To prevent runoff, the slopes of tailings or waste rock piles should be regraded to reduce the slope of the piles and to minimise the erosion during storm events (Hawkins 1995). This provides more stable surfaces, supports revegetation, and thus decreases the pollution of surface streams and ground water. Several of the measures described under water diversion (section 5.7) can be used to prevent surface water from flowing over uncapped waste rock piles (Adams et al. 2000). Sandén et al. (1997) measured the amount of colloids during a storm event 500 m downstream of a remediated mine waste deposit. At this sampling location, the flow in the stream increased 2 hours after the rain started. They clearly proved that the amount of solids started to increase 1 hour later, increased by about 100% after another 30 minutes, rose to a maximum of 150% 4 hours after the runoff water reached the measuring location, and stayed that high for at least 24 hours.

In the Picher Mining District (Kansas/Oklahoma/Missouri, USA), measures were taken to prevent surface runoff from flowing into abandoned mine shafts and boreholes (Tar Creek Superfund Task Force Water Quality Subcommittee 2000a; DeHay et al. 2004). In addition, several areas were covered with impermeable soil to prevent storm water from flowing into the abandoned mine workings, but that measure was not very effective and is not recommended to prevent water flowing into abandoned workings. If there is a need to predict surface runoff, Appendices A and B of U.S. Environmental Protection Agency (2003) describe the necessary measures to predict the amount of runoff by using several graphical and numerical techniques.

Burbey et al. (2000) observed two different effects of rainfall on the discharge water quality. While the main ions in the water, such as Mg, SO_4, and Ca, are relatively uninfluenced by the rainfall, the pH is negatively correlated with increased flow due to rainfall, and the iron concentrations were positively correlated with the rainfall events. They conclude that the rainwater, which contains no iron, flushes the mine water and also lowers the pH.

Sandén et al. (1997) took 47 samples over a period of 24 hours after a 15.5 mm rainfall event during 7 hours. The flow within the stream increased from 0.2 L s^{-1} to 23 L s^{-1} and decreased to 6 L s^{-1} 23 hours after the rain started. Their results were similar to those of Burbey et al. (2000), with an increase in pH, turbidity, and TIC and a decrease of the other trace element concentrations. The main ions and total iron showed two scenarios: they first increased as salt dissolved from the mine waste disposal site, and after 4–5 hours, they decreased again as a result of dilution.

However, there is no general rule of behaviour of mine sites or flooded mines during storm events. At many sites contaminants decrease due to dilution. While the contaminant loads stay relatively constant, the contaminant concentrations decrease with increasing flow (Hawkins 1998).

In conclusion, the negative impacts of surface runoff after storm events or snowmelt can be positively influenced by covering the tailings and waste rock piles, regrading the slopes to reduce the slope angles, revegetationg the surface of the piles, and diversion of waters. Compared to many other remediation measures, especially mine water treatment, these measures should be an integral part of a mine site remediation to limit AMD/ML.

5.5 Reasons for Mine Flooding

5.5.1 General Aspects of Mine Flooding

When the life of an underground or open-cast mine ends, its maintenance typically stops and the dewatering pumps are switched off. Generally, where no drainage adits exist at the lowest levels, the open pits or underground workings will fill with water and the mined voids will be flooded with ground- or surface water (Fig. 15). Eventually, the water will overflow and discharge into receiving water courses. Depending on the open space, the availability of infiltration water, and the type of flooding (uncontrolled, controlled, monitored), inundation can last from several months to more than a decade.

According to the principles of sustainable mining, it is essential that the pre-mining hydrogeological conditions are nearly restored after the mine floods. Furthermore, potential drinking water reservoirs have to be protected, surface streams must not be polluted by the mine water, and the pollutant loads should be reduced to a minimum. Most of those prerequisites can be achieved by reducing the mine's water make, which is the sum of all the water discharging from a mine (Fig. 29). Often the drainage water of a flooded underground mine has a high potential to pollute the surrounding environment and increase the base flows of receiving streams. Therefore, the mine operator nowadays has to ensure that polluted mine water is treated accordingly and that environmental pollution of the water catchment is minimised. To do this, the mine operator has to decide if it is better to accelerate or deaccelerate the flooding process. When mines are flooded quickly, the disulphides are excluded from weathering as soon as possible and the pollutant loads are consequently quickly reduced, according to equation 26. In such a scenario, the water treatment plant can be switched off after a relatively short time. On the other hand, if the water treatment plant will have to run for a very long time, the flooding process should be deaccelerated. Several options for an adapted mine water treatment are available, and will be discussed in chapter 11.

The main reasons for mine flooding are:

- the mine is no longer economical
- all the raw material has been exploited
- accident, war, or political reasons
- geotechnical stability of the abandoned mine workings
- prevention of disulphide oxidation
- safety reasons (prevention of unintentional visitors)

Controlled mine flooding is conducted in areas where people or the environment might be affected by polluted mine water or by raising the water table above a certain level. There, maintenance of pumps or treatment facilities will take as long as deemed necessary by the mine operator or the authorities. In some mines in the former German uranium mining areas of Saxony and Thuringia, controlled flooding was prepared for extensively; several expert opinions about the mine water rebound and the chemical evolution of the mine water quality were sought, taking nearly a decade to draw final conclusions (Jakubick et al. 2002). Controlled mine flooding does not necessar-

Fig. 15. The flooded Reiche Zeche shaft of the Freiberg/Saxony mining field (2 March 2006). The brown colour is a result of ochre precipitates as a consequence of the catastrophic flooding of August 2002. Width of picture approximately 2 m.

ily mean monitored flooding. Of course, in nearly all mines that are flooded in a controlled way, the water level in the mine and the drainage water quantity are measured. Monitored flooding of a mine is a subset of controlled flooding, and includes the installation of physico-chemical probes, regular measurements of the drainage water quality within the flooded part of the mine and at the discharge points, as well as a geotechnical monitoring system. The Prediction Workgroup of the Acid Drainage Technology Initiative (2000) described such monitoring schemes for coal mines while the ERMITE Consortium et al. (2004) did the same for all types of mining operations. Usually, some pumps remain installed to interfere, as necessary, with the raising mine water. Such a scheme was used by the Niederschlema/Alberoda/Germany mine, where the mine water table rose quicker than expected and the mine water table had to be lowered by pumping to avoid the discharge of heavily contaminated mine water (Meyer et al. 2002).

Installations for controlled mine flooding include:

- temperature measurements
- geomechanical installations (geophones)
- conductivity measurements
- piezometers
- flow measurements
- pumps

There are two possibilities for controlled mine flooding. In the first case, enough water in the catchment area of the mine is available to flood the mine. The mine can then be flooded passively, without using water from other catchments. The second case occurs where not enough water is available to flood the open space or where the water that would be available has a chemical composition that would cause problems. Such situations can be found in the Lausitz/Germany lignite mining areas, where the rivers Spree and Neiße do not have enough water to flood all the open pit lignite mines that were closed after 1990 (Kaden and Schramm 2000; Schlaeger and Köngeter 2001; Grünewald 2001). Other examples are salt mines that are flooded with saturated brine to avoid the destruction of nearby mines or surface subsidence due to salt dissolution (Herbert 1989). Because the flooding of abandoned salt mines is different in many aspects from the flooding of abandoned coal or metal mines, their mine water management shall be excluded from the following discussion. The interested reader is directed to the investigations of the central German salt mines by Brendel et al. (1982).

Uncontrolled flooding usually happens in the case of accidental mine flooding, which can occur due to mine water inrushes (section 5.5.4) or where no damages to the ecosphere or engineering construction are assumed. Other examples are mines where the mine budget is unclear, or crisis or wars begin and pumping has to be stopped. Usually, the mine water pumps are stopped at a certain time and neither a geotechnical nor a chemical or physical monitoring system is installed.

For reasons of completeness, it should be mentioned that there are situations where the flooding of a mine or of a mining region would cause or has already caused detrimental impacts on the eco- and anthroposphere. Such situations will occur in the German Ruhr, the Dutch Limburg and the Polish Silesian coal mining areas, where decades of mining lowered the ground water table and cause surface subsidence under the pre-mining ground water level (Bekendam and Pöttgens 1995; Younger 2002b). Flooding of salt or gypsum mines – be it intentional or accidental – can also cause catastrophic situations, as occurred in Polish and German mines (Schwandt 1993; Seifert and Schwandt 1996; Schwandt and Seifert 1999; Zuber et al. 2000).

Flooding prediction is still a highly fuzzy job, needing experience and knowledge of the underground workings. Although several empirical and numerical models exist (e.g. Heidenreich et al. 1991; Younger and Adams 1999; Frydrych et al. 2000; Banks 2001; Adams and Younger 2001; Spiessl et al. 2002; Kalka et al. 2002), the modelling results are still highly uncertain and are only truly satisfactory for simple mine geometries (e.g. Aurada 1970; Brown 1982; Hanzlík and Vydra 1985; Rogoz 1994; Banks 1994; Gatzweiler et al. 1997; Younger 2002d; Whitworth 2002). One of the rare examples, where prediction and reality show a good correspondence, is the Whittle mine of the Shilbottle coalfield. Adams and Younger (2001) investigated the hydrological and hydrogeological situation at this mine site and used the GRAM code to predict the flooding scenario in 1998. The model predicted that the water would reach the 50 m level by around May 2002 and in fact the prediction was a 'precision landing' (Younger 2004a). More details on water rebound will be given in section 5.9.

5.5.2 Surface Mines

Because some surface mines have been or will be converted to underground mines (e.g. Ronneburg/Germany, Chuquicamata/Chile, Palabora/South Africa, El Valle and Carlés gold mines/Spain, Aguablanca nickel mine/Spain), such operations might be hydraulically connected with each other. For those mines it is necessary to have at least an idea of the relevant processes in surface mines. Therefore, this short section gives some general aspects that are important for flooded surface mines.

Surface mines, also called open pit mines, and opencast mines, are used for near surface raw material deposits if the costs for removing the overburden are less than the expected profit from selling the raw material. Surface mines are used to extract large raw material deposits such as coal, gold, copper, or industrial minerals like gypsum, limestone, or hard rock. After mining ceases, they are often flooded, forming pit lakes (Fig. 16). Though open pit lakes are not the focus of this publication, they should be mentioned for completeness because they are unique water bodies that attract different kinds of creatures: bacteria, fish, birds, and scientists. Since the 1970s, the number of open pit mines have increased steadily due to an increasing demand on raw material resources, the increasing technical possibilities to extract ores from low mineralised ore bodies such as copper-porphyry type deposits, and the relatively low costs of surface excavation at some sites. Leading surface mining countries are the USA, Canada, Australia, Chile, Brazil, Peru, Mexico, Indonesia, the Philippines, Papua New Guinea, West Africa, and Kazakhstan (Castro and Moore 2000).

Depths of open pits range from several meters to several hundreds of meters. One of the deepest open pit lignite mines is located in the Western German area around Cologne while the deepest metal open pit mines are the Chuquicamata/Chile and Bingham/USA copper mines. Open pits near Cologne reach a depth of up to 500 m while the Chuquicamata open pit mine has a depth of 850 m and is projected to reach 1100 m by 2013 (Rowan McKittrick, Water Management Consultants Ltda., Santiago de Chile, pers. comm.; Ossandón et al. 2001; Arancibia and Flores 2004). The Bingham copper mine is currently deeper than 900 m (Borden 2003). Besides the depth, the aerial size of the open pits can vary considerably from several hundred square meters to even square kilometres.

In contrast to underground mines, the hydrodynamics of open pit lakes, their geochemical development, and the prediction of open pit filling have been studied by many researchers using numerical, geophysical, geochemical,

Fig. 16. Spreetal Nordost open pit lake in the abandoned Lausitz/Germany lignite mining area. The final water level will be about 8 m above the current water surface.

and tracer techniques (e.g. Prediction Workgroup of the Acid Drainage Technology Initiative 2000; Castro and Moore 2000; von Rohden and Ilmberger 2001; Bowell 2002; Mudroch et al. 2002; Ardejani et al. 2003). A review of pit mine hydrology and geochemistry is given by Bowell (2002) and in a special issue of IMWA's *Mine Water and the Environment* journal (Bowell 2003).

Flooding of surface mines is a common procedure after production ceases or the operations discontinue. In many countries, guidelines for surface mine flooding were developed (e.g. Länderarbeitsgemeinschaft Wasser [LAWA] 2001) and besides the hydrogeological prediction, geochemical predictions of surface mine flooding were published (e.g. Klapper and Schultze 1996; Klapper et al. 1996; Prediction Workgroup of the Acid Drainage Technology Initiative 2000; Geller et al. 2000a). The inflow into an open pit mine can be modelled in several ways, using an equivalent well approach, two-dimensional flow equations, or numerical techniques (Finite Element Method, Finite Difference Method, Boundary Element Method). For an operating mine, such models can also be used to estimate the cone of depression of the dewatering procedures (Singh and Reed 1988).

At such sites, the open pit gradually fills with ground or surface water and a pit lake is established after the dewatering pumps or drainage wells are stopped. Surface waters (precipitation) and runoff water from the pit slopes can be captured and stream water can be diverted or pumped into the pit lake. Ground water can be water from a ground water body connected to the pit or from old abandoned underground mine workings. Water tables usually equilibrate with the ground water table, but they might be higher or lower depending on the local meteorological conditions. High precipitation causes pit lake water levels to be higher than the ground water table, whereas high evaporation lowers the pit lake level. If highly polluted pit water is expected, pumps might be installed in the pit to lower the pit water table and prevent the contaminated pit water from interacting with uncontaminated ground water (e.g. the Aznacollar pit, where the pit lake level is kept 6 m below the ground water table; Schultze et al. 2006; Kurt Friese, pers. comm.).

Most man-made pit lakes are similar to natural lakes and the hydrodynamic and geochemical evolution of such lakes has been studied intensively during the last couple of decades. GeoRef lists more than 400 and the ISI database more than 100 entries concerning pit lakes, but there are few general publications on mining pit lakes. Geller, Klapper and Salomons discuss many aspects of acidification and remediation as well as water quality management of acidic mining lakes (Geller et al. 1998; Klapper and Geller 2001). They also address numerical and empirical models that have been developed to predict water table recovery within and around pit lakes.

Water quality in such pit lakes varies over a broad range, from drinking water quality to highly acidic mine water, and it is known that polluted open pit mine water has killed birds and fish. Besides water quality, the depth and the physical properties of such lakes vary over a broad range. They can be only several meters deep, or reach depths of up to 500 m. Typically, such pit lakes have a low surface area to depth ratio with values below 10 $m^2\ m^{-1}$, compared to natural lakes with surface area to depth ratios above 2000 $m^2\ m^{-1}$ (Doyle and Runnells 1997; Castro and Moore 2000; Bowell 2002). This might cause a stable stratification of the pit water body to develop, such that the depths of the epi-, meta-, and hypolimnion stay relatively constant for a long time. Some studies have suggested that, due to the reducing conditions at the bottom of stratified mining lakes, such lakes could be used as mine waste disposal sites (Doyle and Runnells 1997; Levy et al. 1997).

Not all mine pits are kept open, or are flooded. Alternative uses might include soil reconstruction, forestry, recreation, agriculture, or wildlife habitats (Brenner 2000; Kennedy 2002). Some open pits are backfilled with overburden from other open pits or with reactive material to prevent acid mine drainage. Another option is to backfill pyrite-rich overburden or tailings in the pits to prevent the pyrite from weathering under oxidising conditions. Bowell et al. (1998), and Bowell (2002, 2003) give guidelines on environmental issues that have to be considered and list procedures on how to abandon open pits.

5.5.3 Underground Mines

Underground mines have been used since Neolithic times to extract many different types of raw materials. While the first Neolithic underground mines only reached several meters of depth, Bronze Age underground mines went down more than 100 m, and modern mines, like the planned AngloGold's Western Ultra Deep Levels mine (WUDLs) extend up to 5,000 m (Zschocke et al. 1932; Holgate 1991; Shepherd 1993; Vieira et al. 2001; Lynch 2002; Fig. 17).

While the number of surface mines have steadily increased over the last several decades, numerous underground mines have closed. Yet, it can be assumed that in the future, underground mines will probably be more attractive, because the depths of surface mines are limited, while the depth at which raw material deposits can be found theoretically reaches the thickness of the upper crust. Once the economically extractable near-surface raw material deposits are exploited, underground mining will re-emerge and the knowledge of ecologically safe closure procedures for such mines will become more important than ever. An example of an open pit mine that will be converted to an underground mine is the Chuquicamata mine in Chile, which will start to work underground in the year 2013 (Arancibia and Flores 2004). Río Narcea's El Valle and Carlés gold mines as well as the Aguablanca nickel mine (Asturias/Spain), have been converted to underground mines since 2004 (Paul Younger, pers. comm.; Río Narcea web page).

Fig. 17. Unflooded access Gallery in the Western Deep Level/South Africa underground mine. The Gold Reef comes in from the top right side and dips to the lower left corner. Concrete support helps to support the geomechanical stability in the 2500 m deep gallery.

There are three different types of underground mines: adit mines, incline mines, and shaft mines – and of course combinations of these (Fig. 18). Their hydrological and hydrogeological characteristics differ significantly and the closure procedures for each require different techniques. From an environmental point of view, adit mines should preferably be closed by constructing dams (bulkheads) at the mine entrances to allow the mine to flood, re-establishing a water table that will resemble the pre-mining water table. Unfortunately, this closure technique requires reliable engineering constructions whose stability cannot be guaranteed in all circumstances due to highly acidic drainage waters and extreme water pressures. Also, the rising water table can cause unforeseeable non-point discharges of polluted mine water (ERMITE

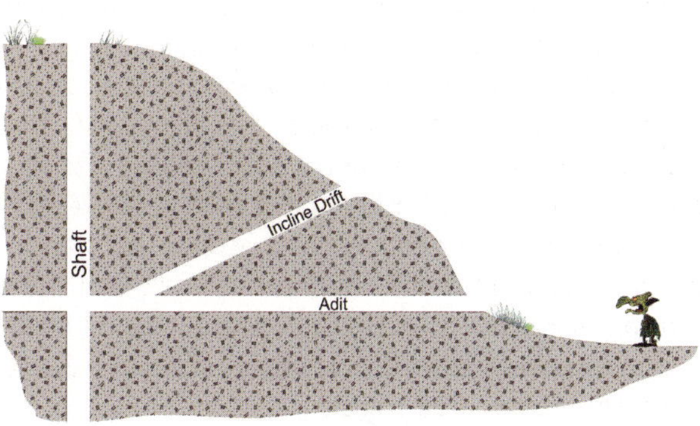

Fig. 18. Principal mine types: shaft mine, incline mine with incline drift, and adit mine. The "Stollentroll" © Walter Moers, Eichborn Verlag, Germany.

Consortium et al. 2004). Therefore, the construction of dams in adit mines always needs extensive investigations to exclude environmental problems, personal injuries, or damage to engineered constructions. If the situation is unclear, the full closure of such mines should be avoided (ERMITE Consortium et al. 2004). More details about such physical barriers can be found in section 5.7.3.

After a mine has been abandoned, the maintenance of the wooden and steel support stops or pillars start to weather and, depending on the physical properties of the support, its stability will gradually decrease. If the rock pressure, caused by the static or dynamic loading, exceeds the supportive forces, the support fails and the workings collapse, eventually giving rise to surface subsidence. In unsupported parts of the mine, especially in the stopes and adits, overburden pressure decreases rock stability and, depending on the extraction ratios, causes roof failures or pillar crushings as well. Alternatively, the buoyancy forces of the rising mine water will cause the pore pressures to decrease and the overburden pressure and the potential for subsidence might be minimised; even uplift can occur. Although such effects have been known since Terzaghi's pioneering soil mechanical investigations (Terzaghi 1925), only minor systematic research has been conducted on this topic (e.g. Oberste-Brink 1940; Goetz et al. 1994; Hunt and Reddish 1997). While Goetz et al. (1994) found proof of positive buoyancy effects, Hunt and Reddish (1997) think that those effects are negligible. Yet, as the mine flooding in Limburg/The Netherlands shows, vertical lift of up to 0.2 m can occur in areas that are affected by the raising water table (Pöttgens 1985; Bekendam and Pöttgens 1995). Oberste-Brink (1940) investigated the surface uplift in the abandoned Witten/Germany hard coal mining area. He noticed uplifts in a mine ranging up to several centimetres.

The geophysical records of flooded mines show that in tectonically weak areas of the mine, the number of micro-earthquakes usually increases during flooding (Flach and Heick 1986; Rudolph 1997; Ogasawara et al. 2002), and normally come to a halt after flooding is finished. Such micro-seismic events within a mine can induce significant changes in the mine water level. One of those events was observed in the Niederschlema/Alberoda uranium mine on 21 November 1995, where a micro-seismic event dropped the mine water level by 34 cm. Even 14 years after the start of the flooding process, the micro-seismic events induced by flooding did not cease, though their magnitudes steadily decreased (Krentz et al. 2004). Flach and Heick (1986) observed that micro-seismic events were clearly linked to the flooding, and when the quantity of flooding water into the mine changed, and that the number of micro-seismic events decreased after flooding. These measurements verify the observations of Goetz et al. (1994) that during the flooding process, areas that were at the limit of their stability will fail and, after flooding, the buoyancy forces will stabilise the mine workings.

One of the main reasons for the flooding of underground mines is the prevention of disulphide oxidation and consequently acid mine drainage. Although not recorded in detail, acid mine drainage causes world-wide damage in the range of several billion Euro. Literature about acid mine drainage is numerous and the reader interested in the details of acid mine drainage is referred to Stumm and Morgan (1996), Younger et al. (2002), Lottermoser (2003), or Jambor et al. (2003) for further reading (see also chapter 4). Disulphides, namely pyrite and marcasite (FeS_2), but also the monosulphide pyrrhotite (FeS) are the crucial minerals for acid mine drainage formation (the chemical formula of pyrrhotite is FeS, yet, most natural pyrrhotites are found to be $Fe_{1-0.83}S$, usually written as $Fe_{1-x}S$, Strunz and Nickel 2001). In the presence of water and oxygen, and catalysed by bacteria (see section 4.4), these disulphides weather to form acid conditions as well as ferric and ferrous iron (Stumm and Morgan 1996). In addition, the acid further accelerates dissolution of other minerals, resulting in an increased trace metal concentration in the mine water. These products of the disulphide oxidation cycle can have major impacts on the environment; the highly acidic mine drainage leaves the drainage adits and if not properly treated, dyes the streambeds, natural wetlands, and flood plains an unpicturesque red to yellow colour. Iron concentrations in highly acidic mine waters can reach several thousands of milligrams per litre and pH values of –3.6 were measured in the Californian Richmond Mine (Nordstrom et al. 2000). Though acid mine water can be treated by several passive or active treatment options (Younger et al. 2002; Brown et al. 2002), the best option is to prevent the formation of acid drainage at its source: in the mine. Mine flooding isolates the disulphides from atmospheric oxygen and therefore significantly reduces the kinetics of acid formation. Due to the low diffusion coefficient of oxygen in water as well as the low oxygen content and redox potentials of ground water, mine flooding reliably limits the formation of acid mine drainage.

Safety reasons are another motivation for mine flooding and – of course – mine sealing. Non-flooded or unsealed underground mines are frequently visited by people who are not aware of the dangers they might encounter underground, and mine owners can be held legally responsible for personal injuries. Mine gases (mainly Rn, CH_4, H_2S, CO_2, CO) are another reason for mine sealing (those gases are usually not much of a problem during the active mining period because of the mine air ventilation). Especially in remote areas of the USA or Australia, deadly accidents are reported frequently, and programmes to close open shafts or adits are ongoing (e.g. Suman et al. 2000). Mining archaeologists investigating the history of abandoned mines, be they professionals or semi-professionals, are usually well equipped and accidents therefore are rare (Leupolt and Hocker ca. 1999). Mine owners therefore should try to work together with local interest groups to find closure solutions that fit the needs of mining archaeologists, environmentalists, and health- and safety regulators.

Economics and the depletion of the deposit are the most obvious reasons for mine flooding. Dewatering pumps, especially in deep underground mines, are extremely energy consuming and therefore add significantly to the costs of underground mining. It is therefore in the interest of the mining company to stop dewatering pumps as early as possible after the productive period. For underground mines involving techniques that do not include surface subsidence, including most metal mines, such a procedure usually presents no danger for the people living in the vicinity. In contrast, longwall mining operations, where the roof is allowed to collapse behind the advancing face, induces surface subsidence, as can be seen in many coal mining areas of Britain, Poland, Germany, Australia, and the United States. Switching off dewatering pumps in those areas, which include very heavily populated regions, can mean the flooding of many well-known industrial zones (Younger 2002b). A compromise approach is to collect and discharge relatively uncontaminated infiltration water to avoid high water treatment costs; this approach has been used during the flooding of the Eastern German uranium mines.

The controlled flooding of an underground mine should include the following precautions:

- removal of potential water hazards in the mine
- backfilling of near-surface structures that might subside or collapse during or after the flooding process
- installation of water diversion systems (see section 5.7)
- grouting of tectonic structures that are known to be connected to drinking water reservoirs
- installation of a hydrogeological and geotechnical monitoring system underground and at the surface
- prognosis of hydrogeochemical and hydrological development of mine water

5.5.4 Mine Water Inrushes

Besides mine gas explosions, mine water inrushes are the most common reason for catastrophic accidents in underground mines and still claims many victims. Though mine water inrushes due to insufficient pumping capacities are relatively rare – compared to historical times – they still occur. Recent cases are the Russian Sapadnaja-Kapitalnaja coal mine (23 October 2003), the Chinese Shentang coal mine (22 June 2002), the Chinese Xuzhou coal mine (1 January 2000, Anonymous 2000) and other Chinese mines (Anonymous 2002; Bei and China Daily 2004). Singh (1986), Sammarco (1986), Job (1987), Vutukuri and Singh (1995), Kroker and Farrenkopf (1999) as well as Younger (2004b) give in-detail-studies of such accidents and some of them also discuss their prevention.

By statistics, flooded abandoned panels or parts of a working mine pose the highest risk of accidental mine water inrushes (Fig. 19, Job 1987; Vutukuri and Singh 1995). Job (1987) investigated 208 water inrushes in British coal mines between 1851 and 1970. He was able to show that 78% of all those accidents were due to contact with abandoned old workings, whereas he identified 11 distinct reasons for such accidents: ineffective barriers, incomplete old plans, absence of plans, incorrect plans, incorrect interpretation of old plans, failure to obtain the abandonment plans, absence of protective boreholes, ineffective protective boreholes, failure to plug boreholes, unknown old workings, and incorrect seam correlation. One of the latest of such accidents was the Quecreek/Pennsylvania/USA accident (24 July 2002) where the barrier pillar between this mine and the adjoining Saxman coal mine breached, entrapping 9 miners, who were rescued 78 hours after the accident (Pauley et al. 2003).

5.6 Remining Abandoned Mines – A Special Type of Remediation

Many abandoned underground mines, waste rock piles, or tailings ponds still contain a great amount of unprocessed ore and acid-producing minerals. This source of acidity may have been left due to the mining method, for example in the pillars used for support, because the concentration of the commodity

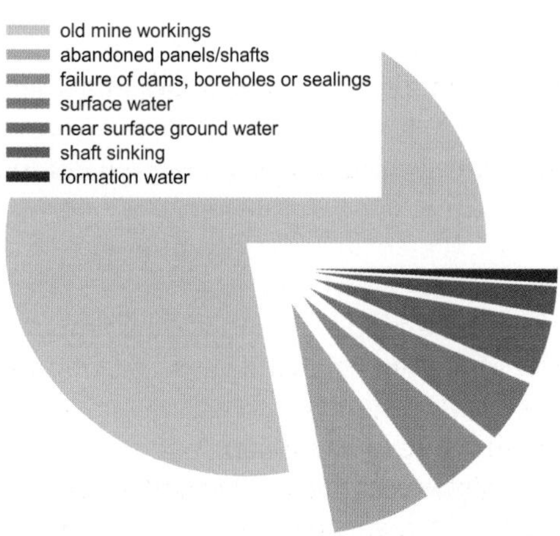

Fig. 19. Reasons for mine inundation (modified after Job 1987; Vutukuri and Singh 1995).

Fig. 20. Remining of an abandoned underground gold deposit in the Kalgoorlie/Western Australia Super Pit by daylighting through Kalgoorlie Consolidated Gold Mines. Some of the old mine workings can be seen in the middle of the photograph. The remining contributes to a reduction of the pollutant loads as potentially acidic strata are excavated, processed and disposed of.

was below an economic concentration, or because no appropriate mining or extraction method was available to extract it at the time of first mining. It is common though to reopen such mines if the economic or technical conditions that caused the closure of the mine change. Though remining, strictly speaking, means the reopening of an existing abandoned or closed mine, this section will also cover daylighting of underground mines (which means that a former underground mine is converted to an open pit mine, Fig. 20), and the reprocessing of mining waste disposals or tailings. Furthermore, as will be shown, remining can significantly contribute to a reduction of the pollution loads and is therefore described here.

Following the definition given by Hawkins (1995), remining would only describe mining operations aimed to improve pre-remining water quality. Another definition is given by the U.S. Environmental Protection Agency et al. (2001): "Surface mining of abandoned surface and/or underground mines for which there were no surface coal mining operations subject to the standards of the Surface Mining Control and Reclamation Act. Remining operations implement pollution prevention techniques while extracting coal that was previously unrecoverable". Neither definition would cover the reopening of non-coal mines. However, in this publication, "remining" will be viewed from a slightly broader context and will cover all of these applications.

Remining was also conducted in the Germany, near Witten/Ruhr, where the history of German coal mining started at the end of the medieval times. At least three mining periods have been recognised there and all of them reflect the stage of technical possibilities at their time (Koetter 2001). Each subsequent period reused the old abandoned workings to mine coal that was uneconomic in the previous period or which could not be extracted because of technological reasons. Remining has been covered by several official publications in the USA and a lot of good guidance books or publications are available (e.g. Smith 1988; Hawkins 1995, 1998; U.S. Environmental Protection Agency et al. 2001).

Unfortunately, at least since the German Rammelsberg court cases (Anonymous 1994, 1995), mining companies refrain from remining abandoned mines because they will become responsible for the historic pollutants caused by the mine that they used for remining (Amezaga and Younger 2002). This situation differs from the Pennsylvanian legislation, where remining (*sensu* Hawkins) started in 1983. There, the mine operator will not be responsible for the pre-remining pollution as long as he does not further degrade the water quality. Should the pollutant loads after cessation of the remining activities be below the previous pollutant loads, he will be released of liability concerning this pollution (Hawkins 1994). Due to this legislative support, Pennsylvania has become one of the most intensively remined states in the USA and possibly on a world-wide scale. Approximately 91% of all remining projects in the USA, a total of more than 300, are located in Pennsylvania (U.S. Environmental Protection Agency et al. 2001). The history of the legislative procedures needed for remining in Pennsylvania is well documented in the earlier cited publication. Remining can be seen as a remediation technology for abandoned mines which – usually – does not to have paid for by the tax payer but by the industry.

In the case of the Tasmanian Mount Lyell copper-gold deposit, the Australian and Tasmanian Governments formed a new company, Copper Mines of Tasmania, to remine and remediate the heavily polluted Mount Lyell mining area (Sustainable Development Advisory Council and Copper Mines of Tasmania 1995; Copper Mines of Tasmania 1995). Yet, even after special legislative measures were taken to regulate the non-responsibility for the pre-remining period, the plan failed, for various reasons, and the Mount Lyell area still endures non-remediated tailings and acid mine drainage.

Nonetheless, many remining projects prove that the environmental impact of acid mine water discharges can be positively influenced or at least generally do not worsen (Perry and Evans 1999). The reasons for this are twofold: remining removes the left-over pyritic strata (very often coal or lignite) and due to the more stringent modern reclamation standards, potential pollution sources have to be removed, covered or reclaimed in other ways. Furthermore, daylighting often exposes and disrupts alkaline strata that provide alkalinity; pH improvements from about 3 to 7 are known (The Pennsylvanian Department of Environmental Protection 1998). Anyway, the conversion from an underground mine to a surface mine will usually produce better water quality because surface mines typically leave less pyritic material and thus have better water quality than underground mines (Fig. 21). The number of remining projects that resulted in higher pollution loads is negligible (Hawkins 1994, 1995; Prediction Workgroup of the Acid Drainage Technology Initiative 2000; West Virginia Division of Environmental Protection – Office of Mining and Reclamation 2001)

After remining, it is comparatively simple to count the hectares of land reclaimed, the numbers of trees replanted or the amount of tailings or waste rock piles covered. Those measures can therefore be checked relatively quickly against the necessary remediation requirements. Conversely, a thorough statistical investigation of the pre-remining and post-remining situation is essential to determine whether remining activities changed the pollutant loads. As in other mine water cases, a thorough understanding of the hydrogeological and hydrological situation of the mine site as well as statistical methods are required. Especially the pre-remining baseline pollution load has to be known and evaluated precisely (Smith 1988). Some of the simpler statistical techniques that must be used are normality testing, exploratory data analyses, and ranked correlation coefficient analyses (Hawkins 1994, 1995). Furthermore, time series analyses, T-tests, or ANOVA tests would be necessary to get statistically significant differences in the data sets (see also section 9.3.4).

From a mine water management perspective, the sealing and elimination of abandoned workings is the most important aspect of remining. During

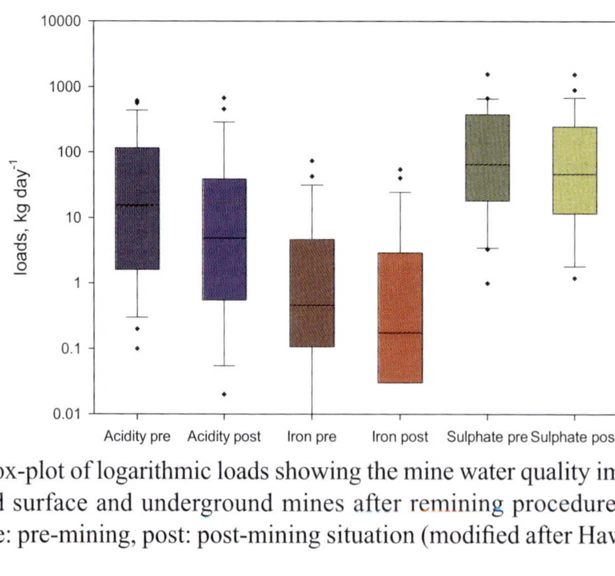

Fig. 21. Box-plot of logarithmic loads showing the mine water quality improvements in selected surface and underground mines after remining procedures were conducted; pre: pre-mining, post: post-mining situation (modified after Hawkins 1995).

remining, old abandoned workings are exposed at the side walls of the open pit. To prevent pyrite oxidation and subsequently the formation of acid mine drainage in these abandoned workings, it is essential to seal them with low permeability material. Yet, care must be taken so that the seals will not be blown out if extreme pressures build up behind the seals. Should this be a concern, it is advisable to install drainage pipes through the seal to reduce the water pressure behind the seal (U.S. Environmental Protection Agency et al. 2001).

From an economical point of view, remining activities are usually cost effective, especially if the operator would have had to treat the discharge water of a previous underground mine. Skousen et al. (1998) summarises some remining projects, including ecological and economical data. The REMINE software package is used by U.S. agencies to calculate the effects of a remining operation (U.S. Environmental Protection Agency and Office of Water 1988). As a conclusion, remining can be seen to be a cost-effective and environmentally sound measure for the remediation of abandoned underground mines.

5.7 Diverting Water

5.7.1 Why Divert Water?

Compared to most geological processes, which might last millions of years, mining is a short term use of our landscapes that can cause substantial change to our environment. Though the changes might be acceptable, or even desirable, pollution of river catchment areas must always be excluded, because according to most national laws (e.g. the European Water Framework Directive), no negative influences on our surface and ground water resources are allowed. It is therefore important during a mine's closure phase to prepare the mine and to guarantee a good status of the environment after the mine's abandonment. Some guidance to minimise post-closure impacts during the closure phase is given below.

Because treatment of mine water might be a long lasting and expensive task, it is advisable to divert surface or ground water from the potentially polluting sources in and around the underground mine. This is already a common practice. Nonetheless, this method is neither practical nor environmentally advisable in the case of huge open pit mines, where the mined area has to fulfil its function as an aquifer after mine closure. In other cases, the diversion of surface or ground water will generally minimise the amount of water that has to be treated but there are cases, where the mixing of highly acidic mine water with neutral or well buffered surface water might have a positive effect on the mine water quality. In such a case, the mixing of the two waters significantly improves the mine water quality, causing the pH to rise and pollutants to precipitate, thus improving mine water quality.

Obviously, the formation of polluted mine water could be prevented if:

- the reactants, oxygen or water, are not allowed to get into contact with the disulphides
- water is not allowed to get into contact with the secondary minerals
- water chemistry is changed in such a way that dissolution processes are hindered
- minerals are removed or covered by inert material

Several methods have been used to exclude the contact of unpolluted water with potential water contaminants. Most of these methods were adopted to avoid ground water flowing into a working mine (Morton and Mekerk 1993), but they also minimised post closure pollution:

- underground dams
- diversion ditches or channels
- diversion wells, pipes, or adits
- grout curtains
- grout injection
- cover drilling
- backfilling of open voids
- backfilling with alkaline fly ash to buffer the acid and seal the ore
- mine entrance sealing

Last, but not least, neither diversion nor mine sealing installations are walk-away options for mine closure. Depending on the circumstances, technical installations might need longer maintenance or control according to a quality control plan. Sometimes, technical measures, such as backfill operations, can offer solutions without maintenance (Potvin et al. 2005).

5.7.2 General Measures to Keep the Water Away

Diversion of water can:

- prevent the contact of acid producing material with the water
- minimise mine water inrushes into opencast and underground mines
- minimise the amount of water that has to be treated
- mix acid mine drainage with alkaline surface water

In the first two cases, the surface or ground water has to be diverted around the potential inflow zones. Such zones could be sinkholes, seeping places, shafts, fracture zones, or boreholes (e.g. Scott and Hays 1975; Ackman and Jones 1988; Plotnikov 1988; Younger et al. 2002). If dams are constructed within the void space before inundation, the dams have to ensure that the water will either be prevented from leaving potentially polluting areas or will be guided around such areas.

Mine diversion should be avoided where it would prevent post-closure mine flooding that could result in a better ecological status of the catchment area. Such decisions must be based on appropriate hydrogeological and hydrogeochemical investigations and a thorough monitoring programme.

If rocks are highly fractured or karstified or there is a significant risk of surface water inflow into the mine voids, diversion can be an extremely useful tool to minimise post closure impacts. This method is therefore highly recommended if a high acid-producing potential in the mine voids is expected during and after mine closure.

Channels, pipes, or ditches can be used to divert surface water from mines. Such systems can be extremely large, sometimes several kilometres, and can affect different water catchments. They have to guarantee that the surface water will not percolate into the mine and that the water is diverted around the mine as fast as possible before it might enter the mine. Therefore, such systems must be securely sealed. The capacities of ditches, channels, and pipes must be evaluated by hydrological investigations or calculations and the destruction of such systems by storm events should be prevented, because their destruction could cause slope failures or sludge movement (see section 5.4.2). Regular inspection and periodic maintenance of such diversion systems is highly recommended – if not mandatory.

Besides surface water diversions, ground water diversions can be installed. Such systems consist of grout curtains, grout injection, or large subsurface concrete walls, as has been constructed in the Eastern German lignite mining areas (Fahle and Arnold 1996), where they protect nearby wetlands from ground water dewatering and prevent ground water flow into the opencast mines. In the Idrija/Slovenia mercury mine, grouting of abandoned mine workings prevents pollution and helped to prevent surface movements (Dizdarevič and Režun 1998). Similar techniques can be used during the mine closure phase if potential pollution sources are encountered. Cover drilling/cementation, normally used for developing new adits or shafts, can also be used to prevent unpolluted ground water from flowing into mine voids (Nel 2000).

From tracer investigations in already flooded mines, it is known that underground dams constructed before flooding can fail. This could be due to pressure differences on the sides of the dam, which could result in the dam's damage during the flooding process (Wolkersdorfer 2001). As the case of the M. Mayerova Mine/Czech Republic shows, grouting after the flooding of the mine might be a successful way to stop drainage of polluted water (Kipko et al. 1993).

Sometimes, the diversion of alkaline surface water might serve as an on-site remediation measure for acid mine water. By mixing alkaline and acidic mine water, the pH of the water can be raised to ecologically reasonable values as reported in the Montana/USA Stillwater River basin, where fresh surface water enters a highly contaminated mountain stream and thus improves the biological diversity (Gurrieri 1998). Although such techniques can reduce the concentrations of iron, toxic trace elements, metals, or semi-metals by hydrolysis and co-precipitation, they cannot minimise conservative water pollutants, such as sulphate, chloride, or sodium. Such a procedure can sometimes reduce the pollution load and improve the ecological status of the water; it is advisable not to apply the "dilution prohibition" rules of the waste directives, because this method must be seen as a passive treatment option for polluted mine water. At abandoned sites, it may be the only affordable option.

In the case of surface water diversions, it is important to know the catchment area of the underground mine so that the mine's water make might be calculated. In those cases where exploration boreholes were converted to observation wells (piezometers), the mine's catchment area can be calculated to a certain degree (Plotnikov et al. 1989). However, if not enough observation wells are available, the mine's catchment area can only be estimated since the inflow from "old" deep ground water can cause uncertainties in the calculation. Though numerical codes are an excellent tool to help the researcher in estimating the mine's recharge area (e.g. Singh and Reed 1988; MacLeod et al. 1993; Lewis et al. 1997; Dumpleton et al. 2001; Witkowski et al. 2003), they are nothing more than another way of producing a conceptual model of the mine's surroundings. Dumpleton et al. (2001) calculated two different flooding scenarios for the Nottinghamshire coal field with surface break outs of the mine water 19–20 and 52 years after the end of dewatering. They clearly state that neither scenario can be "identified as an optimum management proposal", showing that numerical models can be regarded as only a working hypothesis.

The following example shall exemplify how complicated it might be to calculate the catchment area of an underground mine. It is based on a draft project of the German "Wismut" company predicting the flooding of the Niederschlema/Alberoda/Germany uranium mine in the early 1990s.

The mine's catchment area was thought to be 22.6 km², in the river Zwickauer Mulde catchment area. Together with the Schlemabach, Lößnitzbach, Silberbach, Floßgraben, and Alberodabach tributaries, the catchment has a total area of 115 km². Based on geological investigations and hydrogeological observations underground, it was known that most of the water infiltrated through the upper 300 m of rock. A mean precipitation of 825 mm a^{-1} and a mean surface runoff of 38.7% as well as a mean annual temperature of 8.5 °C were reported (Veit et al. 1987). By using the equation of Turc (Gray et al. 1970) for calculating the real evapotranspiration, a total annual mine water make of $1.02 \cdot 10^6$ m^3 was calculated. However, the annual water make known from mine dewatering was $7 \cdot 10^6$ m^3 per year. Using this number to back-calculate the mine's catchment area, an area of 155 km^2 would be needed to get the actual annual water make of $7.0 \cdot 10^6$ m^3. Interestingly, by using Saul's observation, that 20–40% of the annual precipitation above the mined out area can be used to estimate the total water make (Saul 1948), the numbers would be $3.7–7.5 \cdot 10^6$ m^3. It was therefore suggested that stream water is infiltrating into the mine or that the mine affects an area bigger than the 22.6 km^2. Having such a big difference between the observed and calculated mine water makes, and the fact that it is not unusual that stream water infiltrates into underground mines, flow measurements in surface water streams to evaluate potential inflow zones would be necessary. Tracer techniques can be used successfully for such an investigation (chapter 10 and e.g. Kimball et al. 1999). Once it has been confirmed that stream losses are occurring, electrical conductivity can be used to detect the leaky areas and the fractures can be inexpensively grouted (Ackman and Jones 1988).

5.7.3 Physical Barriers

Under normal conditions, all parts of a mine are well connected with each other to guarantee quick, safe, and unhindered movement of material, workers, water, and air (Gertsch and Bullock 1998). Therefore, any fluid (which might be water, oil, or air) will pass through the horizontal and vertical connections in a relatively short time. Unused parts of a mine are commonly back-filled after the ore has been extracted (Landriault 2001; Potvin et al. 2005) but only in the case of cemented or hydraulic backfill will the hydraulic connections between the different parts of a mine be significantly reduced. Usually, the formerly open voids will be a preferential flow path for the mine water because the backfill still has a porosity that exceeds that of the surrounding host rock.

Consequently, the water in flooded mines can flow without restrictions into nearly all parts of the mine workings, and water pollutants can rapidly be transported through the mine by Poiseuille or free and forced convective flow. Additionally, the pore space of the previously dewatered rocks and sediments will be filled with water and Darcian flow will occur. Flow will also occur in fractures and faults.

Physical barriers, which are all types of constructions that manipulate or prevent the flow of fluids in an underground mine, delay or prevent water flow in the horizontal or vertical workings (Skousen et al. 1998). Such installations are commonly called bulkheads or dams and in working mines they are constructed to isolate abandoned workings, to control the water make of the mine, to manipulate the ventilation system, to exclude dangerous gases such as radon or methane, or to prevent unwanted access (Fig. 25). Of course, there are other reasons for constructing dams with no direct connection to proper mining behaviours. Physical barriers could be made of steel, concrete, sand, or wood, or combinations of those materials, depending on their function and the hydrostatic pressure they have to resist. Another method to control water flow into the mine or within the mine is grouting, where different types of cement mixtures can be used depending on the water chemistry or pressure. Only rarely have physical barriers been constructed to control the water flow during and after mine flooding, though their regulative character is obvious and has been studied by several scientific groups. If in-situ treatment of a flooded underground mine is planned (section 11.6), bulkheads should be constructed after a thorough hydrodynamic investigation of the mine (chapter 10). It must also be clear that bulkheads can have negative effects if constructed poorly or at an inappropriate position within the mine.

Bulkheads or open dam walls are constructed for reasons of control, emergency, precaution, or consolidation. Chekan (1985) discusses detailed conventional bulkheads to control the water flow and water level in mines. He distinguishes five types of bulkheads:

- Control bulkheads
- Emergency bulkheads
- Precautionary bulkheads
- Consolidation bulkheads
- Open dam walls

Control bulkheads aim to block abandoned parts of still active mines and prevent water inflow from behind the dam into the worked mine (Fig. 22). Usually, there is enough time available to carefully plan and consider the layout and position of the bulkheads. They have to withstand high water pressures and are usually permanent constructions with no access to the abandoned panels or workings. Often, they include pipes with valves and pressure gauges (in American English: gages). Emergency bulkheads are constructed in case of accidents, to protect the mine from unexpected water inflows. Their position and dimension is determined by the place of the inrush. No access to the workings behind the dam is planned and they also have to withstand high water pressures. Precautionary bulkheads are constructed well in ad-

Fig. 22. An underground dam in the abandoned Königstein/Germany ISL uranium mine to prevent the discharge of highly acidic leachate fluid (8 January 1992).

vance of potential water inrushes at locations that isolate the main haulage roads and mine entries. They can be shut in case of an accident and secure the mine's safety. Water pressures behind such constructions might be extreme. Consolidation bulkheads are temporary constructions that keep water away during grouting or ground consolidation. Open dam walls, finally, partly seal an adit and are a means to store mine water for reasons of treatment, conservation, or reuse. Besides those reasons listed under the five bulkhead types in Chekan (1985), some more reasons for constructing bulkheads for water control are conceivable. Environmental reasons should be added under the control bulkheads and open dam walls. For advance drilling, it might be useful to construct temporary bulkheads mentioned under consolidation bulkheads, and precautionary bulkheads could be used for hydrogeochemical investigations. Bulkheads could also be constructed as flow-through bulkheads to control water exchange within different otherwise separated water compartments. Finally, in the case of expected acid mine drainage, bulkheads have to be constructed using acid-proof concrete (Bagdy and Sulyokm 1990). Yet, this is only a classification about the use of bulkheads. More important is a classification that relates the construction of a bulkhead to its use.

Wolkersdorfer (1996) suggested the construction of vertical bulk heads to prevent the overall convective transport of contaminants for the Niederschlema/Alberoda uranium mine in Germany. Due to a lack of existing experiences, this method was rejected by the mine operator at that time. Heidenreich et al. (1991) used numerical methods to study the flow in a radioactive waste disposal and concluded that physical barriers could positively influence a convective flow regime. One of the rare examples where physical barriers were constructed before flooding is the Ronneburg/Germany uranium mine. There, 120 physical barriers shall separate highly contaminated mine water from low contaminated ones and thus minimise potential environmental impacts and treatment costs (Gatzweiler et al. 2002).

Lewis et al. (1997) made numerical simulations to find the best places for bulkheads in the Michigan/USA White Pine Copper Mine. They used MODFLOW to model the regional hydrogeological regime and found that nine bulkheads would positively influence the mine water's hydrogeochemistry and flow regime. However, their objective was not to influence the mass transport of contaminants in the mine, but to influence the amount of water draining from the mine. Without bulkheads, an annual water make of about $625 \cdot 10^3$ m^3 was calculated. Installing the bulkheads resulted in $405 \cdot 10^3$ m^3 per year, thus reducing the water make by 35%. Their solution is interesting as the bulkheads aimed to raise the water levels in the different mine compartments; because the cones of depression would be smaller, the water infiltrating into the mine would be reduced.

In the M. Mayerova Mine near Karlsbad (Karlovy Vary/Czech Republic), $8 \cdot 10^3$ m^3 of clay-cement grout were injected into the flooded workings to control the water flow in the mine (Kipko et al. 1993). As a result, the flow was reduced from $5 \cdot 10^{-1}$ m^3 min^{-1} to $2 \cdot 10^{-4}$ m^3 min^{-1} and the water temperature decreased from 31 °C to 18 °C, thus proving the positive effects of the grouting procedure. This example shows that physical installations can also be installed in mines that are already abandoned and flooded. However, a proper grout composition must be found for the mine water chemistry. The grout's viscosity and composition has to be selected so that it can flow through the disposal pipes without blocking them and build up in the adits that have to be blocked without flowing away with the water stream.

Another physical method to reduce water inflow into flooded mines is the sealing or divergence of surface streams crossing over an underground mine. Such measures can minimise the influx of oxygen-rich surface waters and consequently the oxidation of disulphides. Furthermore, the amount of drainage water is decreased and treatment operations have to deal with lower water flows. Yet, a wrong composition of the grout or the wrong grouting technique could cause unnecessary costs (Ackman and Jones 1988; Morton and Niekerk 1994; Niyogi et al. 1999). Measures to locate potential infiltration zones are electrical conductivity investigations within the stream segments that could infiltrate into the underground mine, tracer investigations, or flow measurements (Ackman and Jones 1988; Kimball et al. 1999). At the abandoned Teichtalstollen near Königshütte/Harz Mountains in Germany, a tracer test with microspheres proved that there is a good hydraulic connection between a surface stream and the drainage adit, which is used as a drinking water supply (Wolkersdorfer and Hasche 2003).

A different type of physical barrier would be the complete sealing of the surface above a mine, called encapsulation. Plastic liners or low permeable natural material, such as clay, are emplaced above the affected area. This method prevents the infiltration of surface water into the worked veins, seams, or waste material to stop or minimise the formation of acidity by disulphide weathering. In addition to the clayey material, alkalinity-producing material can be used to buffer the acidity. Yet, there are difficulties in doing this. For one, the area to be sealed could be extremely large. At a USA surface coal mine in Alton, West Virginia, $1.8 \cdot 10^5$ m^2 of a combined disposal site for coal refuse and overburden was sealed with a plastic liner. This resulted in a reduction of ground water recharge and a significant diminution in the acidity loads (Skousen et al. 1998). Another problem is locating the correct area of recharge above the underground mine. Even if this can be delimited to a certain degree, the subsurface recharge area (that means ground water inflow into the mine) still stays unclear until isotopic studies or tracer tests define the full recharge area. Those reasons, together with the fact that some parts of the surface cannot be sealed because of slope stability or construction cause such systems to be only partly successful.

All of the techniques described above are designed to control the water flow in or around the flooded mine. Thus, the oxygen-rich air in the mine is replaced by oxygen-poor water and the rate of acid production consequently reduced. As described above, the flooding of adit mines can cause geotechnical problems due to high potential pressures at the water side of the seals. For situations where a full sealing of an adit might be dangerous, partial physical barriers have been developed. They guarantee a free flow of the water, but eliminate the flux of oxygen into the underground mine by insulating the mine against the air, and can reduce acidity by 50–80% within the first months of mine sealing (e.g. Anonymous 1936). Such techniques usually

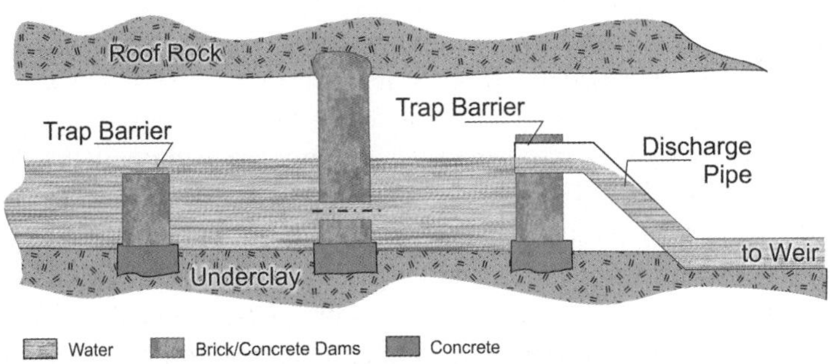

Fig. 23. Partial seal of an abandoned adit mine (after Foreman 1971).

consist of a concrete or brick wall, a U-bended pipe, or a wall combination which allows free flow of water but eliminates oxygen flow into the mine through the adit (Fig. 23). Because of such a physical barrier, the concentrations of radon behind the wall might increase above the MAC-values, and the oxygen content of the mine air will drop under the values necessary for safe working conditions. Such techniques seem to contradict the needs for bat protection, because bats need good quality air for living underground. The construction of such air-barriers should therefore be shifted several tens of meters into the adit if bats are known to live in the mine. Yet, as Foreman (1972) has pointed out for the mine sealing projects in Butler County, Pennsylvania/USA and Schindler et al. (2004) for the Randolph County, West Virginia/USA, "the effect [air] seals have had on water quality is inconclusive".

Mine sealing operations usually exclude future options for in-situ remediation techniques and may shift the mine water problem from a point-source to a non-point source problem. Furthermore, bulkheads or dams are engineering constructions that are subject to chemical weathering or by rock mechanical failure – in no way should such constructions be seen as "walk away options" (ERMITE Consortium et al. 2004). Adit mines should not be completely sealed because the breakout of heavily polluted mine water could cause disastrous damages to human life, engineering constructions, and the environment. Positive cases of mine sealing operations can be found in the literature (Stoertz et al. 2001), but it is not always clear if the positive effect is due to the mine sealing or the natural attenuation of the mine water.

5.7.4 Mine Entrance Sealing

Permanent mine seals are being used more nowadays because such systems overcome several post-closure problems. They:

- prevent drainage of polluted water
- minimise oxidation of disulphides
- stop the emission of hazardous gases
- prevent unauthorised entrance to the open voids
- protect cultural heritage

Yet, the total sealing of mine entrances might contradict with the needs of protecting endangered species, for example bats (which need only small holes to get through), and should therefore be used with care. As discussed above, the construction of such barriers can be shifted several tens of meters into the adit if bats are known to live in the mine. Furthermore, the total mine seal could activate pre-mining water drainage systems, such as karst features or fractures, which were dry during the operation of the mine. In such circumstances, mine water could drain uncontrolled into the environment and mine water treatment, in such cases, could be problematic. Such situations can occur when the water level is allowed to raise in an uncontrolled manner to its near pre-mining level.

In the USA, mine entrance sealing studies with the aim of reducing surface stream pollution date back to the late 1920s; by the late 1930s, over 1,000 coal mines had already been sealed with pollution reductions ranging from about 50% to 90% (Leitch 1935; Fellows 1937). One of the most comprehensive description of pollution prevention and control from abandoned mines has been given by Scott and Hays (1975), though most of the techniques described have been improved since then. Mine sealing, according to this report and other publications, depends firstly on the mine layout (e.g. Foreman 1971). Mines with access through adits or galleries have to be sealed differently than mines with access through shafts or inclines.

Before any sealing is conducted, the void where the seal will be placed should be cleaned of all trash, such as timber, metal, wires, and other material, that could hamper an effective sealing procedure. In addition, wastes that are potentially harmful to the aquatic environment, such as oils, chemicals, or cleaning detergents, should be removed. Most other materials used in mining have a low pollution potential under anoxic conditions. Mine sealing can be conducted to either prevent water from draining into the river catchment or to minimise the amount of oxygen coming in contact with the exposed rock. Wet mine seals can be used for both of the before mentioned options, while dry mine seals are only intended to prevent the entry of oxygen (Fig. 26).

Another method, first described by Taylor and Waring (2001), prevents acid mine drainage by displacing oxygen with reducing gases generated by anaerobic bacterial activity, such as CO_2, CH_4, or H_2S. This GaRD (Gas Redox and Displacement) System can halt sulphide oxidation and acid generation, and may result in the precipitation of secondary sulphides from the accumulated drainage water.

As will be shown in the examples, not all mine sealing operations are successful. Dam failures, though rarely observed, are mostly due to seal blowout or leakage around the constructed dam, or chemical reactions between the dam and the mine water. It is important to note that the hydraulic head above a mine seal can be extremely high, and so mine seals must be well engineered. Most mines would have a hydraulic head exceeding that of the highest water dams (Fig. 24). Another issue is the mine water quality. As several investigations show, mine water quality might change during time and chemical reactions between the water and the concrete could destroy the dam construction. Most of the failures that have been reported can be attributed to insufficient hydrogeological or geotechnical investigations, which would have

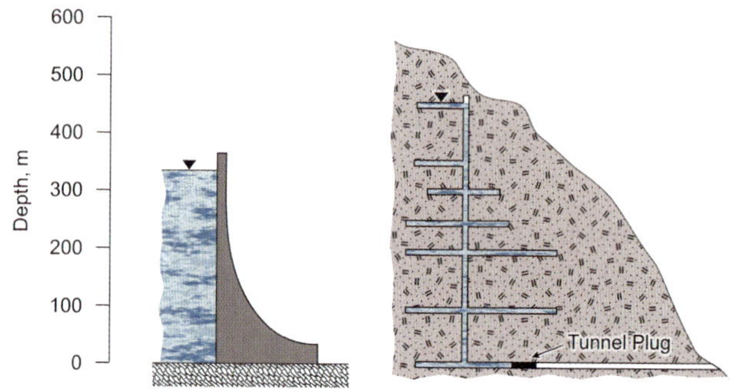

Fig. 24. Comparison of hydraulic head above a mine plug with the world's highest dam (Rogun Dam, Tadschikistan, 335 m).

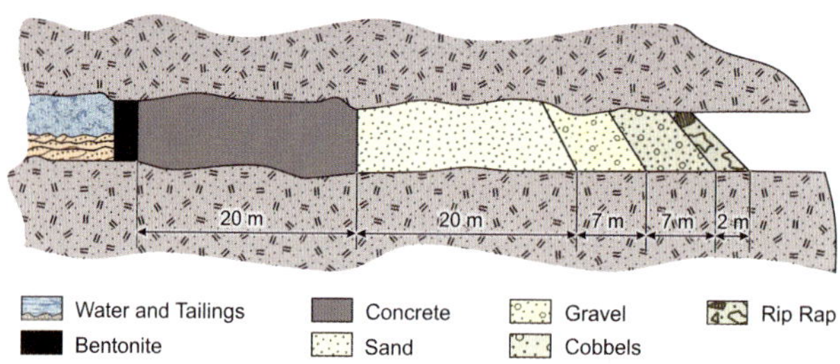

Fig. 25. Example of an underground mine dam (after Lang 1999).

been less cost intensive than the subsequent remediation measures! Some of the possible dam constructions and mine sealing techniques are illustrated in Fig. 25 to 27.

Shafts are usually sealed by either backfilling them with inert or alkalinity-producing rock material or with a concrete seal. Each technique has its advantages and disadvantages which should be discussed within the mine closure plan. For monitoring purposes, a concrete plug should be constructed with monitoring pipes with a diameter of at least 8 cm (3 inches) to allow access for measuring instruments (see chapter 9). Because backfilled shafts could be a pathway for water infiltrating into the mine, the top of the backfilled shaft should be sealed to direct drainage away from the shaft.

Mine adits that do not drain polluted water and are unlikely to drain polluted water in the future should not be permanently closed. Such openings can serve as reserves for endangered animals such as bats. Gates can be constructed safely to allow free drainage of water and free access for the animals. Local environmental protection groups, bat protectionists, or mining archaelogists can help maintain such localities.

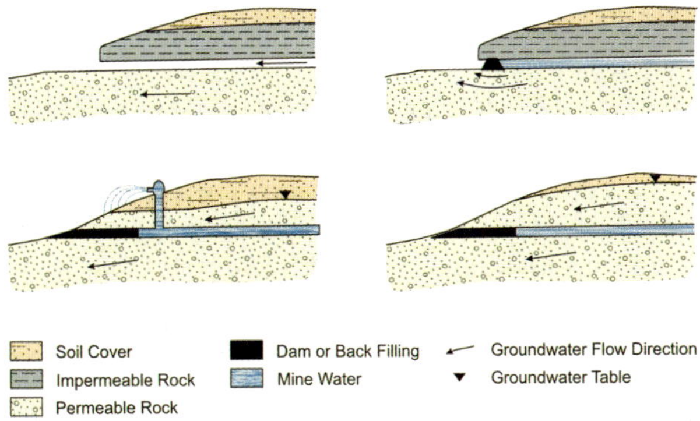

Fig. 26. Possible mine sealing techniques (after Fernández-Rubio et al. 1987).

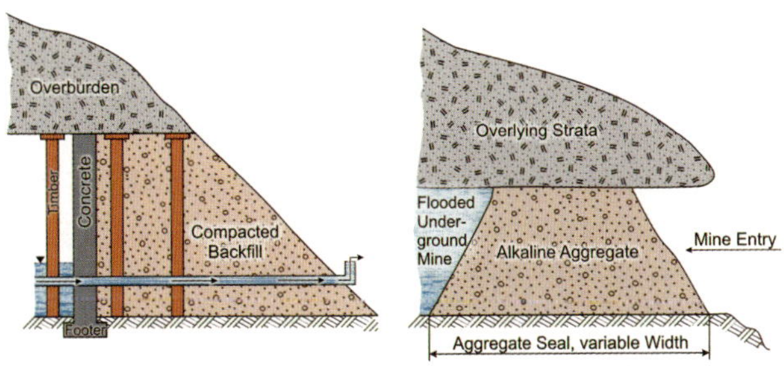

Fig. 27. Possible mine sealings (after Scott and Hays 1975).

To stop the production of acid mine drainage, total or partial mine seals with dams or barriers can be installed. Usually, such underground barriers or dams must just prevent oxygen coming into contact with the disulphides thus minimising the production of acidity. They must not be constructed as water dams withstanding the water pressure from a total mine flooding. Yet, as Barnes and Clarke (1964) state, oxygen is not the only controlling reactant of acid mine drainage/metalliferous leachate and before a mine is sealed, a thorough understanding of the theoretical geochemistry of the site as well as laboratory experiments and field observations are necessary.

A positive mine sealing example is the Beth Energy's underground Mine № 105W near Buckhannon in West Virginia/USA. Two of the three mine portals had to be sealed with five seals to prevent the drainage of acid mine water. At the same time (1984/85), the mine was treated with an in-situ application of apatite rock-dust to the potentially acid parts of the mine and by injecting alkaline substances after sealing in the mine water pool. Even after four water changes, the mine water, that originally had a pH of 3.9, retained a pH of around 7 (Hause and Willison 1986) and the iron concentration ranged from 1 to 15 mg L^{-1} in 1989, after an initial concentration of 144 mg L^{-1} (Aljoe and Hawkins 1993).

Yet, there are cases where a mine entrance sealing had a negative impact on the drainage water quality and the receiving streams. At the Mynydd Parys mine site in Anglesey/Wales, the Afon Goch stream is highly polluted by acid mine drainage with pH values between 2 and 3 (Boult 1996) and elevated concentrations of metals (e.g. Al 81 mg L^{-1}, Fe 431 mg L^{-1}, Cu 39 mg L^{-1}, Zn 45 mg L^{-1}; Younger et al. 2004). In the early 20th century, the Dyffryn Adda drainage adit was sealed with concrete and the mine water, containing elevated amounts of copper, was processed to extract the copper from the mine water. During and after this regular water extraction from the underground mine, which was controlled by valves and pipes through the dam, a mine water body with approximately $270 \cdot 10^3$ m^3 of acidic mine water and a hydrostatic pressure of $3.6 \cdot 10^5$ Pa built up behind the dam. One reason for the development of that mine water pool was the closure of the valves after the *ex-situ* mining was stopped in the mid 1950s. As no data about the dam was available and the dams safety was not clear, $245 \cdot 10^3$ m^3 of acidic mine water were pumped out of the mine to dismantle the dam. After the dam was removed, the mine water quality improved slightly, based on microbiological and geochemical analyses (Bryan et al. 2004; Coupland et al. 2004; Johnston 2004). Consequently, the water quality of Afon Goch South has greatly improved while the Afon Goch North is no more impacted than it was before.

5.7.5 Quality Control Issues

Quality control in diversion of mine waters is equally important as quality control in sampling and remediation measures. Therefore, a diversion quality control plan, which will also help to support decisions concerning the water diversion, should be established. In the diversion quality control plan, the following requirements should be described:

- which specific types of data are needed
- what are the data collection requirements
- which monitoring instruments are needed
- how to deal with failure of instruments
- which measurements of which parameters are necessary

Periodic inspection and maintenance of diversion installations is an important component of a diversion quality control plan. Water must flow without hindering in the channels and ditches, and and any erosion or overflow of the diversion installations must be repaired as soon as possible. Maintenance must include the cleaning of any waste material in the channels or at the measuring devices. There must be references to the maintenance intervals in the diversion quality control plan and control of the diversion systems is necessary at least before and after major storm events to guarantee its proper functioning.

If mine entrances are sealed with dams, the dams have to be inspected and maintained regularly to guarantee their proper functioning. Any leakage of the dam must be reported according to the quality control plan. If physico-chemical parameters of the diverted water courses must be measured, the same quality controls as described in section 10.7.1 are essential.

5.7.6 Examples and Case Studies

Examples of water diversions around mines as well as mine seals are numerous and can easily be found in the relevant literature data bases (e.g. GeoRef, Geobase, FIZ-GEO). Here, four examples with interesting results will be given to show the effects of mine water diversion and mine sealings.

Biaglow (1988) reported on a hydraulic mine seal in the year 1980 at the Mine Complex 88 at Big Four Hollow Creek near Lake Hope State Park in southeastern Ohio/USA, which did not result in the predicted pollution reduction. His study was conducted to determine the effectiveness of the mine seal and to clarify whether flooding the mine complex had an impact on the amount of acidic mine drainage to Big Four Hollow Creek. Comparisons with historical data showed that the mine seal initially resulted in a decreased amount of dissolved sulphate and iron loadings. This decrease was temporary; dissolved sulphate and total dissolved iron loadings returned to nearly the same levels observed before completion of the mine seal. Twenty years later, in 2000, the site was reinvestigated and it could be shown that on a long term basis, the hydraulic mine seal had a small, but positive effect (Stoertz et al. 2001). Since its sealing, the pH in the mine water had increased from 2.7 to 5.3, electrical conductivity had decreased from 2,700 to 600 $\mu S\ cm^{-1}$, and dissolved oxygen was lower than 2% saturation. Oxygen concentrations had been lowered, and this was associated with a decreased production of acid mine drainage. Furthermore, it could be shown that the mine water quality has improved gradually in the two decades since flooding, due to mine sealing, natural attenuation and the effect of the first flush. On the other hand, leakage around the mine seals and an accidental seal blowout resulted in less remediation than expected.

Surface water diversions around active mines are commonly used where the surface water might either be a hazard to the surface or underground mine operations or in cases where the surface water is directly above a productive mining area. In the case of the Meadowgate open pit mine south-east of

Sheffield/UK, the River Rother was diverted between 1977 and 1981. After the coal was mined out, the whole area was reconstructed and in 1983, the Rother Valley Country Park was opened. No real problems occurred during the construction of the diversion but, after the river was diverted back, the alluvial clay planned to be used as a river bed caused handling and storage problems due to its plastic character (Brook and Cresswell 1989).

Subsurface concrete sealing walls were used for the first time in eastern Germany, at the border with Poland. They prevented nearby wetlands (Peitzer Teichlandschaft und Spreeaue) from drying out and also prevented the river Neiße from flowing into the large open pit lignite mines (Jänschwalde Open Pit). These diaphragm walls are up to 85 m deep and 11 km long. Wall construction first started there in 1979 and the last one was finished in 2005. The hydrogeological investigations and more than 20 years of experience clearly show that deep sealing walls can be used as an alternative to surface water diversions to prevent water from flowing into working and abandoned mines (Mielcarek 1988; Arnold and Kuhlmann 1994; Häge 1995; Fahle and Arnold 1996). From a technical point of view, such walls can be as deep as 100 m if there is a low permeability strata with a thickness of at least 2 m. Furthermore, no interaction with another cone of depression must be allowed and no dewatering under the sealing strata must take place. Proper pre-construction hydrogeological modelling of the area is an important requirement to guarantee the long term effectiveness of such walls.

During investigations of the flooded Straßberg/Harz/Germany underground mine in the early 1990s, it became clear that the mine water in the flooded mine was stratified, with a shallow water body of less-contaminated mine water and a deeper one with highly-contaminated mine water (Rüterkamp and Meßer 2000). Therefore, three drainage adits were constructed to divert the less-contaminated mine water and prevent it from becoming more contaminated. However, shortly after the three adit system came fully in force in 1996, the stratification broke down and the highly contaminated mine water entered the catchment area of the Selke River. By that time, neither the mine operator nor the consultant were aware of why this breakdown of stratification occurred. However, the situation was similar to what occurred in the flooded Niederschlema/Alberoda uranium mine, where the effect of convective mixing in such a situation has been described, based on a three year study of the flooding regime (see section 12.11 and 12.9; Wolkersdorfer 1996). The Straßberg example clearly shows the importance of hydrodynamic investigations of a flooded mine and how useful tracer tests can be to optimise a proposed diversion scheme (Kindermann 1998; Wolkersdorfer 2001).

One of the biggest remediation projects in the USA was conducted in the Tar Creek area (in the Picher Mining District), in northeastern Ottawa

County, Oklahoma. In 1983, the area was added to the USA "National Priorities List" and subsequently became known as the "Tar Creek Superfund Site". To prevent surface water from flowing into the underground workings, diversions and dike structures were constructed. Surface water was diverted away from about 600 of the estimated 2,600 abandoned mine shafts, open boreholes, and surface subsidence areas located on the site. Approximately 75% of the annual inflow coming from the surface was thought to be prevented from flowing into the abandoned workings and subsequently into the receiving stream, the Tar Creek. Yet, in 1991, it became clear that the decreasing concentrations of pollutants might be a naturally occurring phenomena since the discharged mine water volume had not significantly decreased and the surface water quality had not significantly improved, so it was decided that the well plugging operations had not been effective. Finally, after nearly two decades of remediation, in 2000 the "Governor Frank Keating's Tar Creek Superfund Task Force" was set up to co-ordinate and complete the clean-up process for the most prominent dangers to the environment and people (Griffin et al. 2000; Tar Creek Superfund Task Force Water Quality Subcommittee 2000a, 2000b; DeHay 2003; DeHay et al. 2004). Meanwhile, an eight-year Watershed Management Plan for the Tar Creek and the Lower Spring River was set up (U.S. Army Corps of Engineers 2004). Investigations in 2002 indicated that the mine and ground water quality had improved but though enormous amounts of money were spent to remediate the Tar Creek catchment area, it must be concluded that from a hydrogeological point of view, most of the improvement can be attributed to natural attenuation. None of the diversion measures was effective enough to significantly improve the mine water quality on a long term basis, though the amount of infiltrating water (and thus the load) on the site was reduced. One of the reasons might be that the Pennsylvanian and Mississippian rocks are too permeable and fractured for diversion measures to be effective.

From the above-mentioned examples, it can be concluded that diversion measures after mine closure are not cost effective enough to be used in the future. In fact, it is not clear if the improvements in ground and surface water quality that occurred could simply be the result of natural attenuation processes. It might be disappointing that reactive transport cannot be influenced by diversion measures so as to protect receiving water courses on a long term basis. However, in all these cases, there was a lack of hydrodynamic and hydrogeological understanding. The upstream catchment areas of underground mines can be extremely large and diverting surface water just above the mined out strata is usually not of much help. In cases where tracer tests can prove a hydraulic connection between the surface streams and the mine workings, surface water diversion should be successful; in other cases, it tends to be a waste of money.

5.8 Mine Dewatering

5.8.1 Introduction

Though mine dewatering is usually not involved when a mine is abandoned, there are cases of controlled flooding where mine water has had to be pumped out of a mine during the flooding process, either for treatment or for environmental investigations (e.g. Anonymous 1997; Nuttall et al. 2002; Jakubick et al. 2002; Gammons et al. 2003; Nuttall and Younger 2004; Coulton et al. 2004; Wolkersdorfer and Hasche 2004). Furthermore, as in the case of the abandoned coal mines in the southern Ruhr area, pumping may have to continue to prevent productive pits from flooding (Coldewey and Semrau 1994) and for other similar reasons (Younger and Harbourne 1995; Younger 2002b). Due to the different mining methods, dewatering of open pit mines differs from dewatering of underground mines. Therefore, the following chapter will focus on mine dewatering techniques, some of which can also be applied to active mining schemes.

Since Neolithic times, dewatering of mines was a major task for the miners, and where no dewatering techniques were available, the mine had to be closed down (Shepherd 1993). Beyond manual water haulage and the use of buckets and leather buckets mounted on water wheels, no real innovation was made in underground mining for centuries (Younger 2004b). The first reliable wooden pumps were invented at the German Rammelsberg/Goslar in the 15th and at Ehrenfriedersdorf/Erzgebirge in the 16th century (Agricola 1556; Kreibich and Eisbein ca. 2003), and it was not until 1749 when the first reliable metal pump was introduced in Schemnitz by Joseph Hell. The first water-independent dewatering machine, the *machina pyraulica* of Thomas Newcomen (the first usable steam-engine to operate in a mine) was installed in 1712 (Kunnert 1974; Younger 2004b) and the first one with a metal pump was installed in 1788 in the German Tarnowitz lead-zinc mine (now Tarnowskie Góry, Poland; Seidl and Stähler 1913). Until then, the depth of underground mines was mainly controlled by the depth of the water table or the dewatering techniques available. Nowadays, the dewatering techniques for open pit mines and underground mines are well established and even mines that are several thousands of meters deep can be dewatered (Vutukuri and Singh 1993; Duane et al. 1997; Fig. 28).

Drainage adits are another long-established method to dewater deep mines, and were also used as water supply tunnels in prehistoric and historic times (e.g. the Turkish Troyan Water Mine; Biswas 1970; Wolkersdorfer and Göbel 2004). Such drainage adits needed the close co-operation of all the miners, land-owners, and authorities involved because the construction of such systems took up to 100 years and reached lengths of several tenths of kilome-

Fig. 28. Mine dewatering pump and mine water scheme control room in the Vazante/Brazil underground zinc-mine.

tres, and, during the construction, no profit was made (Jobst et al. 1994; Gundelwein 2001). To protect the miners of such systems and to guarantee the water rights of the operators after the drainage adit reached its final length, the construction, operation and maintenance of such dewatering systems was regulated by special legislation for such *Wassergemeinschaften* (water communities) and *Erbstollengemeinschaften* (drainage adit communities). Some of the best known German *Erbstollen* are the *Roederstollen* at the Rammelsberg, the *Schlebuscher Erbstollen* near Sprockhövel, the *Rothschönberger Stollen* of the Freiberg Mining area or the *Tiefer-Ernst-August Erbstollen* of the Clausthal and Zellerfeld Mining (Ließmann 2002) areas. Some drainage adits are still in use more than 200 years after their construction. Younger (2004b) lists 38 "important" drainage adits for Britain that were constructed between 1617 and 1897.

In many European countries, mine water – if exceeding regulatory limits – is classified as waste water and is controlled in accordance with the relevant waste water directives. Examples of such countries are Finland and Turkey, where all industrial water, including water emanating from mining operations, are classified as waste water (Mäkinen 1996; Wolkersdorfer and Göbel 2001b). Other countries, such as Germany and Ireland (Government of Ireland 1997; Plinke and Wildhagen 2003), have special regulations for mine water. In the future, when the European Water Framework Directive will finally be set in force in all EU countries, a harmonization of water legislation might occur.

5.8.2 Dewatering Techniques

No comprehensive textbook about mine dewatering techniques has been published in recent years, though Younger et al. (2002) describe the basics of mine dewatering and the problems involved. The most comprehensive books about mine dewatering techniques ("*Bergmännische Wasserwirtschaft*") was published more than 50 years ago in German by Kegel (1950) and eight years later by Sinclair (1958). A thorough discussion of more modern dewatering techniques is, unfortunately, beyond the scope of this publication.

In principle, water management in an operating mine can be divided into water collection, water removal, and water disposal. Most surface ("open pits") and underground mines have to be dewatered to gain access to the raw material and to guarantee safe working conditions for the miners. During exploration and working, pollution reduction is usually not the main focus of the dewatering, though several cases exist where successful dewatering also serves to prevent pollution (Morton and Niekerk 1994). Drainage water is usually referred to as "mine water", but according to Wolkersdorfer and Hasche (2001c), mine water can be classified as follows (though this definition might only be suitable for German authorities):

- flooding water: ground water that is used for the flooding of an open pit/underground mine or water that is discharging after the flooding of the open pit/underground mine
- free or pore water: processing water that is accumulated together with precipitation in tailings ponds
- leakage water: surface water that infiltrates into waste dumps as well as waste rock piles and leaks out at the bottom of waste dumps, waste rock piles, or within the open voids of working mines
- geogenic water: ground water from deeper aquifers that are typical for hard coal mining

Using these definition, only leakage water and geogenic water will be discussed in this chapter. Drainage water amounts can range between several litres to several cubic meters per second (Fig. 29) and can vary extremely in their chemical composition (Younger 1997; Younger et al. 2002; Lottermoser 2003). Some drainage waters might even be used as drinking water (see section 11.7).

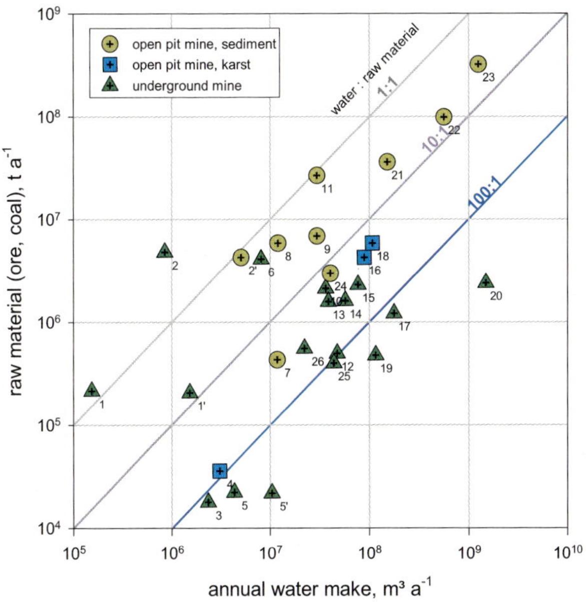

Fig. 29. Selected worldwide ratio of water make and raw material production (modified, updated, and data added after Fernández-Rubio and Fernandez Lorca 1993). 1 and 1' Agnew (Australia, Ni), 2 and 2' Lock (Australia, coal), 3 and 4 Burra (Australia, Cu), 5 and 5' Beaconsfield (Tasmania, Au), 6 German Creek (Australia, coal), 7 Scottish National Coal Board (UK, coal), 8 Bowmans (Australia, lignite), 9 Thorez (Poland, coal), 10 Picher (USA, Pb, Zn), 11 Morwell (Australia, lignite), 12 Fejér Co. (Hungary, bauxite) 13 Chingola (Zambia, Cu), 14 Dorog (Hungary, coal), 15 Tatabánya (Hungary, coal), 16 Pine Point (Canada, Pb, Zn), 17 Konkola (Zambia, Cu), 18 Kingston (Australia, lignite), 19 Bakony (Hungary, bauxite), 20 Kailuna Hebei (China, coal), 21 Atabasca (Canada, tar sand), 22 Rheinbraun (Germany, coal), 23 Lausitz (Germany, coal), 24 China (coal), 25 Vazante (Brazil, Zn), 26 Leonie (Germany, iron).

The following dewatering methods are commonly used in open pit mining, depending on the mine layout and the local geological conditions: filter wells, disposal wells, inverted wells, vacuum drainage, horizontal drains, sealing walls (grout walls), cut off walls, and guard wells (Plotnikov et al. 1989; de la Vergne 2003; Kecojevic et al. 2003). By using such methods, even extremely large and deep open pit mining areas can be successfully dewatered, as can be seen by the German Rhenish lignite mines, which are dewatered to a depth of 350 to 500 m, with a cone of depression of approximately 2,500 km² (Briechle et al. 1998).

In contrast to open pit mines, underground mines that are not dewatered using adits are usually dewatered using sump pumps. A sump at the deepest or at shallower levels collects all of the water seeping into the mine. As a general rule, mines with all of their workings 140–500 m below the surface can be expected to be relatively dry (Saul 1948; Wolkersdorfer 1996; de la Vergne 2003), because hydraulic conductivities at such depths are very low, ranging between 10^{-6}–10^{-9} m s^{-1} (Karrenberg 1981; Stober 1995). Furthermore, it can often be observed that water inflows in non-karstified rocks stop in a relatively short time.

Submersible pumps are installed close to the sump pumping stations to pump the water to the surface. This water can be chemically contaminated or contain suspended matter and so typically needs to be treated after having been pumped out of the mine. While turbidity can be removed using settling lagoons, chemically polluted water has to be treated using appropriate passive or active methods (see chapter 11).

When planning a dewatering scheme within a water catchment, measures have to be taken to ensure that the ecological status of the surface waters will not be adversely affected. The selection of which dewatering systems to use requires careful consideration of factors such as the: hydrogeological and hydrological conditions, lifetime of the mining operation, total water make, availability of dewatering equipment, and drainage water quality. Furthermore, the following aquifer parameters have to be known: type/state of aquifer (confined or unconfined), drainable porosity, transmissivity and storage coefficient, static water level, seasonal variations, depth and thickness, as well as water recharge (Morton and Niekerk 1994).

As several studies have shown, sump dewatering should be given preference against adit dewatering, because after closure, mines with adits usually drain polluted water over a longer life span than sump dewatered mines (Commonwealth of Pennsylvania 1973; Mentz et al. 1975 in Scott and Hays 1975; Hawkins 1994; Younger 1998; Callaghan et al. 2000). In the case of the Shoff and Yorkshire mines, both situated in the Pennsylvanian Bigler Township/USA, the differences between updip (adit dewatering) and downdip (sump dewatering) mining were evaluated. Though the two mines were on opposite sides of Clearfield Creek and were worked in different ways, the geological situation, coal quality, period of operation, discharges, and the hydrological conditions were similar (Commonwealth of Pennsylvania 1973). In 1974, the water from both mines was sampled eight times to evaluate water composition and the water quality of the flooded downdip Yorkshire Mine was significantly better than the water draining from the updip Shoff Mine (Fig. 30). Mentz et al. (1975) also compared the economical differences of both mining methods. Obviously, pumping costs are higher in downdip mines, but if offset against remediation costs, downdip mining has the advantage since the water quality in the flooded shaft mines is better over the long term.

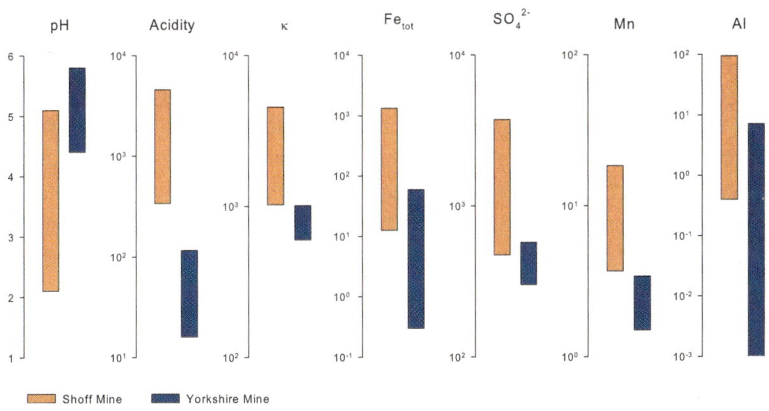

Fig. 30. Up-dip versus down-dip mining: comparison of selected mine water parameters of the Pennsylvanian Shoff (up-dip, adit-drained) and the Yorkshire (down-dip, flooded from shaft collar) mines (after Mentz et al. 1975). $n = 8$; pH without unit, κ (specific electrical conductivity) in µS cm^{-1}; all other in mg L^{-1}.

The reason for this is that sump dewatered mines are usually flooded after mining comes to an end, while mines with adits are subject to oxidation above the water table, thus releasing metals and acidity for a long period of time. Flooded mines, as discussed in section 5.4, insulate the disulphides from the oxygen and thereby limit – to a certain degree – microbiological activity and therefore reduce the release of acidity and consequently pollutants. Examples of adit mines with heavily polluted discharges can be seen at the *Rothschönberger Stollen* in the Freiberg/Saxony mining area (Baacke 2000; Weyer 2003; see section 12.12), the Gernrode/Harz Mountains Hochwarthe Adit, or the Burgfey Adit in the Eifel Mountains (Mair 2002).

5.8.3 Disposal and Treatment of Dewatering Effluents

All mines can encounter polluted waters. One of the most problematic mine waters is acid mine water (usually called AMD [acid mine drainage] or ARD [acid rock drainage]) or neutral zinc-rich water which can already occur during the active mining period. Active treatment plants are usually installed for cleaning highly polluted mine water during the mine's lifetime. Those techniques are well established and the construction of LDS (low density sludge) and HDS (high density sludge) treatment plants (section 11.3) as well as other techniques (e.g. ion exchange, reverse osmosis, nano-filtration, distillation, electrodialyses) are well known. A combination plant has been working at the Polish Dębieńsko coal mine since 1975 (Motyka and Skibinski 1982; Tar

Creek Superfund Task Force Water Quality Subcommittee 2000b; Brown et al. 2002; Turek 2004). An interesting mine water treatment technique is a distillation process described by LaRosa et al. (1971). They used coal refuse together with molten iron to produce energy which is then used to drive either a distillation or a reverse osmosis plant. While the process worked well in the laboratory, it has never been used on site.

After treatment, drainage water can be discharged into surface streams or lakes; the residuals are typically disposed of or pumped back into the underground mine voids. In some rare cases, the residuals can be used as a raw material. One example was published by Hedin (2003), who was able to sell 1000 t of iron oxide recovered from an abandoned coal mine in the Appalachian mountains/USA. Another one is the treatment plant at the abandoned Hungarian Mecsekérc uranium mine (Pécs/Hungary), where uranium is extracted from the mine water and sold on the world market (Gombkötö et al. 2002). Unfortunately, the extraction of pure metals from mine drainage remains economically non-viable.

An alternative to minimise the impacts of mine drainage water are passive treatment methods for polishing already treated waters or for treating less polluted mine water (section 11.4). Those systems can remove different mine water pollutants, at least iron and turbidity and should therefore be installed wherever applicable.

As discussed above, not all mine waters are a source of pollution. Depending on the water quality, drainage water can sometimes be discharged to the river catchment without further treatment. Such waters can be used for irrigation purposes, drinking water, or health resorts (Banks et al. 1997; Annandale et al. 2002). In the German lignite mining areas, several hundred million of cubic meters of geogenic drainage water can be discharged without any further treatment because it is good quality ground water (Regionaler Planungsverband Westsachsen 1998; Landesumweltamt Nordrhein-Westfalen 2002). Nonetheless, unpolluted mine water, when diluted with more polluted surface water, will change the geochemical equilibrium of the receiving water course. This might result in dissolution of previously precipitated contaminants and so must be considered in the mine drainage plan. In the worst case, the surface water quality downstream of the mixing zone can be poorer than upstream of the inflow. Where the problem is the higher suspended load of the mine water, a settlement lagoon can be placed above the discharge point (Fig. 31).

More typically, pumped drainage waters have to be treated according to the appropriate standards, depending on the mining licence. In accordance with the European Water Framework Directive, mine operators should introduce dewatering techniques that minimise the amount of polluted water that has to be treated.

Fig. 31. Settlement lagoon and drainage pumps of an open pit iron mine near Belo Horizonte/Brazil.

An interesting case of acid mine drainage/metalliferous leachate occurs in northwestern Pennsylvania/USA, which is already threatened by several thousands of mine water discharges polluting the surface streams in this area. Acidic, iron-rich water was percolating through gas wells that were abandoned about a century ago as artesian discharges (Hedin et al. 2005). Those wells reach 300–500 m down into gas bearing Devonian sands, which are connected to formation brines. Yet, the acidic water was not saline and during the investigations it was shown that the casings of the wells must be corroded or missing in the shallower aquifers. Those aquifers are connected to shallow coal mine workings and when reaching one of the gas wells, the water upwells to the surface. Because the wells are so numerous, the authors concluded that sealing the wells would not be feasible as the water would just flow to another well or to another place with a lower hydraulic pressure where there would be fewer possibilities to treat that water.

5.8.4 Drawdown Management

Usually, mine abandonment involves the recovery of the water table to near pre-mining conditions. Yet, there can be several reasons to maintain the water level below the potential discharge point to avoid negative economic or environmental impacts. Such a situation can occur when the raising water level is expected to flush tailings or waste rock piles, cause polluted mine water to enter drinking water reservoirs, flush weathered minerals in long-abandoned near-surface mine workings, or if subsided areas might be flooded.
Mine dewatering causes the natural ground water table to drop in the close vicinity of the mining operation. This may lead to the drying out of springs, wells, lakes, or rivers. It is advisable to start the ground water level monitoring well in advance of the mining (or remediation) operation, converse

that exploration drill holes can be converted into ground water monitoring wells (Plotnikov et al. 1989). Water wells have to be constructed in accordance with good practice procedures (Driscoll 1987) so that the drawdown can be precisely monitored during the lifetime of the mine and the remediation process. From a hydrogeological point of view, mine dewatering is a huge steady-state pumping test – if only the data gathered during that test would be used for the closure phase of the mine!

In addition to monitoring, hydrogeological modelling of the expected drawdown is important. The radius of drawdown – and consequently the radius of a potential impact on the water courses – usually depends on how much the water table is lowered by the mining operation, the aquifer's hydraulic conductivity, and homogeneity, and the pumping rate. Simplified, a mine can be seen as a deep well and therefore, the "equivalent well approach" can be used to calculate the drawdown caused by a mine. Up-to-date computer codes are able to model the ground water drawdown and the impacts on the surface and ground water at acceptable precision level. A prerequisite for modelling is an ample experience with such models. Yet, nothing substantially has changed since the basic investigations of Theis, and his principles for calculating the impacts of wells on dewatering are still valid (Theis 1935).

Areas affected by mine water drawdown can reach several thousands of square kilometres, as in the German lignite mining regions. Such large mining operations have to ensure that surface water reservoirs are protected and that natural wetlands are watered during drawdown. Several mines, such as the Western German surface lignite mines, near the Dutch/German border, have installed water re-injection wells to minimise the negative effects of ground water drawdown.

In some cases, diversion pipe lines or channels have to be constructed. Where springs that are used for drinking water supplies dry out, the people affected have to be supplied with water pipelines. Such a situation occurred in the now Polish Silesian mining area at the end of the 19th century. The local authorities, mining authorities, and the different mining companies worked together to drill wells collecting good quality mine water, and construct kilometres of pipelines in 1878 to supply the population of the Silesian industrial area with good quality drinking water (Geisenheimer 1913).

Drawdown management has to ensure that the environmental impacts as well as negative impacts to engineered constructions are minimised. Surface subsidence can reach several meters (Fig. 32), but as long as only small subsidence differences within a single engineered construction occur, no damages to those constructions will arise (Poland 1984). Modern mining and drawdown operations based on high quality hydrogeological and geological investigations can reduce those negative impacts to a minimum. In places

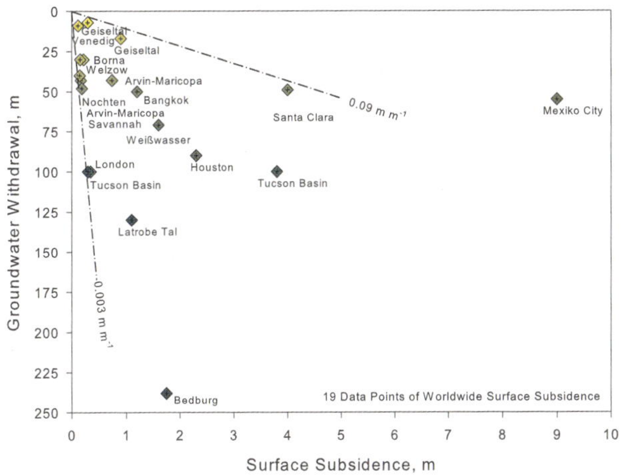

Fig. 32. Surface subsidence of sediments due to mine dewatering and ground water extraction for drinking water (modified after Wolkersdorfer and Thiem 1999).

Fig. 33. Damaged house because of surface subsidence above an abandoned iron ore mine in Roncourt, Lorraine/France.

where subsidence differences and damages to buildings are expected (Fig. 33), buildings may have to be reinforced (Fig. 34) or moved.

Examples where mine water drawdown has to be maintained during and after the remediation phase can be categorised into pollution prevention and flooding prevention conditions. Pollution prevention could either mean that mine water is kept away from potentially polluting material or that polluted water is kept away from drinking water resources. Flooding prevention, on the other hand, may mean that the mine water table has to be kept below the

Fig. 34. Technical solution to secure a private house from destruction resulting from surface subsidence (courtesy Alexander Kattner, 25 March 2003).

surface because the surface subsided during longwall mining, or that urban development has occurred in the area that was kept dry by pumping, so that uncontrolled mine water rebound would flood large, developed or agricultural areas with mine water.

Pollution prevention was and is one of the reasons for maintaining the mine water level below the potential discharge point at the abandoned Niederschlema/Alberoda/Germany underground uranium mine (Gatzweiler and Mager 1993; Hagen and Lange 1995; Rudolph 1997). Because the abandoned near-surface mine workings are within a geological zone of arsenic-rich ores, the mine water was kept about 60 m below the final flooding level. Thus, the quality of the mine water was kept at a reasonable good quality until a water treatment plant could be constructed that could cope with the elevated arsenic concentrations.

Ground water protection is the reason for pumping at the Aznalcóllar/Spain pit, where the pit water table is 6 m below the ground water table (Kurt Friese, pers. comm.). For the same reason, the Horden Colliery/UK is not allowed to flood up to the discharge point, as the polluted mine water might negatively influence the Basal Permian Sands aquifer (Coulton et al. 2004).

A cone of depression is also kept at the abandoned Königstein/Germany in-situ leaching uranium mine. There, the upwelling mine water would impact the 3rd Sandstone aquifer which supplies the cities of Pirna and Dresden with drinking water. Currently, a final flooding level of 140 m a.s.l. (above sea level) is planned, which would maintain the water 170–250 m below the surface until the higher contaminant concentrations of the first flush will have passed (Jakubick et al. 2002; Schreyer et al. 2002). This cost intensive strategy has been questioned by some authors who used modelling tools (e.g. PHREEQC) to suggest cheaper remediation methods, such are diluting the mine water or using the surrounding strata as a passive remediation method (Merkel 2002).

Many mining areas in the world have subsided due to longwall mining. To prevent built-up areas from being flooded or impacted by polluted mine water, the dewatering pumps might possibly run forever. Examples are the abandoned coal mining fields of Durham/United Kingdom, Limburg in the Netherlands, the Aachen and Ruhr areas in Germany, and Silesia in Poland (Bekendam and Pöttgens 1995; Younger 2002b). Since we know no cheaper method to protect these built-up areas, and a resettlement of tens of thousands of inhabitants is unacceptable, continued pumping is necessary.

A final example occurred in Clearfield County/USA. During an investigation conducted in the 1970s, it was recommended that 10 drift openings of the Swank's Mine and 13 of the Shoff Mine be sealed to stop the pyrite oxidation and the subsequent pollution of the Clearfield Creek. Yet, the consultant recommended "a manually operated drawdown system will be installed in the southern end of the mine. This will consist of a pipe to the lower portion of the mine pool with a valve that will act as a safety device to lower the mine pool if required" (Commonwealth of Pennsylvania 1973; see Fig. 23).

5.9 Prediction of Ground Water Rebound

Flooding of an abandoned underground mine ideally means that all mine dewatering pumps will be switched off after the completion of mining and that ground or surface water will be allowed (and sometimes encouraged) to flow into the open voids. The flooding process will stop when the mine water surfaces equals the ground water surface or reaches a point of discharge – ideally a drainage adit. Usually, prediction of the rebound scenario is imperative to allow the planning of certain remediation measures or mine water treatment options. As mentioned before, rebound prediction is a fairly fuzzy job and as so often is the case for mine water, rules of thumb and experience can help a lot to calibrate numerically-based conceptual models (compare chapter 8).

Because conceptual models are a compelling necessity in all hydrogeological investigations, and consequently mine flooding scenarios, the definition of Younger and Adams (1999) shall be given here:

> A conceptual model is an assembly of simplifications about a complex, real system, which achieves a valid representation of that system, including all major features, whilst avoiding unnecessary detail.

Or in the words of Rushton (2003):

> Conceptual models describe how water enters an aquifer system, flows through the aquifer system and leaves the aquifer system

Such a model can start on the back side of an envelope, and almost every hydrogeological project started with a first sketch on a piece of paper the size of this book. Drawing sketches to understand the hydrogeological or flooding scenario is useful in every mine water related research project to assure mutual understanding of the fundamental principles of the investigation area. Of course, colourful graphics and computer printouts are helpful in understanding complicated geometries and numerical results, but they are useless if not based on evolving conceptual models. Finally, everybody should keep in mind that numerical models are – at least in hydrogeology – always approximations of reality (Box 1979) and even when they look very real, they are not reality itself!

Younger and Adams (1999) and Younger (2002d) classified the models available for the calculation of ground water rebound into two categories: semi-distributed lumped parameter models and physically-based, distributed models. From an environmental and a legal point of view, rebound prediction should be conducted well in advance of the mine closure. Such an investigation for underground and surface mines must answer the following questions:

- to which elevation will the mine fill up?
- how long will it take to fill the mine and/or reach a water table equilibrium?
- what will the water quality be during and after rebound?
- is there a potential to impact buildings?
- will mine gases (e.g. Rn, NH_4, H_2S, CO_2) emanate from the flooded mine workings (e.g. Pokryszka et al. 2005)?
- will drinking water aquifers might be impacted?
- could ground subsidence occur?

Based on the results of such a calculation, it might be necessary to backfill parts of the underground or surface mine to prevent the release of acidity or toxic compounds. Alternatively, such a rebound calculation could result in deciding to completely backfill a mine to prevent the accumulation of a large volume of polluted mine water, as was done at the Ronneburg open pit uranium mine in Germany (Zurl et al. 1992; Laubrich et al. 1996; Geletneky 2002; Paul et al. 2002), the Blindwells open pit coal mine in Scotland, and other mines worldwide. Rebound of open pit lakes can very often be estimated using a simple water balance model, or an analytical model. If the hydrochemical situation has to be predicted as well, more complex numerical pit filling codes or coupled ground water flow models are advised (see section 5.5.2; a description of the necessary procedures can be found in Younger and Adams 1999).

In general, the most important question to be solved is the duration of the rebound time, t_r. t_r is a function of many different parameters and can be written as:

$$t_r = \phi(V_m, V_r, r_c, A, P, E, R) \tag{27}$$

with V_m volume of mine voids
V_r volume of matrix
r_c coupling ratio between matrix and mine voids
A area of dewatering
P precipitation
E evaporation
R surface runoff

A central task for the prediction of the rebound time is knowledge about the mined and dewatered volume of rock. Such a calculation is comparatively simple for open pit mines, as the cone of depression, the volume of the open pit and the water make during the mine's lifetime are – usually – well known. Furthermore, the climatic, hydrologic, and hydrogeological conditions have to be specified, as pit filling depends on ground water inflow, surface water runoff, precipitation, evaporation, and seepage out of the mine. An often overlooked fact is that the amount of water infiltrating into the mine changes as the mine water table rises.

Even if all the available data has been considered and the predictions were accurate, technical and meteorological problems can result in catastrophic mine water outfalls, as occurred at the Cornish Wheal Jane tin mine (Hamilton et al. 1994b) on 13 January 1992, when 30,000 m³ of heavily polluted mine water spilled out of the mine (Hamilton et al. 1994a). Furthermore, ecological or economical reasons on a regional scale may not allow all the surface or ground water to be used for pit filling. This situation can be seen in the German Lausitz lignite mining area, where not enough surface water is available to flood the open pits in a reasonably short time because the river Neiße and Spree water cannot be used for pit flooding (Arnold and Baumgarten 1996; Grünewald 2001; Schlaeger and Köngeter 2001).

To calculate rebound times, it is important to consider that the water inflow into an abandoned mine will change as the mine water table rises: it is quick at the start of the flooding and slows down towards its end. This is mainly because the volume of the cone of depression gradually is reduced. Furthermore, the difference of the hydraulic head between the original ground water table and the mine water table decreases so that the total flow into the mine continuously diminishes during the flooding process. Yet, there are several exceptions from this general rule: deep mines within a comparatively impermeable rock with a drainage adit and workings above and below that adit will not encounter much of a change until the water table reaches the level of the drainage adit. The reason is that most of the water in those mines comes from the – usually – less permeable strata above the drainage adit. One such case is the Niederschlema/Alberoda/Germany mine (see section 12.3.3 for details). A comparison of historical, pre-mining data with the current mine water make clarifies that after the flooding process is finished and if the Markus-Semmler drainage adit will be used as the drainage adit, nearly pre-mining conditions can be expected. As the pre-mining water make ($5–9 \cdot 10^6$ m³ a⁻¹) nearly equalled the water make during the active mining period ($8–10 \cdot 10^6$ m^3 a^{-1}), the typical decrease during the flooding process should not occur. The reason for that is that most of the water make comes from the huge dewatered area above the Markus-Semmler drainage adit and relatively little from the interconnected deeper aquifers. Flooding up to the level of the Markus-Semmler adit will therefore not change much of the current situation (Wolkersdorfer 1996).

Banks (2001) differentiates the two characteristic types of inflows into mines; the first explains the reasons for the water make at mines such as the Niederschlema/Alberoda mine:

- A head-independent flow, derived from direct infiltration of precipitation through covering strata, worked veins or mine voids which are open from the surface. This inflow will possibly vary seasonally (or even daily) with rainfall, but in the long term can be regarded as a constant.
- A limited number of head-dependent flows, entering the mine from adjacent saturated ground water reservoirs via permeable horizons, fractures, or abandoned exploration boreholes. The natural ground water head in the reservoir at some distance from the mine drives the inflow from each of these features.

The impact of the flooding of the Durham coalfield on an aquifer that is used for drinking water production was predicted by Younger (1993). Three centuries of coal mining resulted in the dewatering of a large area, which is hydraulically well connected by mine workings. Due to ground water rebound, the water will percolate through the abandoned workings and connect different drinking water resources. From the flooding of adjacent coal fields, it was known that the rising mine water will be enriched with pollutants and will have a high acidity. The following potential consequences were identified by Younger (1993):

- ground water pollution in the adjacent Basal Permian Sands aquifer
- leaching from landfills intersected by the rising water table
- compromised integrity of foundations
- increased flows in old sewers below the new rest water level
- corrosion of other buried services
- surface gas emissions
- long-term subsidence risk from old workings

Therefore, the Horden mine water treatment plant and pumping scheme was put into operation in 2004 to keep the mine water well under the Basal Permian Sands. The plant pumps 150 L s^{-1} from the abandoned Horden Colliery main shaft and treats the water in a HDS treatment plant (Coulton et al. 2004, Fig. 87).

An extensive rebound prediction was also conducted for the Slovenian Mežica lead zinc mine. Based on tracer tests, laboratory investigations, and data from the active mining period, a total flooding time of 1245 days was predicted (Veselič et al. 2001; Prestor et al. 2003). Mine water rebound started in December 1994 and 119 days later, in April 1995, the water drained through the +418 mine level. Veselič et al. (2001) and Prestor et al. (2003) recalculated the data and concluded that the area of the cone of depression and the hydraulic conductivity of the rock matrix were overestimated. However, as will be explained below, it is possible that the researchers did not overestimate those values, but overestimated the influence of the dewatered rock matrix on the rebound course.

Some of the largest abandoned mining regions in Europe are the Aachen/Germany and South-Limburg/Netherlands coal mining districts, where mining ceased in 1992 after 800 years of active mining. During that time, a cone of depression with an area of 400 km^2 built up but since 1992, the mine water has risen by 200–240 m, thus diminishing the dewatered area. The whole ground water rebound was well monitored and controlled, using a step-by-step approach, and so far, at least, has gone well (Heitfeld et al. 2002).

Thus, although several good analytical, empirical, and numerical codes are available for rebound prediction, it is still complicated. Even if all conceivable data and experiences were put into the model, rebound prediction might fail. A comparison of the methodological procedures shows that complicated, numerical models can fail as much as simple empirical models. If the failures described above and in the literature (e.g. Veselič and Norton 1997; Younger and Adams 1999) are considered, it is clear that they are usually due to one or more of the following reasons:

- misinterpreting the cone of depression (usually too large)
- over-estimating the significance of the volume of the dewatered rock matrix
- insufficient data about the volume of the mine voids
- wrong conceptual model

At least the second point can be explained from a hydrodynamic point of view using a mine voids-matrix coupling ratio, c_r. First, mine voids are filled rather quickly while the rock matrix remains unsaturated for a comparatively longer time. After the mine voids are completely filled, the rock matrix re-saturates until the hydraulic regime reaches equilibrium – which, of course, will not be the same as the pre-mining condition. The reason for the slow saturation of the rock matrix can be seen in the fact that the rock matrix consists of three different types of porosities with fast and slow flow occurring in those porosities: flow porosity, residual porosity, and diffusion porosity (Norton and Knapp 1977). Therefore, the volume of the voids is the main factor for the rebound and the volume of the rock matrix accounts only as a minor part. It is understandable that the coupling of those processes has to consider both laminar Darcy flow in the far field of the cone of depression, as well as turbulent non-Darcy flow in the voids and the parts of the cone of depression with high hydraulic gradients – this is what makes the situation complicated (Darcy flow is defined as a laminar flow for a viscous fluid through isotropic porous media).

This section – at first reading – may disappoint those wishing to predict mine water rebound. However, the situation is not that bad and a nearly correct prediction is better than no prediction at all! If all necessary data are considered, a good conceptual model with a first empirical calculation is made, and after enough data are available, a more sophisticated numerical model is used, predictions should, to a certain extent, be accurate. Future research in this field must focus on the rock matrix–mine voids interaction because this mechanism has been largely disregarded in existing models.

5.10 Unexpected Mine Water Discharges During Flooding

As has been seen in the previous chapter, the prediction of ground water rebound can fail and in such cases, mine water may emerge in unexpected or inappropriate places, with unforeseen contamination loads, or underestimated total flows can range from small seepages to catastrophic mine water outbreaks. Some of these failures have already been addressed in section 5.7 or will be described in the case studies (chapter 12). Furthermore, the facts that can be learned from such failures can be found throughout this work without referring to them in all instances. This chapter will focus on unexpected discharges from closed mines during or after the flooding process. Failures or accidents during the operational period of the mine has already been addressed in earlier chapters and dam failures, though one of the accidents with the most public interest, are out of the scope of this book.

One of the more prominent failures during the flooding of an abandoned mine occurred at the Cornish Wheal Jane tin mine in the United Kingdom. In January 1992, at a time when the water was allowed to rise in the abandoned workings due to pump maintenance works, the water level reached the Nangiles adit and 40,000 m³ of polluted mine water discharged into the Carnon River, the Restronguet Creek and Carrick Roads of the Fal Estuary (Younger 2002e). Though the accident found a wide interest in the media and was described as a catastrophe for the region, investigations two years later showed that natural attenuation and the robustness of the organisms were strong enough to overcome the accident and most parameters came back to pre-accident conditions (Somerfield et al. 1994; Hunt and Howard 1994). After the accident, an active and a pilot scale passive treatment scheme were constructed at the site (Hamilton et al. 1994a, 1997; Johnson and Hallberg 2002; Walder et al. 2005; Younger et al. 2005).

Another accident was the mine water outbreak at the Anna Stollen (Anna adit) near Nassereith/Austria, in June 1999. This adit had been abandoned about 50 years ago and the mine water was drained through a concrete pipe through the Nassereith conglomerate. In the course of time, parts of the adit and the pipe became blocked with fine-grained material from the conglomerate. During normal precipitation periods, the system worked fine but after the extremely wet spring of 1999, the system was no longer able to drain the mine water. Consequently, a large mine water pool built up behind the blockage and the system finally failed, giving rise to a huge mine water outbreak and mud flow that partly flooded the nearby valley (Amezaga et al. 2004; Weber 2004; Eckhart and Riepler 2004). Luckily, the mine was a lead-zinc mine within carbonate host rocks and the chemical water quality therefore was of no threat to the local community or the environment. A similar accident had already occurred at the same location in July 1948 and again in June 1955, the reasons again being the blockage of the Anna drainage adit. Interestingly, the 1948 "mine water explosion" could have been foreseen and prevented as the mining authority noticed in 1935 that the adit was blocked 13 m behind the mine portal. In a report to the mining authority, Gustav Hießleitner calculated that the total volume of water that had accumulated behind the blockage was some 30,000 m³ (Mining Authority Hall Zl. 1112/1935; 2075/1948; 1342/1955).

A comparable accident happened in August 2002 when the *Rothschönberg Stollen*, the central drainage adit of the German Freiberg/Saxony mining area collapsed after a catastrophic rainfall event lasting for nearly

4 days. This collapse led to a partial flooding of the famous mining academy's teaching and tourist mine (Fig. 15), as the mine water level rose by about 20 m. The functioning of the drainage system was re-established by intensive remediation and reconstruction (Kugler and Weyer 2002; Weyer 2003).

During the research for this chapter, it became clear that many "unexpected" mine water discharges after flooding are known, but – obviously – people tend not to publish such failures, or try to interpret them in a positive manner. One such case occurred at the abandoned Dresden/Germany Coschütz/Gittersee uranium mine. First, flooding was slower than expected (Gatzweiler et al. 2002) and then the mine water discharged at unforeseen places with an unexpected contamination load (Thieme 2005).

It is important to emphasize – again – that old workings and forgotten connections between modern and past workings are the most dangerous threat in the mining industry. It is therefore essential that mining authorities keep track of mine layout maps and that all kind of investigations are conducted to locate forgotten workings in long abandoned mining areas. The Quecreek accident (Somerset County, Pennsylvania) could probably have been avoided had accurate maps of the abandoned Harrison № 2 been available (Pauley et al. 2003). A recently developed and improved method to identify unknown sub-surface mine water pools and acid mine water discharges are airborne geophysical (remote sensing) investigations. This has been successfully used in the Upper Arkansas River Basin/USA, where natural acid drainage in altered rocks occurs, and in several Pennsylvanian river basins for locating acid mine water discharges of abandoned underground mines (Yager et al. 1998; Williams et al. 2002; Sams and Veloski 2003; Hammack et al. 2003b; Ackman 2003; Hammack et al. 2003a; Sams et al. 2003; Al-Fouzan et al. 2004; Sares et al. 2004).

5.11 Mine Closure Legislation and Guidelines – Some Aspects

Mine closure legislation differs from country to country, though some general conclusions can be drawn from recent investigations of mine closure. Detailed assessments are given in McCourt (1999), Anckorn and del Olmo (1999), Hámar et al. (2002), or Kroll et al. (2002a, 2002b), most of them with a European focus. The topic of mine closure legislation covers a huge area of legal aspects, such as waste, soil protection, water protection, socio-economical impacts, and other relevant legal issues. Furthermore, regulations, standards and laws change quickly, so that an in-depth discussion of those matters would be outdated before the publication of this book and would cover several hundreds of pages. For example, Burchi and Mechlem (2005) collected the most important treaties and other legal instruments about ground water in international law and came up with a 584-page volume! A collection of international and cross-boundary laws and directives related especially to mining has been published by Pring (1999). Though his work is not focused on water, he gives many references to how international water law impacts mining issues, for example the "Ramsar Convention on Wetlands of International Importance".

Though not meant as a legislative piece of paper, the Berlin Guidelines on "Mining and the Environment" (United Nations Department of Technical Co-operation for Development and Development Policy Forum of the German Foundation for International Development 1992) has resulted in a lot of recommendations and international directives, which were partly incorporated in national legislation (Natural Resources Management Unit Division "Protection of the Environment and Natural Resources"1999). As Berlin I did not address mine closure, these recommendations did not much affect mine closure legislation (see appendix III). Yet, Berlin II also addressed abandoned mines and several aspects of mine closure were discussed at this round table meeting, thereunder mine water issues.

On a worldwide basis, most water laws do not explicitly refer to mining. Some exceptions are recommendations and directives of the European Union, and the regulations of Palestine's West Bank, Kenya, Australia, African Incomati, and the Maputo Watercourses (Burchi and Mechlem 2005). In the USA, the "Federal Water Pollution Control Act" (33 U.S.C. 1251 et seq.) contains a paragraph (§107: "Mine Water Pollution Control Demonstrations") which allows pilot mine water treatment plants and allocates a 30 Million $U.S. fund. Furthermore, the "Safe Drinking Water Act" (42 U.S.C. 300h[b]) may be amended "to protect the water quality of the Great Lakes from the harmful effects of acid mine drainage" (110th Congress, 1st Session, H.R. 1842.IH; March 29, 2007). Some states in the USA (e.g. Pennsylvania, West Virginia) and Provinces in Canada also have regulations relating to potential acidic mine waters (MEND [Mine Environment Neutral Drainage Program] 1993). Yet, the rare number of laws or regulations related to mine water does not imply that mine closure or mine water issues are not regulated in other countries; it just means that there is usually no special reference to mine water in the water laws (Plinke and Wildhagen 2003)!

In many countries, mine closure planning nowadays starts with the initial mine management plans; it is compulsory to include mine closure plans in the environmental impact assessment (EAI). This movement towards sustainable and environmentally sound mining is mainly due to the influence of the United Nations Environment Program and the World Bank Group, which connects new lendings for mining enterprises to the principles of environmentally friendly procedures and their guidelines and directives (Strongman 2000) related to:

- environmental assessment (including Environmental Management Plans)
- natural habitats
- dams
- indigenous people
- involuntary resettlement
- cultural property

Sheldon et al. (2002) summarised the future issues of mine closure in the following way:

> The traditional approach toward environmental protection and reclamation has largely been for mining companies to take care of any surface safety hazards and secure the pits, flooding open pits and, where needed, providing for pumping or drainage of water from underground workings nearby and sealing the entryways, shafts, and adits. This is no longer sufficient. Good practice today requires removing unwanted plant and equipment, stabilising and securing waste dumps and impoundments, detoxifying hazardous materials, protecting ground water, addressing any acid rock drainage issues (which in the worst cases can be a severe problem), and reclaiming, rehabilitating, and revegetating land in a manner compatible with local vegetation. Such environmental protection and clean-up apply not only to the mine site but also to all locations affected by the operation, including any rivers, ports, and other transshipment points, some of which may be a hundred miles or more from the mine site. For example, in the case of the Ok Tedi mine in Papua New Guinea, the transshipment port is over 200 miles from the mine, and the area impacted by the disposal of mine waste and tailings extends over 1,200 miles from the mine site to the Torres Straits.

Currently, the Global Alliance, a union of different acid mine water initiatives and mining companies, is preparing the GARD: "Global ARD Guide (GARD guide) – Guide for the Prediction, Prevention and Mitigation of Neutral and Acid Drainage from Sulphide Wastes," which is scheduled to be finished in 2008.

BATNEEC means "best available technology not entailing excessive costs" and is for example used by the Irish Environmental Protection Agency (Environmental Protection Agency 2001; Flynn et al. 2005). Independent from any legal issues, if the BATNEEC principle is used throughout the whole mine closure process, the legislative requirements will be met in most instances on a worldwide scale. As BATNEEC is not a static procedure, but depends on the improvements of existing and challenges of new technologies, the BATNEEC principle guarantees that most environmental aspects are met during mine closure. Yet, it should be noted that it is possible and indeed desirable in some instances to set up more stringent rules than those that could be derived from BATNEEC principles. If, for example, mine water might pollute drinking water reservoirs, an even better and more expensive technology might be demanded by the authorities.

Legal issues of mine waters in Europe have been studied intensively by the PECOMINES and ERMITE projects (Jordan et al. 2004; ERMITE Consortium et al. 2004). Those publications partly cover legal aspects of mine closure as well. A publication that falls into a similar category was developed by the French Bureau de Recherches Géologiques et Minières (BRGM) and covers mainly waste issues from the European mining industry (Charbonnier 2001).

Another aspect, which will become important in the future, is environmental liability (Betlem and Brans 2002; Bernasconi and Betlem 2003). The European Commission has issued a Proposal on Environmental Liability (COM[2002]17); each mine operator will do its best to exclude any future liability that could be traced back to his mining operation. Mining rehabilitation is mandatory in Republic Act 7942 (the Philippine Mining Act of 1995). Remediation and rehabilitation of abandoned/mined-out areas are thus accorded top priority in policy agenda to address negative impacts of mining projects. Several legal aspects of mine closure in Iberoamerica were discussed in the first Iberoamerican seminar on mine closure held in La Rabida in 2000 (Villas Bôas and Barreto 2000). Experts from Mexico, Peru, and Chile reported on experiences with new mine closure legislation with respect to environmental protection.

As can be seen, the regulatory or legislative procedures that are necessary to close working mines are numerous and have been adopted in most developed countries. By international treaties and the role of international organizations, they are also followed in many developing countries. Over and above the efforts of INAP and ICMM, the plans of the Global Alliance to produce "Global Acid Mine Drainage/Mine Leachate Remediation Guidelines" will encourage this trend. however, while this is true for the global players in the mining industry, small-scale mining in all countries remains a real problem for the environment, as Armando Zamora (1999) has summarised. Therefore, the problem of closed or abandoned mines is less associated with current mining, but with past mining periods and small-scale mining (Balkau 1999). As discussed in other chapters, mine water pollution from abandoned mines is a worldwide problem and no matter how highly developed a country is, all of them have yellow or red streams somewhere in their backyard.

6 Do's and Don'ts of Mine Closure

6.1 General remarks

During mine closure, regardless of what caused the closure, the focus should be to close the mine as quickly and cheaply as possible, but to do so in an environmentally sound manner. However, even where a proper mine closure plan is available, people rarely think of all of the problems that might someday evolve, and require that a new or extended remediation project be initiated.

Mine water researchers, mining or industrial archaeologists, and economic geologists desire that shafts or underground workings would not be backfilled or fully closed. Yet, safety for the public and the mine operator obviously require it, and therefore the wishes of the first group cannot be fulfilled at all times. Fortunately, many closed or abandoned mines are converted to visitor mines so that research is still possible, and deeper understanding of flooding processes, the life and work of the historic miners, or the genesis of ore deposits are still possible. Also, from a mining heritage point of view, total remediation of all of a mine's remnant should sometimes be avoided because such remnants can tell us much about our own past (Shepherd 1993; Anonymous 1999b; Carvajal and González 2000; Lynch 2002). Some of the most polluted abandoned mine sites can serve as study areas for mass transport and past mining activities, such as the Spanish Rio Tinto or the German Harz Mountains area (Dobler 1999; Leblanc et al. 2000; Carvajal and González 2000). Many countries, for example Malta or Sweden, have heritage issues included for mine closure operations (Wolkersdorfer and Göbel 2001a; Lindgren and Destouni 2002).

Of course, from the perspective of a mine water researcher, there are many wishes on how mines that are going to be flooded should be equipped, the simplest being continuous monitoring of the temperature or electrical conductivity development during and after the flooding process. Such measurements would help to increase our understanding of flooding processes and thus develop improved strategies for environmentally-safe mine flooding. In fact, there are already many mines where a monitored flooding process was of use in later remediation strategies (see chapter 12). Because many mine operators, understandably, just want to close down their mine, this wish will probably remain a long way off.

Another aspect of mine closure that will not be covered here are the associated social-economic effects. Miners and their families lose their regular income, people move away and formerly prosperous regions can lose most of their citizens. Cases like that can be seen all over the world and when no-one else lives in the former mining place, the likelihood of nobody taking care of the abandoned and remediated mine workings can become a problem. Fortunately, modern mine operations require closure strategies well before the first machinery starts digging, so that community interests can be addressed well in advance of closure (Robertson et al. 1998; Robertson and Shaw 1999; Hoskin 2000; Hoskin 2003).

This chapter will list some things that could be done during a mine closure process that would help later researchers to gain access to the old workings without reducing the safety needs of the operator. Regrettably, this list is far from being complete; all types of suggestions are welcome and will be included in a later edition.

6.2 Shafts

Shafts should not be fully backfilled if it is not necessary for reasons of ground stability or the safety of engineering constructions. Usually, a concrete plug with pipes that allow access for sampling equipment or borehole cameras will suit all the needs concerning safety. The cases where fully backfilled shafts caused ground subsidence above the backfill are numerous (e.g. Skinner and Beckett 1993; Klapperich et al. 2001; Busch et al. 2002; Sroka et al. 2003; Meier et al. 2004), thus proving that this technique should not be called "state of the art". The pipe should have a diameter of at least 8 cm (3 inches), which is large enough for most down-hole equipment. As small pipes tend to twist in such plugs over time or during cementation, larger diameter pipes of at least 20 cm (7 inches) would be better. The current procedure in many areas of pumping concrete into abandoned mines *ad infinitum* does not meet the needs for mining heritage and later scientific investigations of abandoned mines. The "severe danger" that would justify such an extreme measure is unusual; more typically, it is just done because the responsible person knows no other techniques. Sealing and grouting techniques can be used to protect both the mining heritage and engineering constructions on the surface. Also, polyurethane foam (PUF) has been successfully used as an alternative to concrete plugs for shaft sealing (Suman et al. 2000).

Furthermore, for the sake of safety, unclassified material or material that will degrade over time should not be used! In the Picher Mining District/USA, unused cars, railroad ties, and oil tanks were used for shaft sealing (Suman et al. 2000). Some abandoned shafts in the Mulreesh–Portnealon lead-zinc mining area of Islay/Scotland (Wilson and Flett 1921) were covered with corrugated iron and waste material and a small shaft in Tyrol/Austria was just covered with old planks of wood, which soon rotted. And, of course, shafts should never be left open so that animals or people can fall into them (Fig. 35); fencing provides an inadequate and temporary form of protection.

Sampling equipment and downhole probes can be damaged by material left in the mine and especially in the flooded shaft. Wood, steel cables, plastic foils, files, gas masks, plastic pipes, etc. and material used as barriers can commonly be found floating in the mine water and can cause problems that could easily be avoided if all onsetting stations would be secured with structural steel lattices. To avoid high amounts of suspended solids, it is useful to construct 1–1.5 m high dams at distances of 5–10 m before the onsetting stations. Of course, the rotting of wooden shaft furnishings or supports in the shaft compartments cannot always be avoided (Fig. 36), but wire or shot-firing cables left hanging or floating in a shaft are traps for probes and *is* avoidable.

A positive case is the clearing-out and rehabilitation of the Virginia shaft at the abandoned Mechernich/Germany lead mine. This abandoned mine shaft was restored in 1986 to allow access to a drainage adit under a waste disposal site that was receiving waste water; the shaft was secured and access to the underground gallery was made possible again (Benning et al. 2002; Mair 2002).

6.3 Adits

Closure of adits was already addressed in section 3.7.4 but in another context. For safety reasons, all adits have to be closed carefully but they should be closed in such a way that future researchers or bat protectors can get ac-

Fig. 35. Unsecured, open shaft in the abandoned Coolgardie/Australia gold mining field.

Fig. 36. Rotten shaft furniture in a flooded shaft of the closed Schwaz/Austria dolostone mine. The water surface is just a little bit below the left corner of the shaft.

cess to the adit at a later stage. Bat protection and mining research (which covers all kind of scientific research within an underground mine) are not mutually exclusive, but there is often a lack of co-operation between the two groups. Again, the Mechernich/Germany mine is a good example of co-operation between environmentalists and mine researchers (Thies 2003).

Visitor (tourist) mines and mining authorities have found that the closure of an adit with a door causes "door researchers" to test the stability of such constructions. It is interesting to see how many different methods have been used by such groups to gain access, including the use of explosives and sledgehammers. As a general rule, such doors must be as sturdy as possible to prevent unauthorised access. Door hinges and padlocks must be inside the adit and protected against humidity. It is a good practise to leave a sign near the entrance on how access to the adit can be legally attained. Such procedures are routinely followed at many caves throughout the world where a relatively common code of honour already exists between most cave researchers.

If polluted mine water is draining through the adit, huge amounts of iron oxide sludge can accumulate behind the seal. This has been observed at the abandoned Gernrode/Germany fluorspar mine. A low wall was constructed one meter behind the adit's seal to prevent waste from building up behind the drainage pipe (Tamme 2002; Schindler 2003; Wolkersdorfer et al. 2003). Within two decades, iron hydroxide built up for a length of approximately 500 m behind that wall. The adit was inspected by a consulting company and the authorities in August 2001 and December 2003. The individuals who walked in the sludge initiated a sludge avalanche that polluted the receiving stream for a length of at least 3 km. It turns out that a blockage of the drainage pipe caused a rise in the water level and thus the sludge accumulation. This

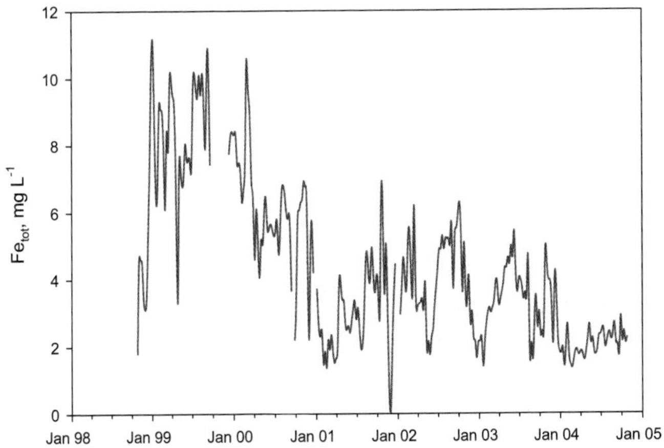

Fig. 37. Seasonal variation of Fe_{tot} in the Glasebach/Germany visitor mine (sampling point A1b; data kindly provided by BST Mansfeld GmbH and GVV). The sampling point is directly under a spoil heap, which shows strong seasonal fluctuations of the total iron concentration (each data point represent the arithmetic mean of three measurements a week).

example shows how important it is that such sealing systems be regularly inspected. This in turn requires that access for inspection be maintained, which is only possible if there is no total sealing.

6.4 Choosing the Final Water Level

A key question for every mine operator is where to establish the final water level of a flooded mine – in other words: where should the underground beach be established? As discussed earlier, the final water level should be as high as possible to minimise disulphide oxidation. Disulphide oxidation preferentially occurs in such mines within the range of seasonal fluctuation of the water table, due to alternating high and low oxygen concentrations as well as redox values within the highest and lowest water levels. Short-term fluctuations occur during storm events or after sustained periods of drought or precipitation. When the water level is low, disulphides are subjected to oxidation and the processes described in section 4.2 will produce acidity, metalliferous leachates, and easily soluble secondary minerals (Fig. 37). In arsenic- or antimony-rich environments, the varying redox conditions could facilitate the release of arsenic (Berg et al. 2001). When the mine water table rises, these secondary minerals will be mobilised. Consequently, acidity ("juvenile acidity", section 5.4) as well as metals are released into the water through these wet-dry-wet cycles (Frostad et al. 2002) until – after decades – all of the disulphide minerals are either exhausted or coated. A storm can increase the contaminant load of the discharged mine water up to a factor of 10 and adversely affect the receiving watercourses (Younger 2000d). The long-term prediction of such discharges after mine abandonment is highly challenging, to say the least (Wood et al. 1999).

Water table fluctuations are usually minimal at adit mines but can reach several decametres in shaft mines being dewatered by pumping. Water table fluctuation will also occur at adit mines that are drained through drainage pipelines because they can only discharge a certain amount of water. A similar situation occurs if sealed adit mines become leaky and mine water trickles through the mine seal. Furthermore, if the mine water discharge has a large seasonal variation, oxidation rates will fluctuate as well, leading to the same type of increased contaminant loads discussed in the preceeding paragraph. Such situations occur in regions with big seasonal differences of precipitation and temperature, for example in Spain or Australia. Observation boreholes should be drilled into the mine water body to allow inspections, and if long-term observations are necessary, they must be constructed in such a way that they function throughout the entire planned observation period.

In karst areas, fluctuations of the water level can cause dissolution of the lime- or dolostone and consequently the formation of sinkholes (Zhou 1997; Li and Zhou 1999). These sinkholes can cause disastrous destruction of engineering constructions and highly dangerous water inrushes (Newton 1984). Thus, water fluctuations must be considered in limestone areas even though the risk of acid drainage is low.

Seismic events can also cause water level fluctuations. In the abandoned Niederschlema/Alberoda/Germany uranium mine, flooding induced a micro-seismic event on 21 November 1995, which caused a water level fluctuation of 34 cm within an hour. During the 18 August 1959 Montana/USA earthquake, three shafts in the Pennsylvanian Wilkes-Barre area experienced water level fluctuations of up to 3 cm (Hopkins and Simpson 1960). Yet, such unexpected and random water level fluctuations need not be considered for mine water management procedures as the duration of the fluctuation is usually short and the effects are therefore insignificant.

Hydrogeological and hydrological data, including precipitation records and ground water recharge and fluctuations (using piezometers), are needed to calculate the final water level,. Although the piezometric water level fluctuations, identified through hydrograph recession analyses, are not the same as the potential fluctuations of the final mine pool, which can be numerically calculated (chapter 8), they do approximate them.

In addition to the hydrogeological data, knowledge about the location of disulphide bearing strata and acid-producing waste rock or tailings is necessary. Using all of this data, the overall mine closure scenario must be to minimise water table fluctuations within the potentially acid producing material of the mine. The fluctuating water level should never be allowed to lie within spoil heaps or tailings disposal sites. If acid-producing material is to be deposited within the abandoned mine or open pit, that material must be deposited well under or above the lowest potential water table to exclude the discharge of polluted mine or ground water. If necessary, technical installations, such as drainage pipes, grouting, or dams can be used to isolate the acid producing material from oxygen and water (Prediction Workgroup of the Acid Drainage Technology Initiative 2000; U.S. Environmental Protection Agency et al. 2001). Fluctuating water levels usually cause worse drainage water quality then totally dry or fully submerged acid-producing material (Perry et al. 1997).

7 Hydrodynamics

7.1 Flow Patterns in Flooded Mines

7.1.1 Introduction

This section describes the types of flow that can occur in flooded mine voids (shafts, inclines, adits, galleries, and stopes). It is based on hydrodynamic theory regarding flow in enclosures and some real-world examples. Hydrodynamic investigations in flooded mines are an evolving area of research and this combination of theory and in-situ investigation is a little unusual.

In hydrogeological investigations, Darcy flow is the most common type of flow considered (eq. 74). This flow is defined as a laminar flow for a viscous fluid through isotropic porous media. If the flow becomes non-linear or turbulent, the flow becomes non-Darcian. The subject of non-Darcy flow, especially when it comes to convective flow, is very complex, and so some basic hydrodynamic understanding will be required. Detailed descriptions of the theoretical hydrodynamic background can be found in the relevant literature (e.g. Gebhart et al. 1988; Landau and Lifschitz 1991; Batchelor 2000; Oertel 2001; Drazin 2002).

7.1.2 Type of Flow in Flooded Voids

Three different types of non-Darcy flow or combinations of them can occur within the open void space of flooded underground mines: laminar or turbulent Poiseuille flow, laminar or turbulent convective flow, or almost no flow except for Brownian molecular forces or diffusion. The dimensionless numbers for describing the transition from laminar to turbulent flow are the Reynolds number Re for the Poiseuille flow and the Rayleigh number Ra for convective and Rayleigh-Bénard flow:

$$\mathrm{Re} = \frac{L \times v \times \rho}{\mu} \tag{28}$$

$$\mathrm{Ra} = \frac{L^3 \times \rho^2 \times g \times \beta \times \Delta T \times C_p}{\mu \times k} \tag{29}$$

with L characteristic length, m
v kinematic viscosity, $m^2\ s^{-1}$
ρ density of fluid, $kg\ m^{-3}$
μ dynamic viscosity of fluid, $Ns\ m^{-2}$
g gravitational acceleration, $m\ s^{-2}$

β	coefficient of thermal expansion, K^{-1}
ΔT	temperature difference, K
C_p	specific heat capacity, kJ kg^{-1} K^{-1}
k	thermal diffusivity, –

According to Landau and Lifschitz (1991), the transition from laminar to turbulent flow in enclosures occurs at critical Reynolds numbers, Re_{Kr}, between 10–100. Many books and articles about hydrogeology refer to Re_{Kr} as being between 6 and 10 in porous aquifers (e.g. Nuttall et al. 2002; Langguth and Voigt 2004; Hölting and Coldewey 2005) but hydrodynamic experiments on the transition from laminar to turbulent flow clearly show that many variables are responsible for Re_{Kr} and that fixed numbers cannot be given with absolute certainty because Re_{Kr} always depends on the fluid and the nature of the flow that is being investigated (e.g. Fedotov et al. 2002; Frigaard and Nouar 2003; Sharp and Adrian 2004). In the simplistic case of water flowing in a horizontal cylindrical pipe without wall roughness, Re_{Kr} is about 2000 (Sharp and Adrian 2004), but can reach values as large as 10^5 (Landau and Lifschitz 1991).

Free convective flow in vertical enclosures only occurs in systems where the Rayleigh number, Ra, is larger than the critical Rayleigh number, Ra_{kr} (Landau and Lifschitz 1991), which is between 658 for two free surfaces and 1708 for no free surface. Small diameter boreholes or wells can therefore not be used for measuring or estimating convective flow in natural systems or in flooded mines!

Furthermore, the flow regime might be characterised by free flow (buoyancy driven, also called natural flow) or forced flow. Free flow develops when density differences, $\Delta\rho$, within the fluid are the only driving forces, while forced flow develops when external forces cause the fluid to flow. However, free and forced flow do not exclude each other and in many cases of practical interest, both effects are of significant magnitude (Gebhart et al. 1988). Such cases, with both flow characteristics involved, are called mixed convective flow. In free convective flow dominated systems, the dimensionless Grashof number, Gr, replaces the Reynolds number, Re:

$$Gr = \frac{L^3 \times \rho^2 \times \beta \times \Delta T \times g}{\mu^2} \tag{30}$$

with	L	characteristic length, m
	ρ	density of fluid, kg m^{-3}
	β	coefficient of thermal expansion, K^{-1}
	ΔT	temperature difference, K
	g	gravitational acceleration, m s^{-2}
	μ	dynamic viscosity of fluid, Ns m^{-2}

In flooded underground mines, all combinations of the above-listed flow types can exist at the same time. Changes of flow type can occur within short distances and short time intervals, too. Even if Darcy flow in flooded mines is of minor importance, such flow characteristics will occur wherever the water flows through backfill, downfall, or goaf material (the latter case was numerically modelled by Jäger et al. 1990).

To describe the flow regime in flooded voids, the Navier-Stokes differential equation for incompressible fluids is applied. An explicit solution of this equation can only be given for very simple cases, but in fact, a mine cannot be called a simple case. For such situations, numerical methods have to be used to model the fluid flow. Unfortunately, the number of cells needed to appropriately model turbulent flow grows exponentially with the Reynolds number, as $Re^{3/4}$, and the memory needed for such calculations is on the order of $Re^{9/4}$. Numerical CFD simulations of flooded underground mines conducted by the author in 1996 as closed cavity simulations (FIDAP 7.52; Sun Workstation) and another simulation by Unger (2002; ANSYS FLOTRAN 2.2, Convex Workstation) clearly proved that no convergence for turbulent flow was reached within an adequate time. Furthermore, the model cases computed in laminar flow situations only converged when calculated in dimensionless matter.

Besides channel flow within the voids, fracture and porous flow through the rock matrix occurs as soon as all the voids are filled with water (Norton and Knapp 1977; Jäger et al. 1990; Wolkersdorfer 1996). Permeability decreases with the mine's depth (Nordstrom et al. 1989; Wolkersdorfer 1996) and also depends on the thickness of the disaggregation zones around the voids, typically ranging between 0 and 3 m with a maxima of 8 m (Müller-Salzburg 1978; Jacobi and Everling 1981; Militzer et al. 1986; Stoll and Bauer 1991). While these fractured zones where the water-rock interactions occur are important in determining the chemical composition of the mine water, they are of minor importance for the flow within the voids.

Water levels in a pipe network are only similar when the density of the water is similar throughout the whole pipe network. Therefore, wrongly calculated weir heights for controlling the water flow within adjacent mine parts (see section 12.11.3) or density differences in interconnected shafts can also cause forced flow within flooded mines. During the Friedenshall/Germany salt mine flooding, Pelzel et al. (1972) observed water level differences of several meters in the shafts. They concluded that those differences were due to the density differences between the saline mine water in the interconnected shafts. The difference in the water level, Δh, of two interconnected shafts with water levels h_1 and h_2, can be calculated according to the following equation:

$$\Delta h = \frac{\rho_1 \times h_1 \times \left(1 - \frac{\rho_2}{\rho_1}\right)}{\rho_2} \tag{31}$$

with Δh head difference between the shafts, m
h_1 height of water column in shaft 1, m
ρ_1 density of mine water in shaft 1, kg m^{-3}
ρ_2 density of mine water in shaft 2, kg m^{-3}

Sander and Herbert (1985) conducted depth-dependent temperature and velocity measurements in several flooded German salt mines. Their investigations focused on the potential isolation of radioactive waste deposits from the anthroposphere. Yet, the results from their investigations are important for understanding temperature and electrical conductivity measurements in other mines as well. It becomes clear that zones with nearly constant physico-chemical parameters are dominated by turbulent flow and zones with gradients in the temperature or electrical conductivity are diffusion dominated (they call those layers 'static layers'). A typical situation is shown in Fig. 38. Zones 1 and 3 are characterized by constant temperature and fluctuating velocity. This is an indication for a turbulent flow situation. Because the water does not discharge from the shaft, there must also be convection occurring in this layer. On the other hand, zones 2 and 3 are characterised by a temperature gradient and no velocity fluctuations, which indicates a more or less static situation. Sander and Herbert (1985) conclude: "from laboratory experiments and in situ tracer tests, it may be deduced that the transport of matter: (a) in static layers takes place by diffusion, i.e. very slowly; (b) in turbulently mixed layers is chiefly a function of flow, i.e. very fast; at the boundaries of the layers the flow reverses; (c) across die boundaries of the layers is essentially due to diffusion, even if each of the two adjacent solutions is in a state of turbulent flow."

In the Niederschlema/Alberoda, Brixlegg, and Straßberg mines, turbulent flow must be assumed from the results of time-dependent temperature measurements. Temperature and electrical conductivity oscillations (Fig. 39) are characteristic of such mines. Wolkersdorfer (1996), Kolitsch et al. (2001), and Johnson and Younger (2002) conducted in-situ measurements of water velocities in flooded shafts, and observed characteristic oscillations of velocity and temperature. Such oscillations with extremely irregular and unsystematic temporal variation in velocity and temperature are characteristic of a turbulent flow regime with high Reynolds or Rayleigh numbers (Landau and Lifschitz 1991). Bau and Torrance (1981a, 1981b, 1983) studied open and closed fluid loops, which can be compared to the situation in flooded un-

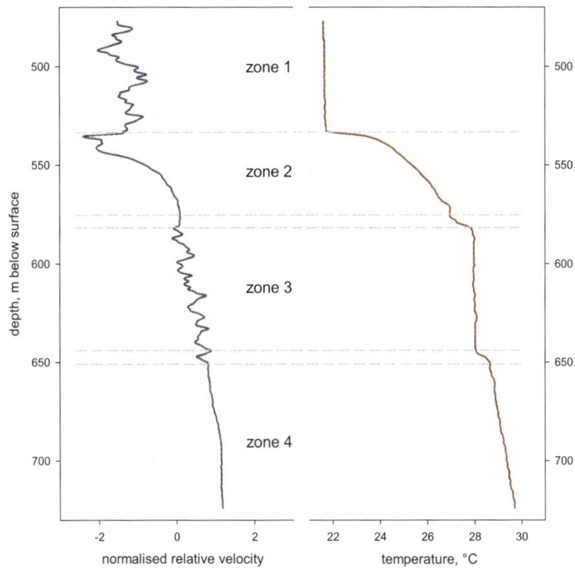

Fig. 38. Depth dependent velocity and temperature measurements in the shaft of an anonymous flooded salt mine (modified after Sander and Herbert 1985). Zones 1 and 3 are characteristic for turbulent convective flow and zones 2 and 3 for diffusive flow (static layer). Between zones 2 and 3 as well as 3 and 4 are intermediate layers that isolate the main zones from each other.

derground mines with two or more shafts. They found out that such systems become chaotic as they become turbulent, and that the flow starts depending on critical heating situations and values as well as perturbations, and eventually can reverse. Furthermore, they concluded that "free convection loops provide a means for circulating fluid without the use of pumps" (Bau and Torrance 1981b).

Results of tracer tests in flooded underground mines showed that the flow in mine voids tends to be nearly always turbulent, caused by the rocks' roughness, the mine support, broken roof, and waste material left by the miners. This explains why the tracer breakthrough usually arrives quicker than predicted. Velocities in flooded mines are reported to be in the range of 0.3–1.6 m min^{-1} (42 tracer tests, 95% confidence interval of all velocities; Wolkersdorfer 2002b; table 19). However, to react in the same way as the fluid, the tracer particles must be small compared to the convection cells, which Landau and Lifschitz (1991) show for general cases of fluid-suspended particles in fully-developed turbulent flow. Unfortunately, according to the suggestions of Mandelbrot (1982; and, of course, in his other outstanding works), the solution of the Navier-Stokes equation must be a fractal – and

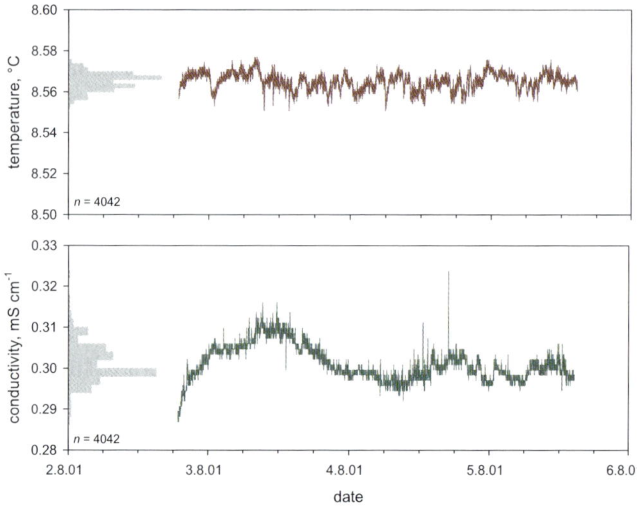

Fig. 39. Time-dependent temperature and electrical conductivity fluctuations over a 68 hour interval in the Georgi Unterbau blind shaft. The histograms on the left indicate the near Gauss normal distribution of the measured values.

consequently of chaotic nature. This explains that "turbulence is one of the oldest, most complicated, and frustrating chapters in physics" (Mandelbrot 1982)!

7.2 Stratification in Flooded Underground Mines

7.2.1 Description of the Stratification "Problem"

Water stratification or layering occurs when water is separated into horizontal layers of different physical or chemical characteristics (Fig. 40). Whilst the stratification might be a consequence of changes in chemical (e.g. contents in sulphate, sodium chloride), physical properties (e.g. temperature, viscosity, density), or turbidity, the stratification itself is always due to density differences of the fluid (Fig. 41). A classical case would be a layer of fresh ground water above a highly saline mine water (Fig. 122).

The stratification might be stable or unstable, depending on the temperature or density differences within the single layers of water (Gebhart et al. 1988). Generally, stratification can be assumed to be stable when low temperature differences or high density differences occur *within* the stratified layer. Because water is a Newtonian fluid, pressure differences can usually be neglected. Stable stratification, written in simple terms, exists when the following conditions apply:

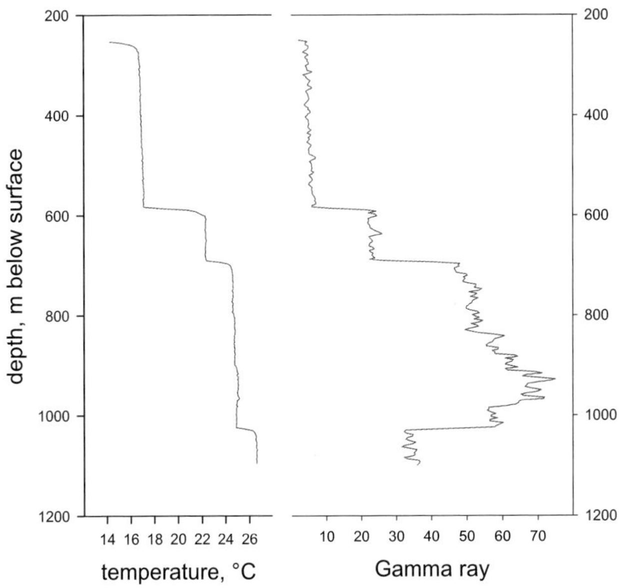

Fig. 40. Temperature stratification in the West Rand Consolidated Mines' Deep Shaft. The gamma ray log shows the temperature stratification and proves that different types of water lie above each other (data courtesy Alan Harley).

$$\frac{d}{dx}\rho_\infty\left(T_\infty, p_\infty\right) \leq 0 \tag{32}$$

with x depth, m
ρ fluid density, g cm^{-3}
T temperature, K
p depth dependent pressure, Pa

To compare different types of stratification, the dimensionless Richardson number, Ri, is used. This number describes the ratio between power and energy and determines whether two layers in a fluid will mix with each other or not:

$$\mathrm{Ri} = \frac{g \times \dfrac{d\rho}{dx}}{\rho \times \left(\dfrac{dv}{dx}\right)^2} \tag{33}$$

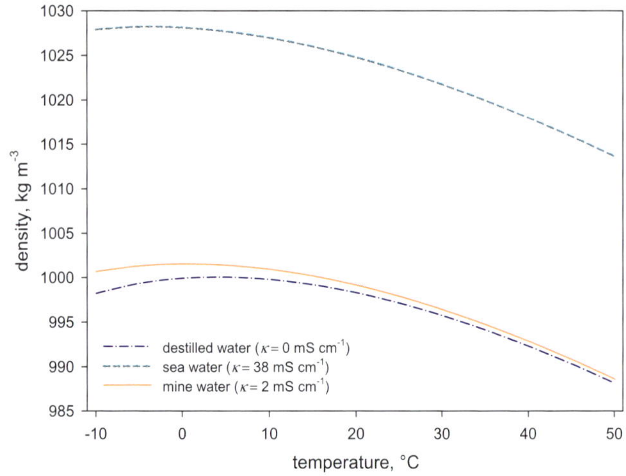

Fig. 41. Calculated density of water as a function of the electrical conductivity κ and temperature T. Calculation is based on Fofonoff and Millard Jr. (1983).

with g gravitational acceleration, m s^{-2}
x depth, m
ρ fluid density, g m^{-3}
v velocity in the fluid, m s^{-1}

Stable stratification, meaning that there is no heat or particle interchange between two different layers, occurs when Ri > 0.25. If Ri < 0.25, internal waves spontaneously occur and break down.

Under such circumstances, the density difference, $\Delta\rho$, between the two different stratified layers are very large and the velocity differences, Δv, are comparatively small, so no exchange will occur between the layers. Examples of such cases are temperature differences above 10 K, total dissolved solid differences of above 3% (static stratification) or big differences in turbidity (dynamic stratification), resulting in density differences of 2 kg m^{-3}, 20 kg m^{-3}, and 200 kg m^{-3}, respectively (Kranawettreiser 1989). Uerpmann (1980) experimentally studied such stable stratifications and the Niederschlema/Alberoda/Germany 371 II b shaft had a stratification that was stable for at least two years (Wolkersdorfer 1996). The extreme situation of the Atlantis II deep in the Red Sea also proves how stable such stratification can be (Blanc and Anschutz 1995). On the other hand, stratification can be broken down by external forces, such as earthquakes, storm events, inflow of fluids with significantly differing density, changes in the overall flow regime, or opening of new adits or voids (e.g. due to a broken dam, construction of a drainage adit, or flooding of a new level during the rebound process).

In some of the above-mentioned cases, an intermediate layer will be established between the two layers. This intermediate layer is a very effective barrier to the vertical transport of dissolved ions, gases, or temperature; in limnology, it is called either a metalimnion or thermocline (Geller et al. 1998). Depending on the thickness, Δx, of this intermediate layer and the velocity difference between the two layers, the upper and lower layers will or will not interfere with each other (Faust et al. 1983; Kranawettreiser 1989). No interaction between the two layers will be observed when:

$$\Delta x > \frac{(\Delta v)^2}{24 \times g \times \frac{\Delta \rho}{\rho}} \tag{34}$$

with g gravitational acceleration, m s^{-2}
x depth, m
ρ fluid density, g m^{-3}
v velocity in each of the fluid's layers, m s^{-1}

Water stratification is a common feature in surface lakes and has been studied there intensivcly (see section 5.5.2). Stratification in natural lakes (e.g. Tietze 1978; Boehrer et al. 2000; Boehrer 2000) as well as mining lakes (e.g. Goodman et al. 1981; Sumioka 1991; Klapper et al. 1996; Murphy 1997; Stevens and Lawrence 1998) is an important component of all limnological surveys, as the exchange of nutrients, pollutants, or gases is controlled by the seasonal variations of the epilimnion, metalimnion, and hypolimnion distributions. An intensive laboratory and theoretical investigation of stratified saline fluids has also been conducted by Barry et al. (2001) investigating dependencies of the turbulent length scale and the turbulent diffusivity for mass. More research is needed to determine if their results are applicable to flooded underground mines.

A feature yet to be investigated in flooded mine shafts is double-diffusive convection, though there are some indicators (e.g. staircase profiles) that double-diffusion occurs in some mines (e.g. Wismut's 296 II b shaft: Wolkersdorfer [1996] page 127). Convective flow in fluids occurs when density variations are present. Double-diffusive convection takes place when two or more components of different molecular diffusivities contribute in an opposite sense to the vertical density gradient. Typical scenarios are the dynamics of magma chambers, growing crystals, or salt water/fresh water systems. Yet, the classic example is heat and salt in water, which causes "salt-fingers" (Brandt and Fernando 1995).

Furthermore, the examples and theoretical contributions given in Brandt and Fernando (1995) suggests that double-diffusive convection cannot be excluded for flooded mine shafts. Double-diffusive convection would explain many features that can be observed in flooded shafts, e.g. a linear temperature distribution and a step-like electrical conductivity distribution (Fig. 44). However, as Kelley et al. (2003) point out, staircase profiles can also be a result of mixing striations or internal waves, as recently investigated in Lake Constance (Boehrer 2000; Boehrer et al. 2000).

Slight horizontal or vertical, jet-like inflows into a stratified water body within a shaft, reservoir, or pit lake can destratify the whole system in a comparatively short time even if no forced flow for the whole system is created. Such systems can develop if different parts of a mine are connected. Small density differences between the jet and the ambient water, exemplified by densimetric Froude numbers, Fr_d, lower than 10, are sufficient for effective mixing and can therefore break down or prevent the stratification within a flooded mine. In that way, a water body with a formerly quasi-linear temperature distribution can be converted into two clearly separated water bodies (Fig. 42; Stefan and Gu 1992).

7.2.2 Stratification in Underground Mines and Shafts

Stuart and Simpson (1961) and Barnes et al. (1964) reported stratification in flooded underground mines of the Pennsylvanian anthracite fields in the USA (Fig. 98). An extensive investigation of mines in this region was conducted by Erickson et al. (1982). Another North American mining area where stratification was found to occur was the Picher Mining District in the Tri-

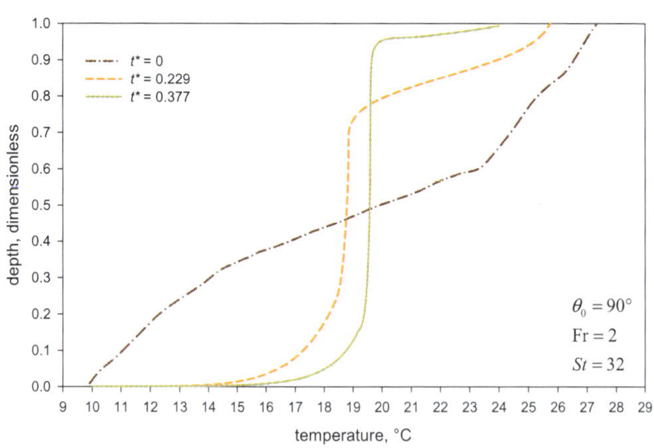

Fig. 42. Temperature distribution in an enclosure prior and after jet-induced mixing. t^*: dimensionless time, θ_0: jet angle, Fr: Froude number, *St*: stratification number (after Stefan and Gu 1992).

State district (Playton et al. 1978). Though mine water stratification was known as early as the 1950s, its potential importance was not known about then. Uerpmann (1980), who measured the stratification in five flooded salt mine shafts, proposed to make use of mine water stratification to secure radioactive waste by implying a special mine lay-out. Similar studies with similar results were conducted by Heidenreich et al. (1991) and Czolbe et al. (1992), who also used numerical models to predict the distribution of radioactive pollutants in flooded mines for radioactive waste disposal. Nearly all published cases with mine water stratification are described in chapter 12.

Stratification does not imply that there is no measurable flow in the mine. Several mine water tracer tests have shown that water does flow within the stratified parts of a mine, as do comparisons of experimental data and field measurements.

A characteristic feature of stratified mine shafts is that the single mine water bodies are separated from each other, usually at the onsetting stations of adits connected to the shaft. Three reasons can be responsible for this stratification: water inflow from the connected adit, fluid loops through interconnected shafts and adits, and flow disturbance due to changing physical properties of the shaft. A FIDAP 7.5 numerical model showed that fluid loops can occur and that the direction of those loops changes rapidly with minimum changes of energy input (temperature changes within the measured range of temperature in the shafts). The same thing was described by Javeri (1995), who investigated potential waste disposal in flooded underground mines using the code TOUGH2.

König and Blömer (1999) found out that with increasing length to diameter ratios, L/d, of a shaft, the number of convection cells and the velocity increases. They modelled a single-shaft closed cavity thermo-convection problem with dimensionless Rayleigh numbers Ra = $0.03 \cdot 10^6$, $6.2 \cdot 10^6$, and $243 \cdot 10^6$ (see equation 29) using the Navier-Stokes equation and a Boussinesq approximation in the software package Hot-Spring.

In the case of the Ronneburg/Germany uranium mine (for details, see below), it is hoped that stratification will occur without intervention (Gatzweiler et al. 2002). Kalka et al. (2002) modelled the density-driven vertical convection in the Ronneburg mine caused by the geothermal gradient and found out that those effects are neglectable there. Paul et al. (2002) measured the temperature-depth and salinity-depth profiles in an observation well of the Ronneburg northwest mining field. Their temperature-depth profiles show a quasi-linear course with a gradient of 42–44 K^{-1}, thus indicating that diffusive transport dominates and vertical convection at the Ronneburg site might in fact be neglected. However, the geothermal gradient is not the only driving force for free convective flow; density differences in the fluid may exist due to effects such as mineralization or turbidity. Therefore, the geothermal gradient cannot be used as the only measure to estimate if free convection will arise or not.

Most of the measurements conducted so far indicate that stratification in mines can clearly be traced by temperature as well as electrical conductivity changes. Similar results were obtained by Erickson et al. (1982), who correlated electrical conductivity and sulphate concentrations, concluding that acid mine drainage contamination can be mapped by continuous electrical conductivity measurements. Those electrical conductivity and temperature measurements in flooded shafts are a first indication of the chemical and hydrodynamic processes occurring in the mine water pool. Generally, it can be concluded that higher conductivities in the lowermost mine parts indicate higher mineralisation and often pollution while lower conductivities in the higher parts of a mine water pool indicate less polluted water or fresh infiltration water.

This dichotomy can be observed in nearly all stratified mines. It can also be observed in the Austrian Nikolaus-Bader shaft, where "pollution" in its usual understanding can be excluded, because none of the exposed rocks or sediments (tillite, dolomite, and limestone) has an acid producing potential or contains significant amounts of potentially toxic metals (Fig. 43). There, the higher conductivities, and thus anion and cation concentrations, in the deeper water are a result of exclusion from an overall convective loop and from mixing with the fresh(er) water in the higher levels of that shaft and the higher mine workings.

During a tracer test in the Niederschlema/Alberoda/Germany mine, suspended material within the 366 b shaft slowly moved to the water's surface and then reversed its motion, flowing downward. Several such convective cells were situated near each other, building a flow scheme similar to Bénard cells for natural convective flow. As Bénard cells are typical for thin water layers with surface tension (Gebhart et al. 1988), it might be assumed that the cell structure seen in the Niederschlema/Alberoda mine was either restricted to the upper surface of the flooded shaft, or the convection cell's symmetry was congruent to Bénard cells. However, the temperature differences in the Niederschlema/Alberoda shafts clearly indicate that Rayleigh-Bénard instability is the likely cause. Numerical simulations of Javeri (1995) also resulted in convective cells like those that were observed in-situ. Independent of whether the cells are Bénard cells *sensu strictu* or not, convective cells in flooded shafts clearly exist. The continuous temperature, conductivity, and pH measurements during the rise of the mine water in the Niederschlema/Alberoda № 383 shaft are an indication of such small convective cells (Fig. 44).

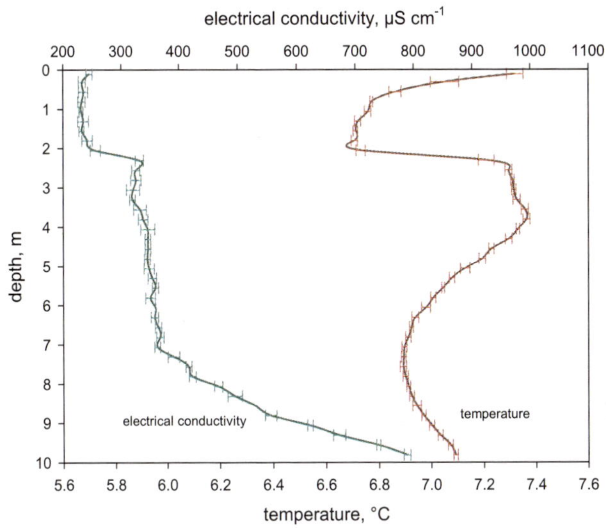

Fig. 43. Temperature and electrical conductivity stratification in the 10 m deep Austrian Nikolaus-Bader shaft, an abandoned small-scale mining site.

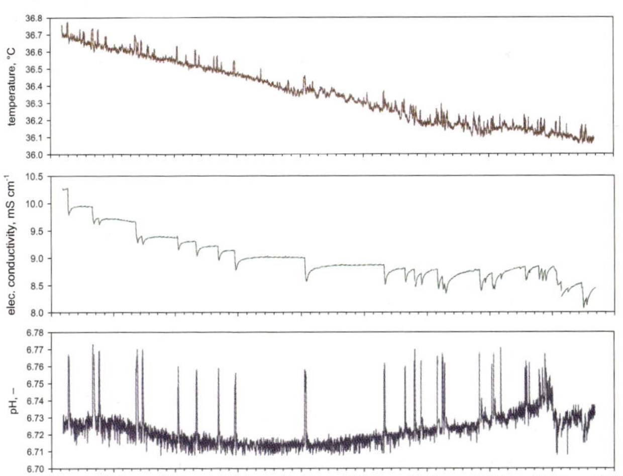

Fig. 44. Continuous temperature, electrical conductivity, and pH measurements in the Niederschlema/Alberoda/Germany № 383 shaft between September and November 1994 while the mine water was rising between levels -945 and -900. Head above probe on 6 September 1994: 42.0 m, 4 November 1994: 60.4 m.

8 Mine Closure Flow Models and Geochemical Modelling

8.1 Introduction

Modelling of flows in flooded underground mines helps one to understand the release of pollutants from abandoned mines. Mine water models can be used to study the release of naturally occurring pollutants, such as metals released from acid mine drainage, as well as anthropogenic pollutants emanating from underground waste repositories for nuclear and non-nuclear waste. Several approaches to modelling the flow in mines were mentioned in chapter 7, without discussing the possible modelling techniques. The geometry of mines is usually more complicated than a "normal" aquifer and even more complicated than a turbine or motor, and so mine models are comparatively complicated. The following sections will give the novice an introduction to mine water modelling, but it is not meant to provide a detailed understanding of the specialty. It should be noted that there is a large overlap between different models, and even conflicting definitions of what a model is. Furthermore, some models have different capabilities and could be grouped into more than one category (see Christopher Wells' "Groundwater Modeling Tools" summary at the InfoMine homepage).

As most modellers recognize, "all models are wrong but some are useful" (Box 1976, 1979). Therefore, one of the first decisions when modelling mine water flow or rebound is: which model shall I use? As described in section 10.1.3, the network of shafts, workings, and galleries can be seen as a karst conduit network and most of the considerations for flow situations from karst aquifers are similar to those of flooded underground mines. For simplification reasons, a flooded underground mine can also be viewed as a network of pipes, which eases the effort necessary to model each feature found in a mine. Collon-Drouaillet et al. (2005) used this modelling approach for the South Basin of the Lorraine/France iron-ore deposits. Because each mine differs from each other, most models are extremely site specific and are adapted to the special requirements of a single mine, and so it would not make sense to discuss site-specific approaches in detail.

There are in fact nearly as many models as model developers and new ones are constantly being developed while established ones are being modified. Many people start by selecting a model and then they compile their data, yet doing things the other way round generally makes more sense. There seems no logical reason for using a sophisticated numerical finite element or finite difference model for a mine and its environment if the data available is insufficient or too fuzzy. Simple model approaches sometimes yield as good results as an intensive x-thousand cell CFD (computational fluid dynamics) model. Note: if you find such terms mystifying and see no reason to learn more, there is no reason for you to struggle through this section; you can jump directly to chapter 9.

This chapter will describe the principal approaches of modelling flow in flooded mines. For each of the possibilities, a short example will be given. Because every mine has its special geometry and unique conceptual model (section 5.9) and special questions to be answered, it is not possible to give general recommendations on which modelling approach to use.

Flow models are generally analytical, empirical, or numerical. Analytical models would be extremely complicated for a real flooded underground mine – if possible at all, and so far, no comprehensive analytical model of a flooded mine has been developed. Thanks to the improvement in computer technologies, 3-D-visualisation nowadays is a convenient means to show the complicated interconnections in underground workings and the chemical or physical developments that might appear during or after the flooding process. Yet, these sophisticated pre-processors must not deceive one into believing that realistic models can be produced by the modeller!

Governing equations of flow in a conduit and matrix network have to include water flow, heat transport, and solute transport, which are in turn derived from the conservation of mass, momentum, and energy laws. Those laws mean that the mass, momentum, and energy in cell one of a model must stay the same in cell two of the model. Furthermore, as temperature, concentration, and pressure are changing in real world mines, the equations of state for density and viscosity have to be included in mine water flow modelling approaches.

In conventional hydrogeological investigations, the governing equations for modelling water flow are Darcy's law and the law of mass conservation. For these standard situations, many analytical solutions and numerical codes are available and have been used successfully (U.S. Environmental Protection Agency 2003; Hiscock 2005). Yet Darcy's law is restricted to flow in porous media, to laminar flow, and small flow velocities. These prerequisites are not valid in flooded underground mines, as they are neither "porous" nor is the flow always laminar and, as can be seen in table 19, the flow velocities can be rather fast. However, several researchers have simplified a real situation so that a Darcian approach could be used, allowing the mine operators or regulators to benefit from numerical codes such as MODFLOW. Some of those investigations are described hereafter. In addition, several software packages have been developed especially for mine water flooding investigations.

8.2 Empirical Models

Empirical models are used when the physical properties or chemical reactions involved are not known in enough detail to analyse them by analytical methods or to set up a numerical model. In contrast to expert systems or expert knowledge, empirical models are not necessarily based on a detailed knowledge about the processes involved. Usually, the data for the empirical model comes from laboratory or field experiments and their systematic connections to the governing equations involved are not clear. In semi-empirical approaches, parts of the model might be empirical, while others are based on a supporting theory (Boudjemaa et al. 2003).

Darcy's law is in fact an empirical equation that is well known by hydrogeologists or hydrologists. In 1856, Henry Philibert Gaspard Darcy investigated the public water supply system in Dijon/France (Darcy 1856). In an attempt to improve water purification through sand filters, he measured the piezometric heads before and after a vertical sand filter filled with the siliceous sand of the river Saone. He conducted 23 experiments with 4 different column heads under atmospheric pressure and 12 experiments with changing pressures at the filter outlet and concluded that "*pour un sable de même nature, on peut admettre que le volume débité est proportionnel à la charge et en raison inverse de l'épaisseur de la couche traversée*" (Darcy 1856, Appendice – Note D: Détermination des lois d'écoulement de l'eau à travers le sable; eq. 74). Darcy's empirical law, if described in modern words, is a simplified averaged momentum balance equation for a Newtonian fluid in porous media, consequently the Navier-Stokes equation. It can also deduced analytically (Bear and Bachmat 1991).

Other empirical models are the discharge equations for thin-plate weirs or for flumes, as described in section 9.2.1. Nearly all of the equations given there, though their physical background is known very well, are empirically deduced from experiments with the equations only valid for the specific weir or flume type.

Empirical models use different statistical, stochastical, or mathematical techniques to reproduce the "real world" parameters (Fig. 45; Box et al. 1978). Their aim can be to describe a physical law or to predict the future or past development of a measured dataset. Such techniques known to most researchers in the mining field are simple or multivariate regression analyses, curve fitting algorithms, parameter fitting packages (see section 9.3.4), or polynomial spline curves. Other methods are Fourier series analyses, wavelets, NURBS (non-uniform rational B-spline), or neural networks.

Splines are functions defining a curve piecewise by polynomals. Though splines might be useful to fit simple existing data with only one variable, their applicability is limited in prediction models or in more complicated hy-

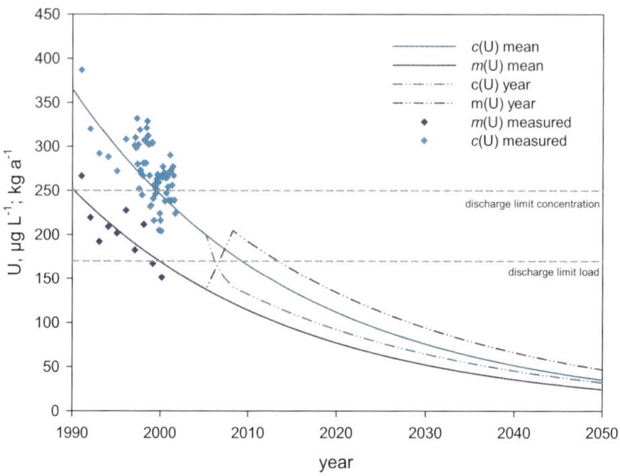

Fig. 45. Forward predictions without advance dewatering boreholes (full line) and with dewatering boreholes (dotted line) for the uranium concentrations and loads of an abandoned uranium mine using dewatering boreholes to lower the uranium concentrations, where *c*(U) = uranium concentration and *m*(U) = uranium mass load. While the concentrations decrease after drilling, the total load is projected to increase. The discharge limits for the uranium concentration and loads are predefined by the mining and water authorities.

drodynamic situations. They are usually used to model discrete data but empirical mine water models based on splines are not known to date. Yet, if the knot position and spline coefficients can be deduced precisely enough and represent the data adequately, most discharge curves or mine flooding predictions could be based on such data.

Fourier series analyses are commonly used in the interpretation of geophysical data sets. Al-Fouzan et al. (2004) describe an airborne mine water investigation using frequency-domain electromagnetic (FDEM) conductivity investigations. To remove the noise in their data, they used Fourier filtering. Yet, empirical mine water models based on Fourier series have not been published so far, though this method can be used to extract seasonal variations or systematic fluctuations. Fluctuations in the head of a mine water discharge in the Ehrenfriedersdorf/Saxony tin mine was analysed using a Fourier series analysis. The hypothesis of the mine operator was that the fluctuations were due to switching the mine water pumps on and off, but the Fourier series analyses could clearly assign those fluctuations to tidal effects.

Wavelets are a modern method to analyze measured time series data according to their scale where the amplitudes or wavelengths of the important cycles are time dependent (Graps 1995; for example, they are used in compressing pictures, according to the JPEG 2000 standard of the Joint Photographic Experts Group). The method originated from signal and image processing, but is nowadays used in a broad range of applications. In contrast to Fourier time series analyses wavelets are used for time-frequency representations with the advantage of a good time and frequency representation. Results of such wavelet analyses are the variabilities of a time series within time (Torrence and Compo 1998; Lord et al. 2000). So far, no wavelet analyses of time series in mine water applications have been done. Yet, they can be used to interpret time series of concentrations, or to find characteristic patterns in those time series that could help to understand the changes in mine water chemistry after a certain – not obvious – treatment option has been applied. One of the most comprehensive commercial software packages, which also has the capability to use Fourier transforms and many other options to analyse measurements, is SeaSolve's AutoSignal (SeaSolve Software 2003).

8.3 Analytical Models

Analytical or semi-analytical models for real underground mines have to find an explicit solution for the general transport equation, which is impossible – at least with the mathematical methods known today. Yet, Grisak and Pickens (1981) have already been presented an analytical solution for comparatively simple cases, such as tracer flow through a single fracture. Pfingsten and Mull (1990) provided an analytical solution for flow in connected fractures. Chan et al. (1987) compared analytical solutions for a single fracture with the numerical 3-D-code MOTIF (model of transport in fractured/porous media) while Reimus and Haga (1999) provided analytical solutions for a dual porosity conceptual model. Further analytical solutions for fracture and multiple fracture flow were presented by Tang et al. (1981) and Tsang and Tsang (1987).

Bau and Torrance (1981a, 1981b, 1983) conducted interesting experiments with open and closed fluid loops that they heated and cooled at different locations along the loop. Though their intention was to find analytical solutions for flow within solar panels, the implications of their results are also valid for flow in flooded underground mines, as mentioned earlier.

Simple mine geometries and consequently simple conceptual models can be used if segments of a mine are being investigated independently. Such approaches have been used by several authors (e.g. Fawcett et al. 1984; Singh and Atkins 1984, 1985). Fawcett et al. (1984) investigated water inflow into coal mines by assuming that a mine acts like a large well. They used the Theis and Jacob approaches for confined and leaky aquifers (Theis 1935, 1940) to estimate the amount of water flowing into a mine.

A novel analytical solution for open pit lake filling was presented by Fontaine et al. (2003). For real cases, they concluded that the lateral extent of the cone of depression was not predicted exactly enough and that neither the time of the maximum extent of dewatering or its temporal effects on springs and seeps could be modelled in detail. For such requirements they concluded that a full numerical code and substantially more data would be necessary.

Małoszewski and Zuber (1984, 1985) investigated tracer transport in fractures. They found an analytical solution for the diffusive mass flux from fractures into the matrix with a Dirac tracer injection. In a later publication, they also provided an analytical solution for a single fracture with flow diffusing into the matrix, which they called Single Fissure Dispersion Model (SFDM; Małoszewski and Zuber 1990).

8.4 Numerical Models

Numerical codes are commonly used nowadays to model mine water rebound, reactive transport, and the hydrogeochemical development of mine waters, or ground water. These numerical models, most of them based on Darcy's law for porous aquifers and laminar flow, have been applied to ground water investigations to a large extent (Deming 2002; Hiscock 2005; Hölting and Coldewey 2005). Commercial and non-commercial numerical codes used in mine water rebound and flow predictions include MODFLOW, VULCAN, FEFLOW, SHETRAN VSS-NET, ANSYS FLOTRAN, MIFIM, KASOMO, GRAM or SPRING; some of these can also deal with non-Darcy flow (Fig. 46). Most of the models are either finite element or finite difference models, with no preference to either; a smaller number of models are based on other numerical methods. Advanced models cover nearly every aspect encountered in a mining environment, and complex models are able to simulate multi-phase flow and multi-species transport. Yet, all those codes have to deal with a major problem in mine water research: the lack of consistent geometrical and hydrogeochemical as well as hydrogeological data. Because all such models require boundary conditions to solve the continuity equation, many starting parameters are based on weak empirical data or in the worst case, have to be guessed.

As described earlier, Darcy's law for flow in porous media, laminar or turbulent Poiseuille flow, laminar or turbulent convective flow, or almost no flow will occur together in a mine environment. Furthermore, the different physical properties of the rock matrix, fissures, and mine workings must be considered. Though some codes have the capability of dealing with Darcy as well as non-Darcy flow situations, the integration of both flow types can become a trial of patience. Exhausting calibration procedures, often based on

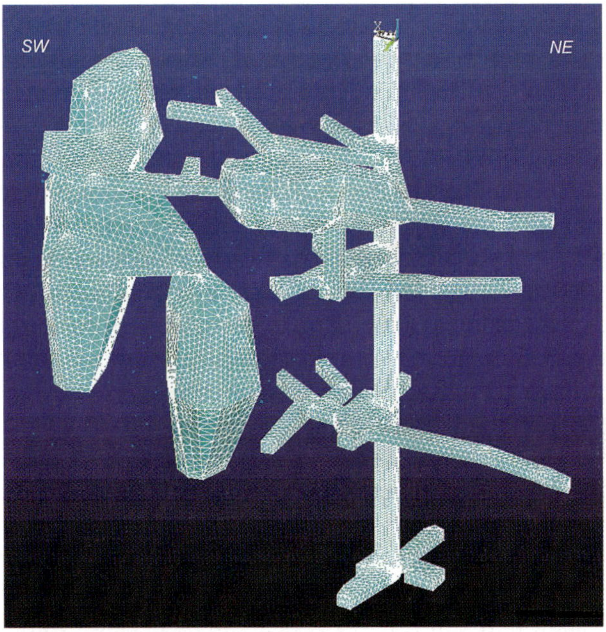

Fig. 46. ANSYS-FLOTRAN finite difference mesh for the numerical model of the Austrian Georgi Unterbau mine (modified from Unger 2002).

trial and error measures, are necessary to obtain trustworthy results. Because mined strata are not a standard condition for most numerical codes, the non-standard use of such codes requires lengthy calibration procedures. To overcome several of these restrictions, black box ("lumped" parameter) modelling approaches have been widely used to solve mine water rebound and flow problems (e.g. Sherwood and Younger 1994). Especially when many input parameters are involved, it is sometimes unclear which parameter should be fitted to get the best result. In a pipe network, for example, where the velocities within the pipes are too large, the head difference, the roughness of the inner walls, or the diameter of the pipes can be modified slightly. A change in any of those parameters would result in lower velocities and it is often unclear which screw is the right one to be driven. In such circumstances, modellers tend to say that the parameters are unidentifiable.

Since numerical codes were first used to model mine water rebound and mine water flow, computer hardware and software as well as the numerical approaches themselves have continually developed. Most models are site specific and adapted to a special situation, and only a few of them have been used for modelling other situations as well. Describing them here would therefore be like shining a spotlight in one direction at random because code development is so quick. Yet, some principles shall be picked out.

To overcome some of the above-mentioned problems, Frydrych et al. (2000) used a mixed-hybrid FEM model to calculate the flow field and contaminant transport of the Stráž uranium mine (Czech Republic). They used the Rothe method for the time discretization, which covered only 7 of the expected 30 year flooding period. A total of 20,000 trilateral elements had to be calculated and they modelled 2560 days with 10-day time steps. Novák et al. (1995) used another numerical model at the same uranium mine and both numerical models resulted in similar TDS distributions in the Stráž block, though showing differences in the details.

Heidenreich et al. (1991) extended a numerical model originally developed to predict the formation of salt caverns to address radioactive waste disposal modelling. The model accomodates free and forced convection, diffusion, heat transport in the fluid and rock matrix, salt dissolution, natural and radioactive heat sources, and radioactive decay. Both, the Navier-Stokes and Darcy equations can be used, depending on the question to be solved. Their model included two key scenarios with one shaft and two levels: one with a blind shaft between the two levels, the other one without the shaft. Their model clearly proved that the blind shaft between the two levels increased the speed of radionuclide transport and should therefore be sealed if the mine was ever used for radioactive waste disposal. Furthermore, Heidenreich et al. (1991) describe the commonly known situation of flooded salt mines where a fresh water lens overlies the brine without a significant mixing between the two layers over time. They concluded that such a situation might be advantageous for radioactive waste disposal and therefore modelled a flooded shaft containing a brine overlain by fresh water. As long as the gravitational forces exceeded the thermoconvective forces, the stratification remained stable for a long time. All together, the numerical model of the authors was compatible with the "multiple" and "single shaft" approach deduced from mine water tracer tests (Wolkersdorfer 2006).

8.5 Models Based on Monte Carlo Methods

Monte Carlo simulations use randomly generated values from a pre-defined range of potential input parameters based on probability distributions. They can be combined with nearly all the modelling approaches described here and are commonly used to evaluate the sensitivity of certain parameters in a given model. This modelling approach is also used to evaluate minimum or maximum results of models through multiple runs of the same model with randomly chosen values. Its scientifically based introduction dates back to 1946, when it was used in the Manhattan project (Metropolis and Ulam

1949), yet Monte Carlo methods were used before then with less sophisticated statistics. Many modelling approaches in the mine water field are based on Monte Carlo simulations as there are many variables that are highly uncertain.

Aurada (1970), for example, modelled ground water rebound after the cessation of pumping in the Mansfeld/Germany mining area. Based on a computer code (*Grubenwasseranstau Mansfelder Mulde*), he set up a numerical simulation where he randomly selected hydrogeological parameters, such as volume of voids, expected precipitation, interconnections within the mine, and the potential discharge point, within a range of observed values. The model consisted of 12 "lamellas," representing the hydrological provinces of the mining area, each of which was filled with water by the code. After lamella 12 filled, the code restarted using another, randomly chosen scenario. From his conceptual approach, his model is very similar to the GRAM model (Burke and Younger 2000).

GRAM (Groundwater Rebound in Abandoned Mineworkings; Sherwood and Younger 1994; Burke and Younger 2000) is an analytical model that uses exactly the same conceptual idea as *Grubenwasseranstau Mansfelder Mulde*. It can also be run with a Monte-Carlo sub-routine for predictions under considerable uncertainty (Paul Younger, pers. comm.). Its first application was predicting ground water rebound in the Durham/United Kingdom coal field, which is very similar in size to the Mansfeld mining area. Based on the investigation of the mine geometry and the hydrogeology of the surrounding area, the mine was subdivided into ponds, which were then filled by the model. All uncertain model parameters were based on ground water levels, mine water pumping rates, precipitation, volume of voids, interconnections, and discharge areas.

8.6 Geochemical Modelling

Since the outstanding work of Nordstrom (1977) on acid mine drainage, geochemical models are regularly used to interpret and predict mine water environments and mine water treatment scenarios. Alpers and Nordstrom (1999) summarised such models and discussed how they can be used in mine water investigations. Nordstrom et al. (1979) compared the results of different numerical models. Nordstrom (2004a) provided an excellent overview of the main concepts and developments in the field of geochemical modelling, including examples. Merkel and Planer-Friedrich (2002) published, in both German and English, an easily understood guide to numerical modelling of aquatic systems using PHREEQC, while Merkel et al. (2005b) and van Berk (2000) described how to use PHREEQC to interpret water-rock interactions in mine waste rocks.

Commonly, such models are referred to as "geochemical models" or "thermodynamic models," but whatever they are called, they use a database of chemical thermodynamic data and a numerical computer code to solve the governing equations, and require an experienced scientist who understands the concepts behind the numerical approach. Geochemical models are a tool to understand the species distribution in an aqueous solution as well as the theoretical mineral assemblage that would result in a water with the given chemical composition and physical properties. Modern codes can calculate the mass balance, speciation, reaction-path, sorption processes, and kinetically controlled reactions (like pyrite oxidation). PHREEQC, nowadays, is the programme used by most hydrogeologists. The most recent version also includes the Pitzer equation for solutions with higher to extreme activities, such as brines. Many researchers who are using numerical codes for interpreting water analyses take the results as 100% correct and representative

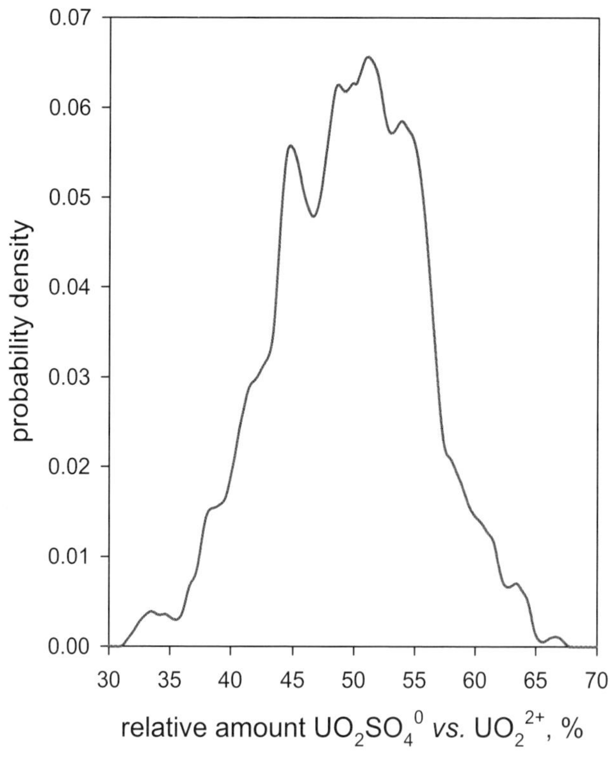

Fig. 47. Results of geochemical modelling for the $UO_2SO_4^0$ *vs.* UO_2^{2+}-fractions based on database uncertainties. Acid mine water at a pH of 2.24; $[SO_4^{2-}] = 3.1 \cdot 10^{-3}$ mol; $[U(VI)] = 4.2 \cdot 10^{-5}$ mol (modified after Nitzsche et al. 2000).

for the water they analysed. Yet, the chemical thermodynamic database is based on experimental results, which have an inherent experimental error range. Nitzsche et al. (2000) or Meinrath et al. (2004) varied different constants in the database within their error range and were able to show the uncertainties of such numerical codes – and their results might be surprising for those who tend to discuss the 2nd or 3rd decimal place in the results of the programmes (Fig. 47). Error bars of up to 20% can be obtained when the uncertainties of the codes are taken into account.

Numerical codes have a wide range of applications in the mine water field. On the one hand, they help us understand the processes observed in the field or laboratory; on the other hand, they are a useful tool in predicting future conditions, based on different scenarios. When the results of a chemical analysis of a mine water are interpreted using a numerical approach, it is possible to understand which minerals the water has contacted and to determine where in the mine the water might have come from. An example of this approach is the mine water discharging from the abandoned Gernrode/Harz Mountains fluorspar mine. Two of the issues that had to be addressed were the complicated mine water hydrogeochemistry and how to treat it. The hypothesis for the chemical composition of the mine water was that "pure" mine water is to a certain degree mixed with highly mineralised mineral water. Because the underground part of the mine was not accessible, no analysis of the water within the mine was possible. Therefore, historical mine water quality data and water quality data for the mineral springs surrounding the mine were used. Based on these analyses and the current mine water composition, a mixing model was set up in PHREEQC; the mine water was mixed with different mineral waters at different mixing ratios. Based on the results, it was concluded that the mine water discharging from the drainage adit comprises 15–30% of highly mineralised water and that isolating that water and the mine water would ease the burden of treating the mine water discharging from the drainage adit (Schindler 2003).

No numerical modelling of aqueous speciation can ever substitute for an experienced scientist. There are many publications discussing how a numerical code calculated the oversaturation of some 'nice' iron phases, and how the authors were expecting precipitation of those phases, but tunable to identify them in the sediment or water! Everybody in the mine water management field, along with researchers, regulatory authorities, stakeholders, and politicians should know that the results of geochemical modelling can only be relied on if the results are properly interpreted by an experienced modeller. Furthermore, modelling should be based on observations in the field or in the laboratory. Croxford et al. (2004) and Coulton et al. (2004) gave an excellent example of how an active mine water treatment process was optimised using geochemical modelling. Based on the results of the model, laboratory and field experiments were conducted and the final treatment plant was thus optimised.

To overcome the problems with missing kinetics, Hamm et al. (2005) used a simplified approach of just modelling the predominant reactions that they identified through kinetic laboratory tests. Furthermore, they coupled the kinetics to a simplified pipe network model for investigating the impact of backfill material on the water quality of an abandoned iron mine in the Lorraine/France.

9 Monitoring and Sampling

9.1 Introduction

The objective of this chapter is to show how important reliable measurements of hydrogeochemical and physico-chemical parameters of mine water are for the design of mine water treatment facilities and remediation actions. There are numerous scientific investigations which go into all the details for a specific mine site. Such investigations are essential for understanding the compartment interactions and the influences of the mine water discharges on the aquatic environment or the water catchment. Yet, those detailed investigations are beyond the goals of semi-routine remediation or monitoring programmes by the typical mine operator or regulator.

Mine water can significantly affect the environmental quality ("good status") and ecological potential of water catchments and therefore, the measurement of chemical and physical parameters should receive special consideration. A baseline site assessment that collected all relevant data for the mine sites within the catchment should be a prerequisite for any catchment monitoring programme.

All the techniques used for analysing mine water discharges evolved from ground or surface water monitoring. There are several pitfalls in sampling and monitoring mine water discharges that this short guide will help you to avoid. Sampling programmes should avoid acquisition of unrepresentative data that might lead to inappropriate treatment or remediation design.

It is essential to obtain as much information about the hydrogeochemical characteristics of a mine water discharge as possible. In addition to the flow measurements and physico-chemical measurements, chemical analyses of the mine water will have to be conducted over a sustained period to account for seasonal variability and storm events. Mine water discharge qualities can alter significantly during the period between sample acquisition and analysis. Concentrations of the relevant elements can decrease or increase. Some critical parameters must be determined on site. Other parameters require special handling or preservation in the field so that lab analyses are accurate. Depending on the parameters being analysed, mine water samples typically need to be filtered or acidified in the field, and they need to be transported to the laboratory as quickly as possible in cool (ca. 4 °C) and dark containers.

In addition to obtaining chemical data on the discharge, it is advisable to consult people with knowledge of the mine and local residents who often have information about extreme situations or other data that could be relevant for the mine water investigation. This information supplements the experience of the consultant.

When establishing a monitoring programme, a step-wise approach is recommended. In a preliminary screening investigation, simple parameters (e.g. electrical conductivity, pH, total Fe) can be used to indicate the worst discharges in the catchment. Based on these results, further water samples can be taken and the number of monitoring stations can be increased or decreased accordingly.

This chapter will give some hints and list some experiences gained over several years of mine water research. For in-depth studies, it is highly recommended that you consult national or international guidelines (e.g. DIN, ISO, ISO-EU, EPA, DVWK) or the relevant literature, which varies from country to country and is regularly updated.

Since each mine water is generally different in some way from any other mine water, no specific recommendations for a monitoring programme can be given. It is the obligation of the mine owner, its consultant, and the competent authorities to set up a priority list for the parameters to be measured. A first indication of the parameters that should be measured can be found in the appendices of the European Water Framework Directive (in principal, the inorganic parameters), the regional Water Directives and, in the case of uranium mining, the relevant radiological parameters have to be considered. At a a minimum, the following parameters should be measured: (hot) acidity (base capacity K_B), alkalinity (acid capacity K_A), pH, electrical conductivity, temperature, redox-potential, Fe, Mn, Al, sulphate, and flow (Figs. 48–62).

Furthermore, though sampling, monitoring, or taking photographs at most mine sites is comparatively uncomplicated, the situation can change significantly inside coal mines. Coal mines, and occasionally other mines, such as salt mines, are a source of methane and therefore measurements inside coal mines can only be conducted with intrinsically safe instruments. Most countries are very restrictive concerning the use of electronic equipment in potentially methane bearing mine sites and the relevant regulations always have to be followed!

9.2 Flow Measurements

9.2.1 Direct Flow Measurement

In many investigations, only concentrations (mass of a substance per volume or mass) of contaminants are provided, but equally, if not even more important, are the total mass loads, which are calculated by multiplying the flow and the concentration:

$$M = c \times Q \tag{35}$$

with M mass load, g s^{-1}
c concentration, g L^{-1}
Q flow, L s^{-1}

Low concentrations of pollutants at high flow conditions can consequently be as contaminating as high concentrations at low flow conditions. This fact is often overlooked, as people are used to comparing the concentrations of parameters only.

Flow measurement is essential for calculating the contaminant mass loads in a water catchment area with polluting surface or underground mine water discharges. Some guidance documents were published on flow measurements at mine sites (Minister of Public Works and Government Services 2001; Science Applications International Corporation 2001), but are not commonly available. Precise flow measurement in open and closed channels is a science of its own (Durgin 1981; Herschy 1995; Miller 1996; Bohl 1998; Brassington 1999; Bonfig 2002), and even simple techniques have several restrictions. These restrictions must be considered, as mine water discharges can significantly change: dry streams can become wet and vice versa due to changing hydrologic conditions. Therefore, a given flow measurement technique could be suitable at one time and unsuitable at another time. Each mea-

Fig. 48. The Dirac-impulse injection of Na-fluorescein (dark red when it flows out of the bottle, lighter greenish to yellow in colour when it mixes with the stream water) used to measure flow at the inlet of a lake that receives polluted mine water (photograph by Andrea Berger).

surement technique has special limitations that influence its accuracy. To select the correct technique, it is essential to know the expected flows and the limitations of each approach.

In general, the maxima, minima, and mean flow conditions, including seasonal variations, should be recorded at the highest precision obtainable. In reality, most investigations are planned for one hydrologic year – or even less – and due to the irregular characteristics of precipitation and hence surface flow, the data gathered may not be representative for a longer period of time. Another problem is that all of us tend to prefer fine weather for sampling trips, which often leads to inadequate sampling of storm events. Long-term gauging stations that are located close to the mine water discharge can often be correlated to a certain degree with existing data, allowing one to allow for missing flow and precipitation data. Precipitation data for a special location can be determined by applying one of the following methods: arithmetic mean, Thiessen Polygon, isohyetal method, hypsometric, or multiquadratic methods (Shaw 1994).

Most researchers and practitioners in the mine water field look upon flow measurement as being a simple task. This is not the case. Therefore, to avoid misinterpretations due to a lack of basic understanding, this chapter will describe some of the necessities when conducting flow measurements.

For each method, many international and national standards (e.g. ISO, DIN, ČSN ISO, Ö-NORM, BS, EN) are available. The International Organization for Standardization (ISO), for example, lists more than 100 different methods for flow measurement. These standards describe in detail the limitations of each method and under which conditions the method is applicable. Furthermore, some standards include equations to calculate the errors and uncertainties of the flow measurements, which have to be considered when comparing different measurements. In the following paragraphs, only the methods commonly used in mine water research or remediation projects are described. They assume a channel or pipe flow with a uniform laminar or turbulent flow at the cross section of the flow measurement.

Usually, the national hydrologic stations within a catchment area are insufficient to calculate the total loads of mine water discharges. It is therefore important to evaluate the already existing measurement network and – if necessary – install new flow measurement gauges that fit the needs of the planned treatment or monitoring project. Linsley et al. (1988) provides details on how to interpret catchment hydrology and evaluate the data.

Two options are used for flow measurements: continuous (electronic or drum recorders) and discontinuous measurements. Usually, discontinuous flow measurements will be good enough, but cheap and reliable methods are now available to measure and store flow rates continuously. Data storage devices have became gradually smaller and cheaper and can therefore be installed at many sites without a major investment. Nowadays, fear of vandalism is generally a greater concern than the initial cost. However, because high flow conditions are usually the limiting events in any remediation or mine water treatment operation, the continuous measurement of flows is generally preferable.

Mine water flow can be measured using several methods. The techniques can be divided into primary and secondary flow measurements, whereas primary means a direct measurement of the flow with one variable (e.g. water level: V-notch weirs; pressure: pitot tubes) and secondary involves two variables (e.g. water level and travel time of sound: ultrasonic flow meters). Nearly all of the primary methods can be combined with secondary methods. Here is a selection of the most common flow measurement methods in the mine water field:

- bucket-and-stopwatch (up to 500 L min^{-1})
- impeller current meter (Fig. 52)
- V-notch weirs (Fig. 49)
- H-flumes (Fig. 56), Parshall flumes (Fig. 58)
- Venturi tube
- pump run-times and ratings
- water meters (Fig. 59, Fig. 60)
- pressure meter measurements of boreholes
- transit-time acoustic (ultrasonic) flow meters

Where the flow is measured with weirs, flumes, and tubes, equations relate the head of the water level above the measuring device to the flow through that device. The water levels are usually measured automatically or manually with level gauges or by a technician using contact free instruments (e.g. ultrasonic methods), floats, or gauge meters. All flow equations relate the flow to the water head at a given distance from the measuring device. Therefore, it is essential to install the level gauge or to measure the water level at a proper upstream distance from the installation. The distances are either given by the manufacturer or, for weirs, located beyond the upstream water level influenced by the drawdown.

The best option to measure the water flow depends on the site-specific situation, including the available man power and the water quality. Heavily contaminated water can damage some measuring devices or could change the diameter of a channel or weir (by gypsum or iron hydroxide precipitation). Some mine waters might be contaminant or iron free, but could contain high amounts of carbonates, which can also interfere with the measuring devices. If contaminated mine water is to be measured, H-flumes or contact free methods are generally better.

Fig. 49. V-notch weir measuring the flow of the Straßberg/Harz/Germany mine water treatment plant. Ochre precipitates are regularly removed from the weir plates.

In some circumstances, the flow at a chosen location can be estimated using regression or estimation techniques. This requires a certain number of gauging stations and precipitation values from which the flow at another location might be deduced. It is advisable to use geostatistical methods in addition to the standard regression and estimation techniques (e.g. Davis 1986; Linsley et al. 1988; Lerner et al. 1990; Shaw 1994; Simmers 1998; Burbey et al. 2000).

Though flow measurement in underground mines is very important for calculating mass transport, weirs or channels are often unsuitable for a proper flow measurement in mines. The main problem is that the rough conditions underground (Fig. 50) prevent suitable installation of such devices (Thomas 1959; United States Department of Agriculture 1979). Therefore, such technical installations must be calibrated using independent flow measurement devices. Here is a list of the mistakes commonly found: weirs are constructed directly in the gullets and the water just flows through the weirs without overflowing them (quasi-submerged flow); the blades are mounted in the wrong direction; channels are not cleaned of sediment; metal parts are bent or are covered with precipitates; the weir or flume is of poor design due to a lack of hydrodynamic understanding. Theoretically, submerged weirs or flumes are covered by equations to calculate flow, but their accuracy is low and a lot of calibration measurements at different flow conditions are needed to get a usable head/flow relationship estimation; most 'submerged flow measurement devices' were originally meant as free flow weirs and were poorly designed or installed! A submerged weir is not good practice for flow measurement and must be avoided whenever applicable.

Fig. 50. Badly maintained channel for flow measurements in an underground mine. The channel inflow is not laminar and the channel itself is filled with sediment.

Another problem is that even where a simple method exists to cross check flow measurements, it is not often done – even in scientific investigations. This is a mistake, as will be illustrated using the example of an abandoned Austrian Silver mine. The mine was investigated at the turn of the 21st century and several of the flow measurement channels were badly maintained (Fig. 50). The flow was measured 10 times at 5 different sampling locations and the mine water coming from sampling locations 1–4 flowed through sampling location 5. The total flow ranged between 22 and 60 L min^{-1} and the relative difference between the sum of locations 1–4 and location 5 ranged between 1 and 30%. The apparent error range did not appear to correlate with the total flow; 50% of the differences were in an acceptable range of ±10% and the rest exceeded that limit. There was no data on the accuracy or the uncertainty of the measurement. As we will see, it would have been a simple task to add all the single flow measurements and compare the results with each other.

To detect the "wrong" flow measurements, you should measure both the flow and the electrical conductivity of the water, which is easily done, even in uncomfortable underground situations. If the flow measuring cross-sections are not too far apart from each other and if no precipitation or dissolution of minerals occurs in the area being investigated, electrical conductivity functions as a conservative parameter for the mine water. Table 11 shows how such an accuracy check might be done (data has been partially adapted to illustrate the method). In the first three lines, the parameters of the February measurements are given. This data were accurate within the expected uncertainties. The situation was different in May. The difference in the total

Table 11. Example of a flow measurement in a closed underground mine showing how a wrong flow measurement can be identified and corrected. SL: sampling location; Diff: difference; electrical conductivity in µS cm^{-1}; flow in L min^{-1}, "load" in µS L cm^{-1} min^{-1}.

Date	SL 1	SL 2	SL 3	SL 4	SL 5	Sum 1, 2, 3, 4	Diff	Diff %
22 February								
flow	3.50	0.83	0.25	16.50	21.80	21.08	0.72	3%
electrical conductivity	318	466	619	374	388			
"load"	1113	387	155	6,171	8,458	7,826	633	7%
30 May								
flow	1.16	2.80	0.30	54.40	45.50	58.66	-13.16	-29%
electrical conductivity	859	163	521	167	204			
"load"	996	456	156	9,085	9,282	10,694	-1,411	-15%
relative "load" difference	11%	-18%	-1%	-47%	-10%			
30 May								
flow, corrected	1.16	2.80	0.30	44.40	45.50	48.66	-3.16	-7%

Table 12. Example of a flow calculation by using equation 39. Data modified after Unger (2002).

Parameter	Upstream	Inflow	Downstream
κ	377.7 µS cm^{-1}	328.3 µS cm^{-1}	352.4 µS cm^{-1}
Q measured	1.97 L s^{-1}	–	–
Q calculated	–	2.06 L s^{-1}	4.03 L s^{-1}

flow was 29%, suggesting an error of some sort. By calculating the "conductivity loads" and comparing them with the data from February, it became obvious that there was a mistake at sampling location SL 4. After recalculating the data, the error apparently was caused by a typographical mistake in the flow. Instead of 54.4 L min^{-1}, it should read 44.4 L min^{-1}. After correcting that mistake, the error became -7% and again was in the acceptable range.

You can also use this method to calculate missing flow data. In the abandoned Austrian Georgi Unterbau mine, there is a huge inflow where the drainage adit crosses the Grünwies fault. The amount of water flowing into the mine can become extremely high after rainfall events and thus cannot be measured. Yet, it is possible to measure the upstream water volume in the drainage gullet and the electrical conductivities in the upstream, downstream and inflowing mine water. The missing flows can therefore be calculated, as follows:

$$\kappa_d \times Q_d = \kappa_i \times Q_i + \kappa_u \times Q_u \tag{36}$$

with κ electrical conductivity, µS cm^{-1}
Q flow, L s^{-1}
d, i, u downstream of inflow, at inflow, upstream of inflow

There are two unknowns, Q_d and Q_i, but because

$$Q_d = Q_i + Q_u \tag{37}$$

equation 36 can be written as follows:

$$\kappa_d \times (Q_i + Q_u) = \kappa_i \times Q_i + \kappa_u \times Q_u \tag{38}$$

with only one unknown variable left. Now, after multiplying and rearranging, the equation becomes

$$Q_i = Q_u \times \frac{\kappa_u - \kappa_d}{\kappa_d - \kappa_i} \tag{39}$$

and the inflow and the downstream mine water volume can easily be calculated (Table 12). In the same way, temperature or the concentration of conservative water constituents such as Cl or Br could be used instead of the electrical conductivity.

Bucket-and-stopwatch

One of the simplest methods for measuring flow is the frequently used bucket-and-stopwatch approach. To conduct such flow measurements, the mine water must flow through a pipe or a channel and it must be possible to fill the container without having it overflow, which can sometimes be a problem under high flow conditions. Using this method, you can measure flows over a range of about 500 L min^{-1} down to 0.1 L min^{-1}. You will need a container and a stopwatch; the appropriate container volume can range from 50 mL to 100 L, and the time should be measured in ⅟₁₀₀ths of a second. You simply measure the time needed to fill the bucket or graduated cylinder, and then use eq. 40 to calculate the flow. At least three measurements should be made, but as many as 10 could be necessary under changing flow conditions. Commonly, this method gives mean errors of 5–10%, but under certain flow conditions, better results are obtained.

$$Q = \frac{V}{\Delta t} \tag{40}$$

with Q flow rate, $m^3\ s^{-1}$
V volume of container, m^3
Δt time to fill container, s

Table 13 gives an example of a bucket-and-stopwatch measurement, and includes the uncertainties of the time measurement (0.1 seconds) and the measurement of the bucket, which had 1 L marks. It is assumed that the volume of the bucket can be determined within about 0.5 L during the filling process. By using those uncertainties, 3 flow volumes can be calculated: the mean flow calculated from the measured volume of the bucket and the measured filling time, the minimum flow from those numbers, and the maximum flow. Interestingly, the standard deviation of the mean measurements and the

Table 13. Example of a bucket-and-stopwatch measurement of flows at a mine water discharge. Flows in the mining sector are often measured in L or m^3 per minute and therefore, the results are given in that unit. Uncertainties are calculated in the last 3 columns using the minimum and maximum flow, based on the accuracies of the volume (0.5 L) and time measurements (0.1 s). Bucket volume: 10 L; $\overline{x}$: mean; s: standard deviation.

time, s	flow, L min^{-1}	min flow, L min^{-1}	max flow, L min^{-1}	uncertainty, L min^{-1}	uncertainty, L min^{-1}	uncertainty mean, L min^{-1}
9.76	61.5	57.8	65.2	3.7	3.7	3.7
10.25	58.5	55.1	62.1	3.5	3.5	3.5
10.20	58.8	55.3	62.4	3.5	3.6	3.5
8.98	66.8	62.8	70.9	4.0	4.1	4.1
10.17	59.0	55.5	62.6	3.5	3.6	3.5
$\overline{x}$	60.9				uncertainty	3.7
s	3.5	5.7%			uncertainty	6.0%

mean uncertainty from the minimum and maximum values are within the same range. This is not always the case, but the fact that it does suggests that the researcher worked very accurately.

Such tables and error calculations are rarely found in the geosciences and in mine water research. A more detailed description of simple error calculations can be found in section 9.3.4 and the following text.

Area-Velocity Method

Another simple method to estimate the flow is the area-velocity method. To measure the velocity of the stream, a floating object or a strong dye are placed in the stream and the time needed for the floating object or dye to travel a set distance is measured. As the velocity in a given stream profile is not constant but increases from the stream's bottom to the surface, the ideal floating object would have ⅔rds of its diameter submerged. One can then calculate the flow using the following equation:

$$Q = A \times \frac{\Delta x}{\Delta t} \times f \tag{41}$$

with Q flow rate, $m^3\ s^{-1}$
A characteristic cross section of stream, m^2
Δx measure distance, m
Δt time for object to float the distance Δx, s
f factor for average velocity in area (0.66–0.9)

The dimensionless coefficient f can be calculated as follows:

$$f = \begin{cases} 0.3 \le d \le 1.8: & f = 0.64 + \dfrac{0.2 \times d}{3} \\ d \ge 1.8: & f = \left(1.4 - 0.06 \times \sqrt{d}\right)^{-1} \end{cases} \tag{42}$$

with d water depth in middle of stream, m

Another common method for determining the mean velocity are the Manning and Chezy equations, which include the roughness coefficients n and C depending on the type of channel (Gray 1970; Herschy 1995). Because their tables are rather long and give details for different naturally-occurring stream conditions that are not representative for mine water discharges, their method will not be further discussed here.

In some cases, you need to measure the flow in partly filled pipes, half-pipes, or ditches. In such situations, only the water depth, d, and the width of the water, l, can be measured (Fig. 51). The equation for calculating the area A of the circular segment is:

$$A = \frac{1}{2} \times \left(\frac{l^2 + 4 \times d^2}{8 \times d} \right)^2 \times \left\{ 4 \times \arctan\left(\frac{2 \times d}{l} \right) - \sin\left[4 \times \arctan\left(\frac{2 \times d}{l} \right) \right] \right\} \quad (43)$$

$$d \leq l$$

which, luckily, can be simplified to

$$A \approx \frac{d}{6 \times l} \times \left(4 \times l^2 + 3 \times d^2 \right) \quad (44)$$

with maximum errors of -0.1–9.4% for all cases where $d < l$ and where

- A characteristic cross section of stream, m^2
- l width of stream or water surface in pipe, m
- d water depth in middle of stream or pipe, m

If a dye is used, the dye must be quickly poured into the water to avoid dispersion (see section 10.5.1 for details of tracer injection). The mean length of the dye cloud should be used to measure the time and the factor f in equation 42 can be set as 1. If the mean length of the dye cloud cannot be evaluated exactly enough, the first and last appearance of the tracer cloud should be measured instead and the mean of those two times used as an estimate.

Current (Flow) Meters

Current meters are also commonly used to measure flow in the field (Fig. 52). The different kinds of current meters have various physical characteristics but they are described together here because the results obtained with

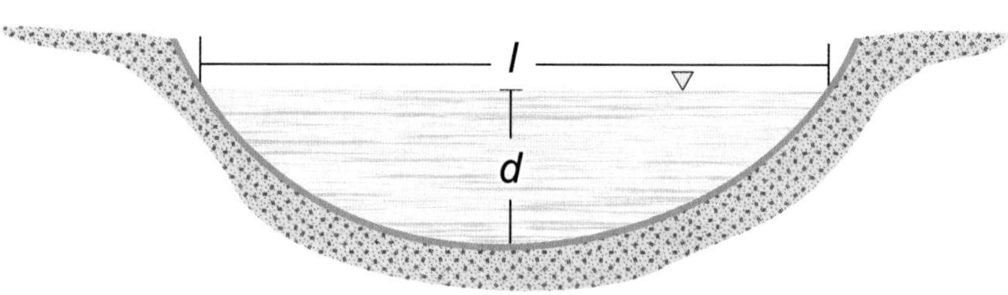

Fig. 51. Necessary parameters for calculating the cross section A of a partly filled pipe, half-pipe or ditch after equations 43 or 44.

them are evaluated in the same way. For the reader wishing to learn the details of closed channel flow measurement, I recommend Miller (1996).

Several types of current meters are used to measure the velocity at a predefined point in the water stream:

- impeller current meter
- heat pulse
- electromagnetic velocity meters
- Doppler velocity meter

Their characteristics and use in the field differ significantly, but any properly calibrated current meter can be used in most mine water related investigations. If uncertain which technique is the most reliable for an actual investigation, you should compare different instruments to make sure that the flow measurements are accurate. The area-velocity method can be used to provide an estimated value for comparison

A prerequisite for current meter measurements is that the flow within the cross-section of the stream must be laminar. To calculate the flow in a given cross section, the cross section is divided into several laminae; within each lamina, the stream velocity is measured at several depths, which vary depending on the specific method used. Both the number of laminae and the number of velocity measurements per lamina depend on the width and the

Fig. 52. Flow measurement with an impeller current meter in the discharge channel of a natural lake receiving polluted mine water. The left operator records the time and the right one holds the impeller current meter and the electronic device for counting the turns of the impeller.

depth of the stream to be measured (Fig. 53). Usually, where mean mine water discharges are being evaluated, three laminae and three velocity measurements at three depths are a good choice. The U.S. Department of the Interior – Bureau of Reclamation (2001) lists eight different methods to measure and calculate the flow in a stream:

- two-point method
- six-tenths-depth method
- vertical velocity-curve method
- subsurface method
- depth integration method
- two-tenths method
- three-point method
- one-point continuous method

You should consult this reference for details about each of these methods. For most of them, the total flow is calculated from the sum of each flow in a single lamina:

$$q_i = \bar{v}_i \times A_i \tag{45}$$

$$Q = \sum_1^n q_i = \sum_1^n \left(\bar{v}_i \times A_i\right) \tag{46}$$

with Q flow rate, $m^3\ s^{-1}$
q_i flow in lamina i, $m^3\ s^{-1}$
$\bar{v}_i$ mean velocity in lamina i, $m\ s^{-1}$
A_i area of lamina i, m^2
n number of laminae

The only problematic variable is the mean velocity, $\bar{v}_i$, which has to be estimated in the above-mentioned methods. One of the simplest methods to do this is to calculate the"simple average," as shown in Fig. 53 for the two-point velocity method. The velocity is calculated in the middle of each lamina in the two-point method as:

$$\bar{v}_i = \frac{v_{0.2} + v_{0.8}}{2} \tag{47}$$

with $\bar{v}_i$ mean velocity in lamina i, $m\ s^{-1}$
$v_{0.2}$ velocity at $0.2 \times d_i$ in lamina i, $m\ s^{-1}$
$v_{0.8}$ velocity at $0.8 \times d_i$ in lamina i, $m\ s^{-1}$

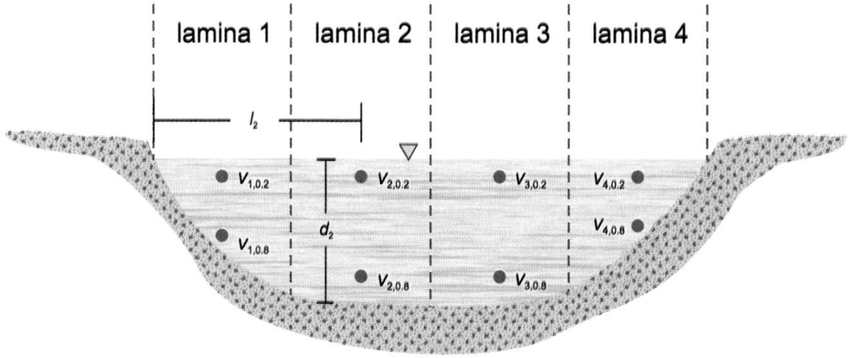

Fig. 53. Calculation of flows in a stream or channel according to the two-point method with four laminae.

Then, $q_{i,(i+1)}$ is calculated as follows:

$$q_{i,(i+1)} = \left(\frac{\overline{v}_i + \overline{v}_{i+1}}{2}\right) \times \left(\frac{d_i + d_{i+1}}{2}\right) \times \left(l_{i+1} - l_i\right) \tag{48}$$

with $q_{i,(i+1)}$ flow in lamina $i,i+1$, m³ s⁻¹
$\overline{v}_i$ mean velocity in lamina i, m s⁻¹
d_i depth in the middle of lamina i, m
l_i distance from stream edge to middle of lamina i, m

The appropriate way to estimate $\overline{v}_i$ varies based on the stream depth, velocity, and its width. Sometimes, the best method must be found by comparing two or more methods with each other. However, if more than two velocities are measured in a lamina, the different velocities have to be weighted; one cannot use an arithmetic mean except when using the simple average method (equation 47). Current meters should not be used for non-laminar flow, which can occur behind pipes, bends or obstacles; the salt-dilution method should be used instead (section 9.2.2).

Weirs

By far, the most common method used for making repeated flow measurements in mine water environments are thin-plate weirs. They can be installed relatively easily under many different flow conditions, their maintenance is comparatively easy, and the flow can be simply calculated by measuring the head above the crest of the weir. Some of the weir types are portable weir boxes and can be used in channels or gullies, others are portable weir plates that can be put into either pre-installed weir holders or hammered into loose soil. They can be easily cleaned from ochre precipitates and any malfunction can usually be quickly identified. One can easily calculate the flow over a standard weir using head-discharge tables or one of the available equations. Though correction methods for submerged flow (also called non-modular or drowned flow) over sharpcrested weirs exist, the water must usually flow freely over the notch as such corrections methods can only be used after careful calibration under different flow conditions. A good summary of the prerequisites for installing weirs is given by U.S. Department of the Interior – Bureau of Reclamation (2001).

However, the flow measuring devices most commonly installed or used incorrectly are also weirs. Weirs or flumes should only be used when the relation between the velocity of the discharged water and the depth of the channel or gully does not exceed a critical value. This dimensionless value is called the Froude number, Fr:

$$\mathrm{Fr} = \frac{\bar{v}}{\sqrt{g \times d}} \tag{49}$$

with $\bar{v}$ mean velocity in channel upstream of weir, m s^{-1}
g gravitational acceleration, m s^{-2}
d mean depth of channel upstream of weir, m

If Fr is greater than 0.5, flow measuring structures lose their accuracy. Thus, in a channel with a depth of 0.2 m, for example, the stream velocity must not exceed 0.7 m s^{-1} or the results will not be reliable.

In free flow conditions, which is the normal working condition of weirs, there must be air between the overflowing water and the weir plate. If there is no air, the pressure condition for which the weir was calibrated is no longer valid and the flow measurements are therefore inaccurate. Another inaccuracy will appear if you measure the head at the wrong position upstream from the weir. For most weirs, the head must be measured upstream of the weir at a distance equivalent to 4–5 times the expected maximum overflow, as the water starts its drawdown at approximately 2 times the head away from the weir. Also, the water level above the crest of the weir must not be too low (see appendix II).

Accumulated sediments must be cleaned from the weir basin behind the weir regularly as they also alter the flow conditions. Furthermore, the flow to the weirs should be laminar; therefore, they should only be installed after a flow section that is long enough to calm turbulent flow situations. The channel sides should be at a distance of at least twice the expected head over the weir to minimise the influence of the channel borders and the downstream water level should always be at least 6 cm below the notch.

The U.S. Department of the Interior – Bureau of Reclamation (2001) lists 12 prerequisites for correctly installing a weir:

- The upstream face of the weir plates and bulkhead must be plumb, smooth, and perpendicular to channel flow.
- The weir crest needs to be level for rectangular and trapezoidal shapes. The bisector of V-notch angles must be plumb.
- The edges of the weir opening must be located in one plane. The corners of the weir opening must have proper specified angles.
- The top thickness of the crest and side plates should measure between 1 and 2 mm.
- All weir plates need to have the same thickness over the entire overflow crest boundary. Plates thicker than 2 mm should be bevelled to an angle of at least 45° on the downstream edge of the crest and sides. The downstream edge of V-notches should have a 60° angle to prevent water from clinging to the downstream face of the weir.
- Upstream edges must be straight and sharp. Burrs and scratches should be removed by machining or perpendicular filing; abrasive cloth or paper should not be used.
- The bottom edge plates and upstream fastener projection should be located at least the distance of two measuring heads from the crest. All upstream faces must be free of oil and grease.
- The overflow sheet or nappe touches only the upstream faces of the crest and side plates.
- The weir head measurement is the difference in elevation between the crest and the water surface at a point located upstream. The upstream point is at a distance of four to five times the maximum expected head on the weir.
- The maximum downstream water surface level should be at least 2–6 cm below the crest elevation.
- The weir head measurement, depending on the weir type, should be at least 2–6 cm to prevent the overflow sheet from clinging to the downstream weir face.
- The weir approach should be kept clear of sediment deposits and other debris.

In general, 3 types of weirs are used: V-notch (Thomson) weirs, rectangular notch (Poncelet) weirs, and trapezoidal (Cipolletti, sometimes also Cipoletti) weirs (Fig. 54). All of them have to be sharp-crested, and the sharp edge has to be on the upstream side. Several researchers investigated different experimental set-ups of full width and contracted weirs, resulting in a set of equations (Ackers et al. 1978), of which Kolupaila (1961) has listed some 100 references.

Kindsvater and Carter (1959) analysed which parameters influence the discharge over a thin-plated weir and came to the conclusion that at least eight variable parameters are necessary to describe the flow over such a measuring structure:

$$Q = \phi\left(L, B, P, h, \gamma, \rho, \mu, \sigma\right) \tag{50}$$

with Q flow rate
L width of the notch
B width of approach channel
P crest height of weir above mean bed level
h gauged head related to weir crest
γ specific weight of fluid
ρ density of fluid
μ dynamic viscosity of fluid
σ surface tension of fluid

They summarised several of the above parameters to the coefficient of discharge C as follows

$$C = \phi\left(\frac{L}{B}, \frac{L}{h}, \frac{h}{P}, \mathrm{Re}, \mathrm{We}\right) \tag{51}$$

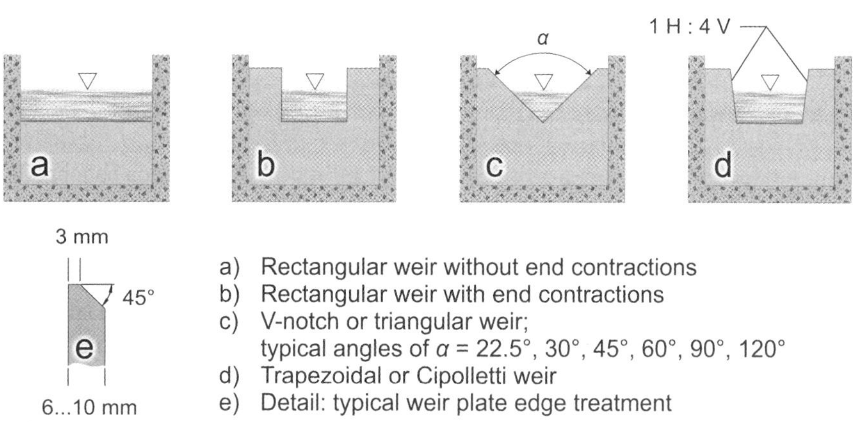

Fig. 54. The three different weir types commonly used in mine water research: rectangular notch weir, V-notch weir, and trapezoidal (Cipolletti) weir (modified after Minister of Public Works and Government Services 2001).

with Re Reynolds number (measures the relative influence of viscosity)
We Weber number (measures the relative influence of surface tension)

Because the influence of Re and We for "large" heads on "large" weirs can be seen as negligible, the coefficient of discharges can be simplified to:

$$C = \phi\left(\frac{L}{B}, \frac{h}{P}, h, L\right) \tag{52}$$

and the general equation for calculating the flow over all sharp-crested weirs, according to some 250 experimental set-ups by Kindsvater and Carter (1959) becomes:

$$Q = C_e \times L_e \times h_e^{\frac{3}{2}} \tag{53}$$

with Q flow rate, $m^3\ s^{-1}$
C_e effective coefficient of discharge, $m^{½}\ s^{-1}$
L_e effective weir length = $L + k_b$, m
h_e effective head above weir crest = $h_1 + k_h$, m
L measured length of weir crest, m
h_1 head measured above the weir crest, m
k_b correction factor to obtain effective weir length, m
k_h correction factor, $9.144 \cdot 10^{-4}$ m

Though this general equation looks rather simple, different solutions are given in the literature and the only way to get a precise head-discharge dependency is to calibrate each notch before it is used. Another option is to install manufactured, pre-calibrated weirs. In most investigations, a rough estimate of the mine water discharge is sufficient. A comparison of the equations given in the literature shows that the errors are usually very big for high flows because the equations do not consider the coefficient's dependency on the head above the weir crest. Because mine water discharges can reach several cubic meters per minute, this inaccurate coefficient might result in an underestimation of up to 13%, compared to the real flow. Therefore, the simplifications given in several publications for measuring the flow discharged by a mine or a mine water treatment scheme should not be used (e.g. U.S. Department of the Interior – Bureau of Reclamation 2001; Younger et al. 2002).

Depending on the weir type, equation 53 can be solved in different ways, of which some are presented in appendix II. For most purposes, a simplification of the equation fulfils the needs of mine water researchers, but all of the simplifications have an error of ± 1–5%. This error is mainly a function of the head above the crest of the weir and normally increases to both ends of the 'boundary' conditions for each weir type. It also increases if weirs are installed that do not meet the restrictions for its use or if weirs are installed in non-linear flow conditions. If high precision measurements are needed, the equations in appendix II should be used and – if necessary – be fitted to the site-specific conditions, as described in U.S. Department of the Interior – Bureau of Reclamation (2001). All of the following SI unit-based equations were computed using the original equations; discharges for different flow conditions were computed and, based on those results, the simplified equations were fitted to the results using a state of the art curve-fitting model (TableCurve 2D and 3D). Before using these equations, it is recommended that another flow measurement technique be used to quantify the measurement uncertainties and to evaluate the real discharge coefficients.

The following nomenclature is used for equations 54 to 59 (see also Fig. 126):

h	gauged head related to weir crest, m
b	crest breath, m
P	crest height of weir above mean bed level, m
Q	flow over weir, m^3 s^{-1}

The unit for all of the coefficients is $m^{0.5}$ s^{-1}.

V-notch (Thomson) weirs (90°, ½ 90°, ¼ 90°; h > 0.1 m; Fig. 49, Fig. 55)

$$Q_{90} = 1.369 \times h^{2.5}; \alpha = 90° \quad (54)$$

Fig. 55. A V-notch weir in the Felsendome Rabenstein/Germany underground visitor mine showing that the sharp-crested side has to be at the upstream side of the weir.

$$Q_{½90} = 0.684 \times h^{2.5}; \alpha = 53°8' \quad (55)$$

$$Q_{¼90} = 0.348 \times h^{2.5}; \alpha = 28°4' \quad (56)$$

accuracy 90°: –4–0%
accuracy ½ 90°: –6–0%
accuracy ¼ 90°: –10–0%

Compared to a 90° V-notch weir, the ½ 90° and ¼ 90° V-notch weirs have a flow that is half and a quarter of the flow over a 90° V-notch weir at the same gauged head, respectively.

Rectangular notch (Poncelet) weirs (full width), accuracy: 0–3%.

$$Q = 1.773 \times \left(1 + 0.148 \times \frac{h}{P}\right) \times b \times h^{1.5} \quad (57)$$

Rectangular notch (Poncelet) weirs (contracted), accuracy: –4–1%

$$Q = 1.758 \times b \times h^{1.5} \quad (58)$$

Trapezoidal (Cipolletti) weirs, accuracy: –1–1%.

$$Q = 1.869 \times b \times h^{1.5} \quad (59)$$

Flumes

Flumes change the dimensions in an open channel to accelerate the water flow by converging the channel sides or by raising the channel bottom. If standard flume dimensions are used, the flow can be calculated from the water head upstream of the flume at a predefined distance from the critical depth. Flows between 0.3 L min^{-1} and 5,600 m^3 min^{-1} can be measured with flumes.

Flumes can be categorised as short- or long-throated flumes, whereas short and long refers to the linear parts in the calibration curve, not to their constructive length (Fig. 57). In contrast to short-throated flumes, long-throated flumes can simply be designed with computer programmes, and it is easy to adjust their dimensions to nearly every flow and constructive situation. Short-throated flumes, according to the suggestions of the U.S. Department of the Interior – Bureau of Reclamation (2001) should be purchased rather than constructed on site due to their complicated calibration and their design considerations. Detailed descriptions for different flumes are available in Ackers et al. (1978). Up-to-date flumes are rarely used in working or abandoned mines because they need more careful installation than weirs. If installed appropriately, their use is as easy as that of weirs.

H- and Parshall-flumes are short-throated measurement structures and each needs its own calibration table. Due to hydraulic restrictions, the dimensions of a given flume cannot be up- or downscaled by multiplying its properties to fit the new flume size. Both flume types have the advantage of coping with high sediment loads, and due to their construction and the velocity increase involved, they have a certain "self cleaning" capacity. Their disadvantage is that the flow measurement is not as accurate as with long-throated flumes and that they need special construction considerations that are not always possible in the restricted space of underground mines or surface channels.

H-flumes

H-flumes (Fig. 56) were invented by the U.S. Soil Conservation Service to continuously and accurately measure small flows as well as higher flows in small catchment areas (United States Department of Agriculture 1979; Hudson 1993). They are named "H" after the eighth experimental setup for testing new flume types. Three different types were investigated for three dif-

Fig. 56. 2-ft-H-flume near Mettmann/Germany (courtesy Umwelt-Geräte-Technik GmbH Müncheberg).

ferent flow conditions ranging from about 0.3 L min^{-1} to 199 m^3 min^{-1} and a total of seventeen construction dimensions, ranging from 0.4 to 4 feet are available (the numbers here must be given in feet, because each of the fourteen calibration tables is especially made for one type of H-flume. The equation given in table 3.2 of Younger et al. (2002) cannot be used for all types of H-flumes (see Fig. 57). They need free fall out at the end and must not be submerged, though a calibration for submerged H-flumes is available as well. Table 14 shows the wide range of flow conditions in which H-flumes can be used. As in the case of V-notch weirs, data loggers and monitoring devices can easily be installed. Ackers et al. (1978) give detailed equations and coefficients for calculating the flow through H-flumes. Those equations and the according coefficients depend on the upstream heads and submerged or free flow conditions. If manufactured H-flumes cannot be installed, the references in chapter 2 of United States Department of Agriculture (1979) must be followed, as erroneously constructed flumes will result in wrong discharge measurements.

One advantage of flumes over weirs was demonstrated at the Shilbottle/UK passive treatment site, where large amounts of iron precipitates were expected. An H-flume installed there worked reliably and demonstrated some self-cleaning capability. On occasions when it became clogged (usually with plant litter in the autumn), it could be manually cleaned in seconds by pushing the clogging matter out through the open "jaws" of the flume (Paul Younger, pers. comm.). This is a big advantage over weirs, which sometimes require laborious digging out behind the plate.

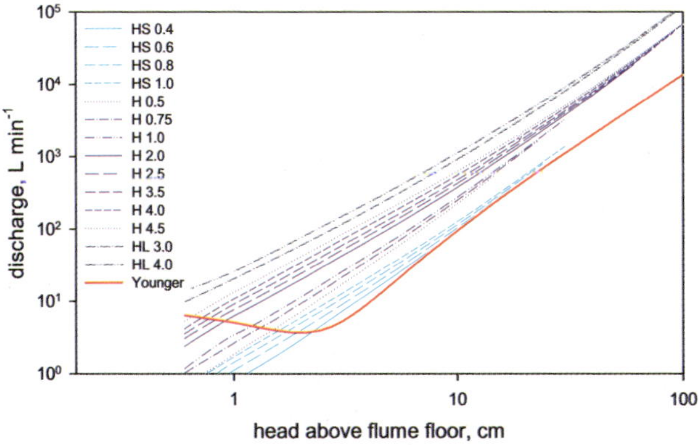

Fig. 57. Graphical comparison of the 14 different HS-, H, and HL-flume types used for flow measurements. The simplified H-flume equation of Younger et al. (2002) is given for comparison. It can only be used with heads over about 6–7 cm and for HS-type flumes.

Table 14. Selection of the commonly used H-flume types for flow measurements (after United States Department of Agriculture 1979). Each H-flume needs its own calibration table. GPM: U.S. gallons per minute. GPM and metric units are rounded. For comparison, the minimum height of a V-notch weir is given at the same maximum flow. As V-notch weirs should not be constructed with heads above 30 cm, these data are given for comparing reason only. The 3.0-HL-flumes are not recommended by United States Department of Agriculture (1979); H-flumes should be used instead.

H-flume type	Width, feet	Flow, GPM	Flow, L min^{-1}	V-notch, cm
HS	0.4	0.07 – 36	0.3 – 140	8
HS	0.6	0.10 – 99	0.4 – 370	11
HS	0.8	0.13 – 205	0.5 – 780	15
HS	1.0	0.17 – 360	0.6 – 1,400	20
H	0.5	0.18 – 149	0.7 – 560	13
H	0.75	0.27 – 430	1.0 – 1,600	21
H	1.0	0.31 – 862	1.2 – 3,300	28
H	1.5	0.49 – 2,392	1.9 – 9,100	42
H	2.0	0.62 – 4,928	2.3 – 18,700	56
H	2.5	0.81 – 8,617	3.1 – 32,600	69
H	3.0	0.94 – 13,644	3.6 – 51,600	84
H	4.5	1.39 – 37,699	5.3 – 142,700	126
HL	3.0	2.60 – 25,637	9.8 – 97,000	108
HL	4.0	2.24 – 52,510	8.5 – 199,000	144

Parshall flumes

Like H-Flumes, Parshall flumes (named after Ralph L. Parshall who developed them in 1922) are short-throated measurement structures, and are regulated by ISO 9826-1992 (E). Being a modification of a Venturi flume, they constrict the flow into the inlet and have vertical side walls (Fig. 58). They are described here because they have been installed in some mine water applications and can still be found. Yet, as mentioned earlier, they should be avoided due to their constructive restrictions. There are 22 variations of Parshall flumes with throat widths between 2.5 cm and 15.2 m, which were calibrated in the laboratory and can be used in mine water applications for flows between 5 L min^{-1} and

Fig. 58. Parshall flume in the flooded South Crofty/Cornwall/ UK tin mine.

5,600 m^3 min^{-1} (Ackers et al. 1978; Herschy 1995). It is not possible to use other dimensions unless the new dimensions are also calibrated. The head difference between the upstream and downstream part is usually low and like an H-flume, they are not as much affected by precipitation or high sediment loads. Therefore, they can also be used in mine water applications where the gullets have a moderate gradient and, due to their design, they can be submerged up to 70% without the need of corrections. This fact is an advantage over Venturi flumes (Science Applications International Corporation 2001).

The head measurement has to be conducted exactly at the position at which the flume originally was calibrated. A shift in that position causes erroneous readings and consequently wrong flow measurements.

Pump Run-Times and Ratings

Usually, in underground mines that have to be dewatered, the pump run times are recorded relatively accurately. This data, together with the pump characteristics, can be used to estimate the expected flow rates (Fernández-Rubio and Fernandez Lorca 1993). It might be acceptable to use the manufacturer's information about the pumping capacities, but wear and tear or cavitatation might have altered the factory parameters and so an independent flow measurement should usually be conducted before that data are used. Furthermore, fluctuations in the power supply slightly change the pump capacity and must be considered when using pump run-times to estimate the mine water flow. Clogging in the pipe network due to ochre precipitates can also introduce errors in the estimated flow. Finally, it should be noted that flow estimates that are based on pump run times are often over-estimated by around 20%, and in some rare cases, up to 50% (Paul Younger, pers. comm.).

Water Meters

Measuring flow with water meters is a very reliable method to quantify mine water discharges. To use such instruments, the mine water must flow through a closed pipe which usually must be 100% full. The flow through the pipe is either measured mechanically with a propeller (Fig. 59) or an electromagnetic velocity meter (Fig. 60). Propeller water meters are available for pipe diameters ranging between 1 and 183 cm (0.5 – 72 inch) and a flow of up to 10 m^3 min^{-1} with an accuracy of 1–10%. They can cause problems if water with high sediment loads has to be measured or when abrasion of the mechanical parts occurs. Magnetic-inductive flow meters are available from 0.1 to 800 cm diameter and typically have an accuracy of 0.2–3% (mainly depending on the pipe's diameter); if maintained properly (coating of sensors), they are one of the best flow measurement devices in mine water research.

Laser-Doppler-anemometers can precisely measure the velocity profile in a pipe or a channel, but they are currently too expensive for regular flow measurement in the mine water field. They require regular maintenance and service, and gases as well as sediments can influence the acoustic and laser signals (Durgin 1981; Zosel 1997). Other methods are acoustic flow meters or pitot tubes.

Fig. 59. Mechanical, propeller based water meter at the abandoned *Dienststelle Marienthal* bunker near Marienthal/Germany.

Dilution Method

Dilution techniques are tracer techniques used to measure flow. Due to their high accuracy – if performed by experienced personnel – they can be used to calibrate other flow measurement techniques. Furthermore, they are the methods of choice for turbulent flow conditions or in streams or gullies with no defined cross sections, as there is no need to measure channel geometry or floating times (Herschy 1995; Käß 1998; U.S. Department of the Interior – Bureau of Reclamation 2001; Science Applications International Corporation 2001). However, backflow eddies or seeping sections should be avoided as they negatively influence the results. All types of conservative, water-soluble tracers, which are described in section 10.2.5, are suitable but to reduce the costs of tracer analyses, a halide (e.g. sodium chloride) or sodium fluorescein are recommended because both tracers can be detected cheaply with on site measurement techniques: electrical conductivity and mobile fluorometers. Spence and McPhie (1997) successfully used that method for mine water impacted streams in the Canadian Northern Territories, where many streams are shallow, braided, and rocky. They analysed the tracer in the laboratory and used potassium chloride instead of sodium chloride because the natural potassium concentrations are usually low compared to sodium. Having compared a variety of stream flow measurement methods, they concluded that the "dilution method compares favourably to other measurement techniques both in accuracy and operational ease".

Two different methods can be used: the "constant rate injection method" and the "sudden injection ('gulp') method". In the first case, the tracer is injected at a constant rate with a pre-defined concentration; in the second case,

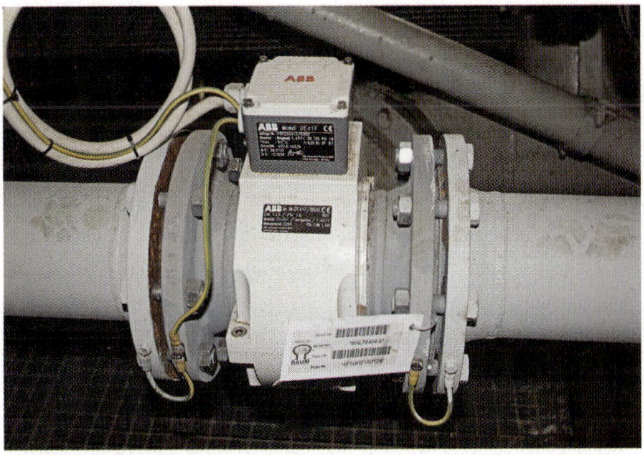

Fig. 60. Magnetic-inductive flow meter in the *Schwarze Pumpe*/Germany mine water treatment plant.

a predefined mass of tracer is suddenly poured into the water. Both methods have their special field of application and Kimball et al. (2001; and other publications of that research team covering exactly the same subject) for example, showed how the "constant rate injection method" can be used to measure the flow of mine drainage into a mountainous stream. The "gulp method" has been used at the Gernrode/Germany fluorspar mine, for example, to measure small inflows into Steinbach Creek and to compare and calibrate the results of current flow meter measurements (Fig. 48, Fig. 61).

Because the constant rate injection method requires a technically sophisticated installation, it is mostly used for scientific-based catchment investigations where different inflows have to be quantified. Either Mariott bottles or constant rate pumps are used to inject the tracer (Käß 1998). The flow calculation is comparably simple because the whole breakthrough curve does not have to be measured; instead, one only has to measure the plateau that develops after enough time, and the distance between the points of tracer injection and detection. The discharge can be calculated (U.S. Department of the Interior – Bureau of Reclamation 2001):

$$Q = q \times \frac{c_1 - c_2}{c_2 - c_0} \tag{60}$$

with Q flow rate, $m^3\ s^{-1}$
q discharge of tracer into the stream, $m^3\ s^{-1}$
c_0 background concentration of tracer, $g\ L^{-1}$

Fig. 61. Flow measurement with the salt dilution method in the turbulent Gernrode/Harz/Germany Steinbach Creek. The salinity probe is hanging in the middle of the stream and the data logger records the electrical conductivity every second (photographed by Andrea Berger).

c_1 tracer concentration of injected tracer, g L^{-1}
c_2 tracer concentration in stream after full mixing, g L^{-1}

If only the flow of a mine water discharge has to be measured or a technical flow measurement device calibrated, the gulp method with tracer integration can be used instead. A given amount of tracer is quickly poured into the stream, channel, or gullet. In a distance far enough away that full mixing can be guaranteed, the tracer cloud is measured from the time of the tracer injection until all of the tracer has passed the measuring cross section. A breakthrough curve is obtained and the concentration-time-integral must be evaluated. By dividing the amount of tracer injected by the integral, the discharge Q can be calculated:

$$Q = \frac{M}{\int_{t_0}^{\infty} c_t \times dt} \tag{61}$$

with Q flow rate, m^3 s^{-1}
M mass of injected tracer, kg
c_t concentration of tracer at time t, kg m^{-3}
t time, s

Before conducting a halide tracer test, the electrical conductivity probe must be calibrated with the salt used in the tracer test and the stream water. This procedure is necessary to determine how the tracer concentration is related to the electrical conductivity of the water. Each type of salt reacts differently and even the same type of salt in the same stream might produce different results at different times. This calibration can compensate for drifts of the electrical conductivity probe and chemical reactions between the salt and the stream water. As for most tracer tests, there is no straightforward way to calculate the amount of tracer needed for the dilution method. For most situations in the mine water sector, 0.5–2 kg of rock salt, depending on the electrical conductivity and the flow of the water, should be adequate.

In a typical flow situation, the breakthrough curve will resemble Fig. 62. Should the background concentration of the tracer used change significantly during the investigation, the graph will deviate from the ideal curve and you will need to repeat the test.

9.2.2 Mine Water Tracing

Until recently, mine water tracer tests were complicated and the results were sometimes unsatisfactory. Monitoring of flows within a flooded mine by tracer tests has therefore been uncommon. However, mine water tracer tests

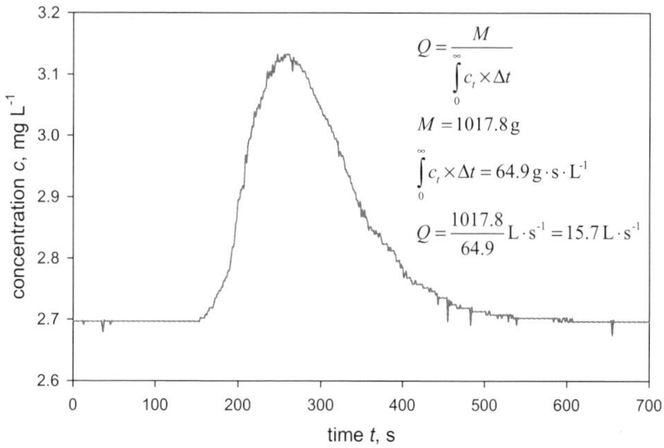

Fig. 62. Breakthrough curve and evaluation for a salt-dilution tracer test in the German Gernrode/Harz Steinbach creek. The distance between the tracer injection and the sampling site was 53 m and the complete mixing was tested in advance using a dye tracer.

have proven to be a valid decision-support tool for treatment or remediation measures or diversions within a mine (for details, see chapter 10). Although tracer tests are site-specific tests, they can be used for catchment-based decisions where it is necessary to decrease the contaminant load in the entire catchment area.

Before a mine water tracer test is conducted, it is essential to monitor the hydrogeological regime of the mine. The overall flow situation must be known as precisely as possible because tracer tests will only be successful if the injection and sampling sites have been carefully chosen. Nothing is more frustrating than conducting an extensive and expensive tracer test without getting usable results!

Of the range of artificial (introduced) tracers used so far, microspheres (polystyrene beads of 0.2–100 μm diameter) seem to be one of the best options (section 10.2.5.2). Most of the other artificial tracers are either not stable in acid mine water (e.g. Na-fluorescein), are too expensive (e.g. bromides), their use is restricted (radioactive tracers), or the amount of tracer that has to be injected is too large (e.g. halide salts). Besides artificial tracers, natural (environmental) tracers, such as isotopes or rare earth elements have been used successfully to trace the surface and underground flow regimes of mines. Always, the background level of the tracers in the waters to be traced must be evaluated before the start of the test.

Before the tracer test starts, at least 2–4 months of work are necessary to investigate the hydrogeological situation, to get the necessary permits and

Fig. 63. Tracer Injection Apparatus (TinA, left) and *Lycopodium* Apparatus (LydiA, right). The diameter of LydiA is 120 mm.

to conduct laboratory experiments with the potential tracers. Successful tracer tests in mines need 2–8 weeks of time to make sure that as much of the tracer as possible is detected. After the tracer test, 1–2 months must be allowed to process the samples and to present the results. All together, 4–8 months are a realistic time for a well based tracer investigation in a mine.

Two different tracer injection apparatuses have been developed for tracer investigations in flooded underground mines. The first one, LydiA (*Lycopodium* Apparatus), was invented for injecting a tracer into large diameter shafts. The other one, TinA (Tracer Injection Apparatus), can be used to inject a tracer through mine plugs or boreholes down to a diameter of 3″ (Fig. 63, Fig. 78). In conjunction with a special auto sampler (MeFiSTo: Multiple Filter Storage Tool), particle tracer tests can be conducted more reliably and more cheaply than in previous years (Wolkersdorfer et al. 1997a, 1997b).

Stable isotopes should be given special consideration as natural or artificial tracers as they provide potentially powerful methods to deduce flow paths and residence times, and will help you to understand the results of your hydrochemical and hydrodynamic investigations. However, due to the com-

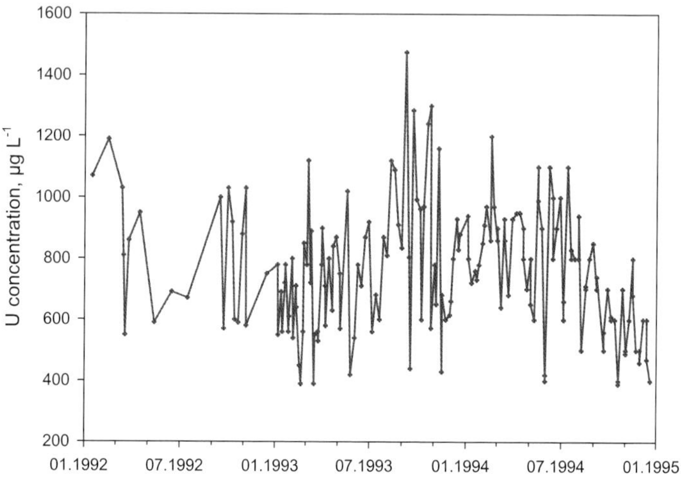

Fig. 64. Changing uranium concentrations at the measuring point m-107 of the abandoned Niederschlema/Alberoda uranium mine (data from Wolkersdorfer 1996).

plexity of stable isotope studies they will not further discussed here (for details see Clark & Fritz 1999); case studies can be found in: Allen and Voormeij (2002); Allen and Lepitre (2004); Davies et al. (2000); Helling et al. (1998); Jiries et al. (2004); Knöller et al. (2005); Melchiorre et al. (2005); Pezdič (1998); Razowska (2001); Rees and Bowell (1999); Tröger et al. (2005). But no tracer test is simple, and in fact, tracer tests should only be considered trustworthy if they are conducted by experienced consultants or research groups. Some examples of successful mine water tracer tests are described in chapter 12.

9.3 Physico-Chemical Measurements

9.3.1 Sampling Procedures

Sampling procedures for ground waters are described in detail in the literature (e.g. Lloyd and Heathcote 1985; American Public Health Association et al. 1998), but the special prerequisites for mine water sampling are only rarely summarised in commonly available publications (e.g. Ficklin et al. 1999). We are still far away from regular process analyses (Kellner et al. 1998) and mine water diode or sensor arrays and mine water chips as are available for wastewater treatment (Bourgeois et al. 2003), are not even under development. Therefore, this chapter will highlight some of the necessities but will also describe standard procedures for collecting good quality mine water samples. Readers who are interested in the details of analytical chemistry or the instruments used are referred to the general literature (e.g. Schwedt 1995; Kellner et al. 1998; Otto 2000) or the literature cited in the following paragraphs and sections.

The chemical composition of mine water discharges usually change over time (Fig. 64), except when it has been discharging for many decades. As discussed in section 5.4, the first flush occurs when water from a flooded abandoned mine reaches the surface; the concentrations of water constituents decrease thereafter. However, changes in mine water quality are not restricted to the time after the flooding of a mine; they also occur during the active working time or after the end of the first flush. In the case of the flooded Freiberg/Saxony underground mines, which are dewatered by the 50 km long drainage adit *Rothschönberger Stollen*, Fe_{tot} was about 2 mg L^{-1} during the active mining period, rose to 170 mg L^{-1} in 1969 (start of mine flooding), came down to about 20 mg L^{-1} in 1975 and has presently reached 1 mg L^{-1}. The flow, in 1997, ranged between 2 and 6 m^3 min^{-1} (Merkel et al. 1997; Baacke 2000). Therefore, both the mine water quality and flows must be measured for at least one hydrogeological year.

In the first set of samples for an unclassified mine water discharge, as many elements as possible should be analysed to identify potentially hazardous constituents. Because mine water usually does not contain large amounts of organic pollutants, it is normally adequate to measure only the inorganic parameters. Nevertheless, some mines discharge organic substances that pollute the receiving stream. If there is any concern or evidence to suggest that this might be the case, the sum parameter TOC (total organic carbon) as well as other organic substances such as BTEX should be analysed. Depending on the results of the analyses, further investigations may have to be considered.

At a minimum, the major ions (Na^+, K^+, Li^+, Ca^{2+}, Mg^{2+}, NH_4^+, Cl^-, SO_4^{2-}, HCO_3^-, NO_3^-) as well as ferric and ferrous iron, manganese, zinc, and aluminium should be measured. Other analytes should be added based on the geology and mineralogy of the mine site as well as any former use of the land property or mine (keep in mind potential underground waste disposal). As most modern laboratories have ICP-MS devices, it may be advantageous to have a screening of the full range of elements that can be measured using this method. The outcome of such desk and laboratory studies will be the basis for the future monitoring programme.

Anions like Cl^-, SO_4^{2-}, and NO_3^- are typically determined by ion chromatography (IC). Cations can be also be analysed using this method, but it is recommended that you determine the cations by ICP-AES, Ca^{2+} and Mg^{2+} by conventional titration, and Na^+ and K^+ with flame photometry. Although the major ions are (except Na^+ and Cl^- in salt mining) usually not considered as pollutants, those parameters are essential for thermodynamic modelling.

At least two water samples should be collected at every monitoring point: one for the main ions and the other for the trace elements (often referred to as "heavy metals" – but this term is even more confusing than "trace elements", because many "heavy metals" are neither heavy nor are they metals at all; Duffus 2002). All water samples being analysed for trace elements should be filtered through a 0.2 µm filter, though most investigations and recommendations use a 0.45 µm filter. The reason for filtering the sample with 0.2 µm is twofold:

- Water contains particles, colloids, and solute constituents (Ranville and Schmiermund 1999). Usually colloids range in size from 0.001 to 1 µm; solute constituents have a defined diameter of less than 0.001 µm (Fig. 65). To separate solute constituents from non-solutes, a filtration with 0.001 µm pore size is required. Since this is very difficult in the field, 0.2 µm pore size is an operational measure (replacing the former 0.45 µm operational measure). However, a 0.1 µm filter is strongly recommended if aluminium is to be determined, since aluminium tends to form colloids that are smaller than 0.2 µm in size.
- Water contains bacteria with an average size of 0.2 to 6 µm; by using 0.2 µm pore size filters, microbiological reactions are minimised in the filtered samples.
- However, not all samples should be filtered. Filtering samples that will be analysed for major cations and anions can change the concentrations of gases, of which CO_2 is of particular concern since it affects the distribution of inorganic carbon species.

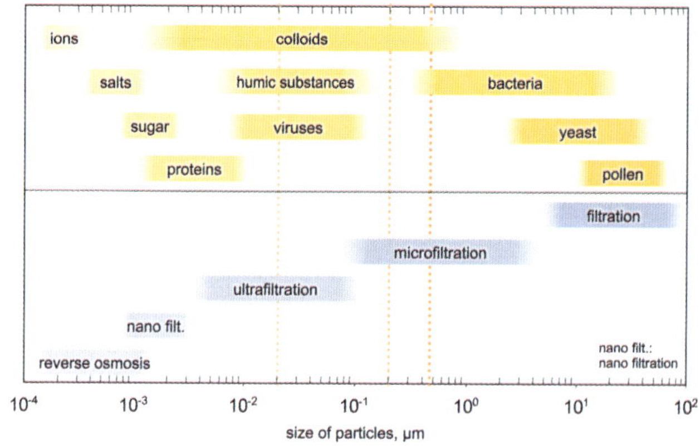

Fig. 65. Sizes of common water constituents and filtration techniques. The dotted lines gives the 0.02, 0.2, and 0.45 µm filter sizes used for filtrating water samples.

Ranville and Schmiermund (1999) showed that the 0.45 µm filtration has consequences when conducting geochemical modelling with PHREEQE, as some phases are supersaturated in the < 0.45 µm filtrate and under saturated in the < 500 Dalton (approximately 0.005 µm) fraction (the latter can be seen as the true "dissolved" fraction of mine water). They conclude that geochemical modelling should be restricted to those constituents that are not significantly attenuated by the filtration process. Shiller (2003) compared results of 0.02 µm filtrations with those of 0.4 µm and 0.45 µm filtrations. He used a field applicable filter tool for syringe filtration and noted some significant differences in the results based on the filter sizes he used. The differences were associated with both the different filtration techniques and the analysed element. Interestingly, the discrepancies between the 0.4 and 0.45 µm filtered results were sometimes larger than those between the 0.4 µm and 0.02 µm filter results.

After filtration, the sample that will be analysed for the trace elements must be acidified according to laboratory instructions or the QA/QC manual. Acidification is necessary to prevent chemical reactions (precipitation or adsorption) during transportation and storage (Lloyd and Heathcote 1985). Usually, 2–3 drops of HNO_3 (ultra-pure!) per 50 mL of sample are used to stabilise the trace constituents by maintaining a pH < 2. For speciation analysis, or for analysis of Sb, Ru, and Sn, HCl (ultra-pure) has to be used instead (American Public Health Association et al. 1998). All containers (made of laboratory glass, PE, or PTFE) must be clean and pre-treated, labelled properly with the place name, name of the sampling point, date, name of operator, and treatment (e.g. filtered, acidified).

Mine water is usually monitored at its discharge point into the surface hydro- or anthroposphere. However, mine water quality does not only change over time, but also with depth. Many mines have stratified water columns, and incorrect treatment technologies have been constructed at some of these sites because this stratification was disregarded. This occurred, for example, at the Straßberg/Germany mine. Similarly to the physico-chemical properties, stratification can change significantly within very short times – even hours! Nuttall et al. (2002) showed that the reasonably good mine water quality changed to a highly contaminated mine water during a pumping test (section 12.7).

Such observations can also be made in wells. Wells interfere with the aquifer or the shaft and disturb the natural situation with regard to the flow field and the chemical composition of the ground- or mine water. Furthermore, wells show stratification similar to what is observed in mineshafts (Lloyd and Heathcote 1985). However, this stratification, though similar to that in a shaft, has a different physical background. It is affected by the length–diameter ratio in a well, which is comparatively large and causes wall roughness to play a more important role than in a shaft.

The usual procedure when sampling wells is to pump out three times the volume of the well before taking a water sample. Yet, it is more appropriate to pump until the on-site parameters remain constant. Those parameters are temperature, electrical conductivity, pH, or the oxygen content, which have to be measured in flow-through chambers. Also, if the well has a steel casing, the mine water in the well may have been altered by a reaction with the steel. Samples taken in such a situation are not representative and must therefore be avoided, as the steel-water reactions significantly change the physico-chemical parameters of the water (Lloyd and Heathcote 1985). Many abandoned shafts are backfilled and equipped with cased wells. The sampling procedures required for wells also have to be applied to such shafts.

Samples taken must be representative of the water or sediment that has to be investigated, because treatment and remediation measures are based on the assumption that the results of any monitoring programme are representative for the catchment. To ensure representative samples, the geological, ecological, and hydrological situation in the catchment area must be known precisely. During the implementation process of the EU Water Framework Directive, numerous European catchments were studied and so a lot of representative data is already available for researchers there. Besides the real sampling points in the catchment area, which represent the regional part chosen, there should also be blind sampling points for quality control purposes (see section 10.7).

Herbert and Sander (1989) explained that sampling highly saline mine waters of salt mines (brines) requires the following precautions to ensure that the water sample is representative:

- avoid evaporation during sampling
- avoid contamination and an alteration of the sample by the mine air
- avoid particles in the sample
- samples should be stored at the same temperature and pressure conditions of the solutions in the rock matrix

During and after the Hope/Germany flooding experiment (section 12.14), Herbert and Sander (1989) developed a patented sampling tool that is able to take samples and determine on-site parameters under "real" in-situ conditions without altering the sample before or during the measurement.

For every dataset, it is important to record the name of the sampling point, dates and times of the sampling, the technicians or scientists who collected the samples, and the weather conditions. The brand of all meters, filters or any other tool should also be documented so that the cause of any potential mistakes can be identified. Finally, all chemical data should be tested for accuracy to reveal unreasonable values. Experienced hydrogeologists normally have a feeling for whether data is reasonable or unreasonable (for example, if pH in a carbonate aquifer is significantly deviating from near neutral to basic conditions). If one cannot count on an experienced hydrogeologist, statistical methods should be used (e.g. Kellner et al. 1998); several procedures are available to compensate for any missing data (e.g. Lloyd and Heathcote 1985).

One of the standard methods to check the accuracy of a water analyses is a cation-anion balance calculation or a test of the electro neutrality of the analyses. Both methods are very similar and either can be used. Some authors use the expression "ionic balance," which is simply twice the electro neutrality and equivalent in meaning:

$$\text{electro neutrality} = \frac{\sum \text{cations} - \sum \text{anions}}{\sum \text{cations} + \sum \text{anions}} \times 100\% \qquad (62)$$

$$\text{ionic balance} = \frac{\sum \text{cations} - \sum \text{anions}}{0.5 \times \left(\sum \text{cations} + \sum \text{anions}\right)} \times 100\% \qquad (63)$$

with cations and anions given in equivalents and the electro neutrality and ionic balance expressed in %.

If both anions and cations are missing in the same amount, this method will not detect the error in the analyses. Therefore, ion balancing should always be accompanied by other statistical methods.

One of these other methods is comparing the electrical conductivity with the *measured* total dissolved solids (TDS). Depending on the water chemistry, the following relations exists:

$$\text{TDS} = \kappa \times 0.6 \ldots 0.725 \qquad (64)$$

with TDS total dissolved solids (calculated), mg L^{-1}
κ electrical conductivity, µS cm^{-1}

Some hand-held probes, such as the Ultrameter 6P (MYRON L Company, Karlsbad/USA), *calculate* the TDS depending on the temperature and the electrical conductivity. Such TDS values obviously cannot be used to evaluate the accuracy of the electrical conductivity measurement.

Some chemical-thermodynamic codes calculate the electrical conductivity based on the chemical analyses. Researchers very seldom use these results though they provide a very easy way to check the accuracy of a water analysis. Another mathematical method that even works for mine waters with complicated chemical compositions compares the ion mobility, *u*, and its concentration, c, in the water (Rossum 1975; Coury 1999; according to a pers. comm. by Peter T. Kissinger, Chairman and CEO Bioanalytical Systems Inc., the second part of Coury's article never was published). This method can easily be used in every spreadsheet or database. The electrical conductivity of a solution with strong and weak electrolytes can be written as:

$$\kappa = F \times \sum_{1}^{n} \left(|z_i| \times u_i \times c_i \right) \tag{65}$$

with κ electrical conductivity, S cm^{-1}
F Faraday constant, $9.6485 \cdot 10^4$ C mol^{-1}
z_i charge of species i, no unit
u_i mobility of species i, m^2 s^{-1} V^{-1}
c_i concentration of species i, mol cm^{-3}

Because the mobility is concentration dependent, this equation cannot be directly used in the calculation without further rearrangement. In experiments, it was found that the concentration dependent parameters composing the equation can be calculated using tabulated "limiting ionic molar conductivities" (also called equivalent ionic conductance), the ionic charge of the species, and its molar mass. Using a small program that can be programmed with spreadsheets or in Visual Basic™ add-ins, chemical analyses can quickly be checked for their accuracy. For details on the method, the limitations of the approach, and relevant literature, please see appendix I.

9.3.2 On-site Parameters

On-site parameters are water parameters that can change quickly during transportation or sample handling. These parameters are meta-stable, redox- or pressure-dependent and include pH, redox potential, temperature, ferrous/ferric iron contents, base and acid capacity (in the mine water literature usually called acidity and alkalinity), dissolved oxygen, and specific electrical conductivity. Therefore, those parameters have to be measured *immediately* after the sample has been taken, in other words: on-site (Fig. 66) or in flow-through chambers. Only in cases where a short time between sampling and laboratory analyses is guaranteed can specific electrical conductivity, acidity, and alkalinity be measured reliably in the laboratory (Cravotta and Hilgar 2000).

A quality assurance/quality control (QA/QC) plan must define which parameters will be measured in filtered or unfiltered water and which filter will be used (see section 8.3.1). Temperature, electrical conductivity, pH, redox potential, total base capacity (acidity), total acid capacity (alkalinity), and

Fig. 66. Mine water and tracer sampling in an underground mine.

oxygen content are *always* measured in the unfiltered sample, whereas ferrous/ferric iron should be measured in both filtered and unfiltered samples. For iron, it must be clear what the parameters "dissolved iron" and "total iron" mean. Authorities and even consultants often use the terms interchangeably to mean the sum of the ferrous and ferric iron (that is, total iron). Similarly, the expression "total metals" should be restricted to the sum of the species! The reason for the confusion is that the relevant regulations (DIN 38406-E1-1 May 1983; EN ISO 11885 November 1997; DIN 38406-32-1 May 2000) are unclear. Only EN ISO 11885 clearly – but unfortunately not precisely from a chemical and physical point of view – defines "dissolved" metals: "The fraction of metals in a water sample that passes a 0.45 µm filter". Concerning aluminium, DIN ISO 10566 April 1999 clearly writes "filterable aluminium" instead of "dissolved" aluminium. Future version of those regulations should become more precise on this point (Table 15).

On-site parameters may never become constant because the physico-chemical properties of the mine water can change rapidly, in seconds or minutes. To overcome this problem, it is advisable to take at least three readings within short time intervals (minutes) and calculate their mean value (Table 16). At this point, all readings must be recorded with their corresponding units and without any field calculation results. Furthermore, the readings should be recorded in full, since all the digits of the result (e.g. pH 7.10 instead of 7.1) are a measure of the precision (not accuracy!) of an instrument.

Table 15. Expressions to be used for analysing iron in mine waters. The same expressions should be used for other metals and semi-metals. The kind of filter (0.02, 0.2, 0.45 µm) must always be recorded.

Constituent	Sampling Procedure	Term	Common Term	DIN/ISO use
Fe^{2+}	Filtered	Fe^{2+} filtered	ferrous iron filtered	–
Fe^{3+}	Filtered	Fe^{3+} filtered	ferric iron filtered	–
$Fe^{2+} + Fe^{3+}$	Filtered	Fe_{total} filtered	total iron filtered	dissolved iron
Fe^{2+}	Unfiltered	Fe^{2+} unfiltered	ferrous iron unfiltered	–
Fe^{3+}	Unfiltered	Fe^{3+} unfiltered	ferric iron unfiltered	–
$Fe^{2+} + Fe^{3+}$	Unfiltered	Fe_{total} unfiltered	total iron unfiltered	total iron

Most of the on-site instruments used were developed for laboratory use and were adapted to conditions outside the laboratory. Although good outdoor equipment is now available, most instruments work best with water that is less contaminated than mine water. Furthermore, they are equipped with plugs or probes that are extremely sensitive to the rough field environment. While the calibration procedure of a probe is simple in a laboratory, it can be difficult on-site – especially when working in potential methane-prone mines. Some companies provide excellent and easy to handle on-site equipment and modern "pencil"-probes can also be used to get reliable on-site parameters.

The mobile on-site instruments that can be calibrated (e.g. pH, O_2, electrical conductivity) have to be calibrated regularly, at least every day when in use and in the case of dissolved oxygen, before every measurement. The instruments require regular maintenance since membranes can be poisoned or damaged and precipitates can cover the probes (e.g. ochre, bio films). Plugs can get dirty or batteries low. All instruments should be checked and serviced before every field trip!

A major problem with mine water is that it often consists of ground or seepage water from waste rock piles or tailings that might contain elevated concentrations of carbon dioxide or low amounts of dissolved oxygen. When this water reaches the surface, it contacts the low carbon dioxide concentration and high oxygen concentrations of the atmosphere. Depending on the

Table 16. Example of a series of on-site measurements with calculated means. The redox potential is not corrected to standard H_2-electrode and 25 °C! The redox potential is not stable and should have been measured for a longer period. From the other parameters, it might also be assumed that the total chemistry and the water temperature of the water was changing during the sampling, which took no longer than 5 minutes.

Parameter	Unit	#1	#2	#3	#4	#5	mean
O_2	mg L^{-1}	0.34	0.30	0.35	0.31	0.34	0.33
temperature	°C	18.2	17.4	16.8	16.6	16.4	17.1
electrical conductivity (25 °C)	µS cm^{-1}	2648	2687	2704	2730	2750	2704
total dissolved solids	mg L^{-1}	1934	1975	1991	2012	2027	1988
redox potential	mV	115	101	92	79	66	91
pH	–	7.19	7.10	7.08	7.04	7.02	7.09

flow regime (laminar/turbulent) it might take a few minutes or hours to achieve a new equilibrium. Since oxygen and carbon dioxide are key parameters, the master variables like pH, redox potential, and dissolved constituents (e.g. iron) might change dramatically and consequently the entire water chemistry will change. Redox reactions in particular are also catalysed by microorganisms. Thus, it is unlikely to see thermodynamic equilibrium in mine waters. This should always be considered in the evaluation of data that are based on the assumption of thermodynamic equilibrium (e.g. redox potential).

Absolute temperatures can be determined with an accuracy of about 0.1 K and relative temperatures with a precision of about 0.01 K (Lloyd and Heathcote 1985). After the temperature equilibrium between the medium and the thermometer is reached, the temperature reading can be taken. Because electronic thermometers are usually very small, such readings can be taken very quickly. In contrast, mercury- or alcohol-filled thermometers may take as much as a minute to reach temperature equilibrium.

Most probes for electrical conductivity do not require special treatment (and storage) nor calibration. However, the accuracy of the probe should always be tested with a standard solution of known electrical conductivity.

Some high quality probes can also be calibrated within the range of use. Readings usually are automatically recalculated to a temperature of 25 °C and can be taken after a short time interval. If the indicated value on the gauge is varying, this normally indicates a malfunction of the probe or the amplifying device (Fig. 97).

Nowadays, pH is commonly measured with pH probes. These field probes are easy to use and should have automatic temperature compensation, automatic buffer recognition and be battery powered. The following stages are necessary to achieve a representative field measurement:

- the pH probe has to be calibrated before any use at least once a day with two, or preferably three or four standards (4, 7, and 10 for normal water; 1.68, 2, 4, and 7 for acid mine water)
- the pH probe must be stored in a 3 molar KCl solution if not in use (or whatever is recommended by the manufacturer)
- the measurement has to be performed in a flow-through chamber to guarantee both a slight movement of water along the probe and absolutely no contact with the atmosphere
- readings can be taken after 10 to 15 minutes with an accuracy of 0.01–0.05 pH units

If pH probes are not available, colorimetric pH measurements with either indicator dyes, litmus paper, or modern multi indicator strips should be used. Yet, for dilute waters, dyes and indicator strips can give very erroneous results (Arthur Rose, pers. comm.). Nordstrom et al. (2000) has reported on extremely acid mine waters with pH values as low as –3.6. In the unlikely event that you encounter negative pH values, you should know that normal pH probes and buffers cannot be used. Furthermore, the pH has to be defined using the Pitzer method for the calculation of the theoretical pH of the standard solutions and the calibration of the glass membrane electrodes (Nordstrom et al. 2000).

In addition, redox potential and dissolved oxygen measurements have to be performed in flow-through chambers to avoid degassing or uptake of gaseous components from the atmosphere. Redox probes are subject to surface poisoning; if this occurs, the platinum electrode can be mechanically cleaned with corundum powder. Redox probes cannot be calibrated but should be tested using standard solutions (e.g. Zobell solution). Since achieving equilibrium can take several hours (obtainable reproducibility about 50 mV), it is recommended that you use at least two probes, one for oxidising solutions and another for reducing conditions. Plotting continuous time series will allow you to decide whether stable values were reached. Lloyd and Heathcote (1985) present time series of pH- and redox

Table 17. Coefficients to be used for the redox calculations according to equation 66.

Sensor Type	Coefficient *a*	Coefficient *b*
Mercury calomel KCl	67798	324
Ag/AgCl KCl 1 mol L^{-1}	62612	279
Ag/AgCl KCl 3 mol L^{-1}	50230	295

measurements showing that neither parameter reached stable values before 30 minutes of measuring. If redox measurements are conducted in shafts or boreholes, the eigenpotential of the system must be considered (section 12.1).

Another factor, often overlooked, is the necessary conversion of the redox potential. The international basis for the redox potential is the standard-H_2-electrode. Such electrodes cannot be used in the field, so the redox potential is usually (but not always) measured with an Ag/AgCl-electrode (3 mol L^{-1} KCl). The measured redox potential then has to be converted to the potential of the standard-H_2-electrode. The redox potential is also temperature dependent, so the temperature must also be measured. Instead of using tables, the following simplified empirical equation and the coefficients in table 17 can be used, where the first two terms correct for the temperature and the square root term converts the reading to what would have been measured on a standard-H_2-electrode (accuracy of equation: ± 0.5 mV between 5 and 65 °C):

$$E_{0(25°C)} = E_t - 0.198 \times (T - 25) + \sqrt{a - b \times T} \quad (66)$$

with E_0 converted redox potential, mV
E_t measured redox potential at temperature T, mV
T measured temperature, °C

Dissolved oxygen (often abbreviated DO) is usually determined by a potentiometric electrochemical probe based on the Clark-sensor (Clark et al. 1953). The sensor has to be replaced regularly (see manual) and calibration for 0 and 100% O_2 has to be completed before each measurement, whereas temperature and pressure calibrations are usually done by the instrument itself (Schwedt 1995). Readings can be taken after a few minutes with an accuracy of 0.1 mg L^{-1}. Common electrochemical O_2-probes have an operation-range from 0.5 mg L^{-1} to O_2-saturation; thus, readings below 0.5 mg L^{-1} are meaningless. Because the sensor consumes oxygen during the measurement, the sensor has to be lapped around during the whole measurement or kept in a flow-through chamber with a flow of about 10 cm s^{-1} (Lloyd and Heathcote 1985). Continuous oxygen measurements in mine waters are problematic due to sensor drifts, temperature and pressure dependency, accumulation of biofilms, and Fe-oxide precipitation.

New ceramic and fibre optic optode-based O_2-probes have been developed (Klimant et al. 1995). These luminescent dissolved oxygen (LDO) sensors are coated with a luminescent material. Light from an light emitting diode (LED) is transmitted to the sensor surface, which excites the luminescent material, which upon relaxation emits light of a shorter wavelength. Both, the intensity of the steady state luminescence and the average relaxation time are inversely proportional to the dissolved oxygen concentration and are therefore measured and correlated to the dissolved oxygen concentration. To eliminate sensor drifts due to ageing of the luminescent material, internal references are used (Rommel 2004; Jackson and Hach Company 2004; Hach Company Product Information; Van Essen Instruments Company Information). Although those sensors have proved to be suitable for wastewater treatment, their reliability for mine water investigations has not yet been verified. Körtzinger et al. (2005) were able to show that such innovative oxygen probes can be used in extreme conditions of high salinity in the ocean and up to a pressure of 20 MPa. It might therefore be assumed that such probes can also be used to measure depth profiles of flooded mine shafts.

Two other parameters that are essential in mine water research, especially for the design of a treatment plant, are the base and the acid capacities (acidity, alkalinity) of the mine water. Different definitions and instructions on how to measure those parameters are known and can be used (e.g. Hedin et al. 1994a; Cravotta and Hilgar 2000): hot acidity with hydrogen peroxide (American Public Health Association et al. 1998; International Institute for Environment and Development 2002) cold acidity with hydrogen peroxide (Fishman and Friedman 1989), or cold acidity without hydrogen peroxide (Ott 1986, 1988). Acidity and alkalinity are usually evaluated by acid-base titration. In normal surface or ground waters, the dominant species contributing to acidity and alkalinity are the species of the carbonate system: H_2CO_3, HCO_3^-, and CO_3^{2-}. Therefore, the measured alkalinity and acidity, strictly spoken, are related to those species. If much mineral acidity exists, the concentrations of the bicarbonate system cannot be evaluated from the alkalinity and acidity titrations, but have to be replaced by the TIC measurements. A detailed discussion of net alkalinity and net acidity can be found in Kirby and Cravotta (2005a, 2005b).

Furthermore, if mineral acidity is present, it is possible to estimate the total base capacity (acidity) in mol L^{-1} from the pH and the concentrations of Fe, Mn, Zn, and Al (equation modified after Hedin et al. 1994a):

$$\text{Acidity} = \gamma_{H_2O} \times 10^{(3-pH)} \times \sum_{1}^{n} \frac{z_i \times [Me_i]}{M_i} \tag{67}$$

with z_i charge of species i, no unit
pH pH-value, –
$[Me_i]$ concentration of species i (e.g. Fe, Mn, Zn, Al, Co, Cu, Cd), $g\ L^{-1}$
M_i molar mass of the species i, $g\ mol^{-1}$
γ_{H_2O} activity of mine water, no unit

Compared to other forms of this equation, it is written here in its generalised form for all the species that contribute to the acidity. Furthermore, the activity of the water is included here, because calculated acidities (Hedin et al. 1994a; Cravotta and Hilgar 2000) regularly overestimate higher acidities, which might be due to the missing activity of the mine water and possibly complexing of ferric iron and Al by OH^- and HSO_4^- (Kirby and Cravotta 2005b).

Since the different methods give different results, it is essential to measure both parameters using a pH electrode and recording the titration end point (Fig. 67). Acid and base titrations cannot be performed in the field without the sample being in contact with the atmosphere. Thus, the properties of the sample can change due to degassing of carbon dioxide or to chemical and physical reactions between the sample and the atmosphere. A recommended alternative method is taking a sample in the field, storing it in a gas-tight container and determining the total inorganic carbon (TIC) in the laboratory. This

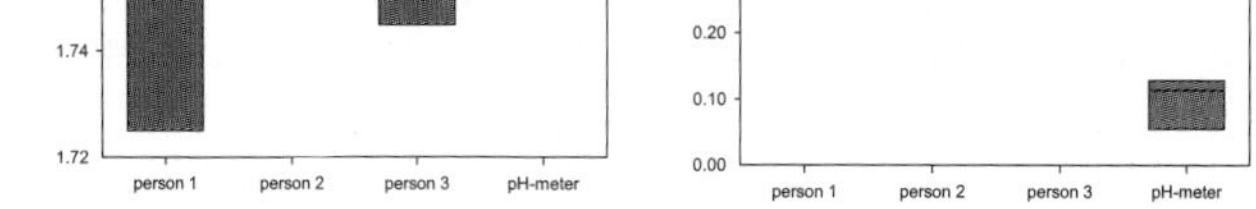

Fig. 67. Repeated measurements of acid capacity $K_{A(4.3)}$ and base capacity $K_{B(8.2)}$ of mine water (Drainage Adit *Rothschönberger Stollen*, Freiberg/Germany). 1–3: titration by an experienced technician using indicators to determine the titration end point; 4: titration using a pH meter to determine the titration end point. Each titration was repeated three times.

can be achieved by lowering the pH to 2 and purging the CO_2 from the sample before the detection via non-dispersive IR-photometry in the gas phase. The TIC values can then be used to calculate the acidity and alkalinity.

During the field studies of the German Gernrode/Harz mine water, it became apparent that the acid/base titrations that were being used for evaluating the acid capacity K_A and base capacity K_B (alkalinity, acidity) of the mine water seemed to be faulty because the calculated bicarbonate was either too high or too low (Fig. 68, Fig. 69). Consequently, four different mine drainage waters (Straßberg/Harz, *Fürstenstollen*/Halsbrücke, *Rothschönberger Stollen*/Rothschönberg, *Verträgliche Gesellschaft Stollen*/Freiberg) were tested to find the reason for this observation.

Each mine water was titrated in the field with HCl (end point pH 4.3) and NaOH (end point pH 8.2) three times with two different methods (organoleptic test with indicator, pH-meter) and by three persons. Furthermore, TIC (total inorganic carbon) was determined for each mine water and compared with the field results. It became clear that in some cases, the different methods used for evaluating the acid and base capacities produce significantly different results. Furthermore, the laboratory TIC determination of $K_{A4.3}$ differed significantly from the results of the field determination, whereas the acid titration with indicators and the pH meter did not show significant differences. Acidity based on equation 67 could not be calculated because no metals were analysed.

The situation was reversed in the case of $K_{B8.2}$. TIC-determined $K_{B8.2}$ is close to the value determined with the pH meter, while the organoleptic determination significantly differs from the value determined with the pH meter.

From the results of this study, it was concluded that:

- $K_{B8.2}$ and $K_{A4.3}$ have to be determined in the field using a pH meter instead of indicators
- further studies have to be conducted comparing TIC-determined acid/base capacities and field determined ones.

If all the above-mentioned parameters are measured on-site, and samples for main and trace elements be taken, a 30–60 minute sampling period must be allowed. An additional 10–30 minutes must also be allocated for precise flow measurements.

9.3.3 Presentation of Results

A lot of information is collected and stored in various forms during a mine water study, e.g. graphs, tables, or drawings. All this information has to be compiled to understand the hydraulic, hydrological, hydrogeological, hydrochemical, and geological conditions investigated. Usually, this information is presented in a written report that describes the investigation, results, conclusions, and finally the recommendations. Therefore, writing a final re-

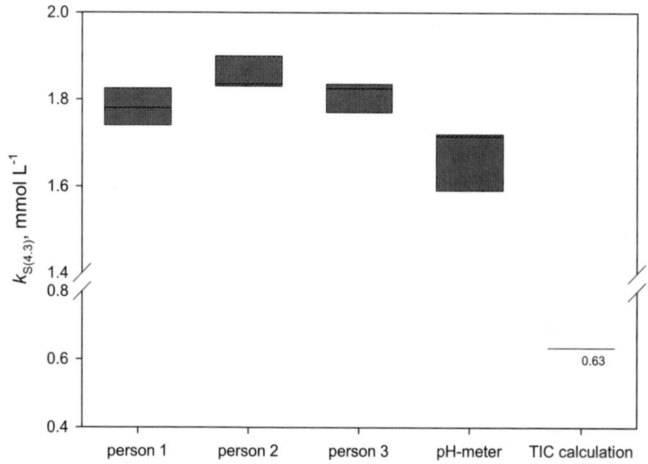

Fig. 68. Results of acid capacity measurements $K_{A4.3}$ (alkalinity) with three different methods: 1–3: persons 1, 2, 3 using indicators; 4: titration end point using a pH meter; 5: TIC (total inorganic carbon) calculation: determination in the laboratory, $n = 3$, except for TIC (data from Lieber 2003).

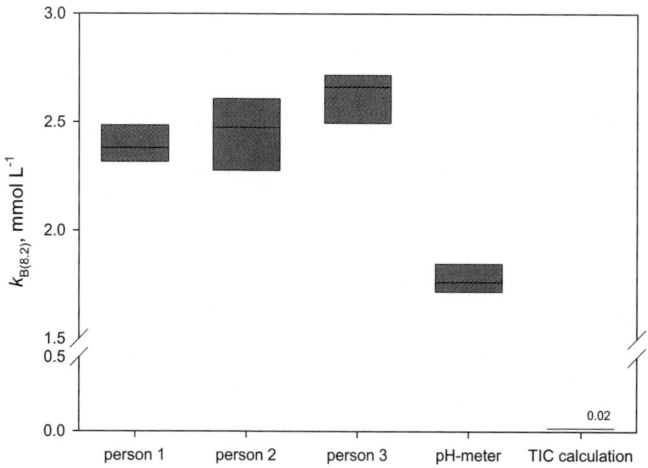

Fig. 69. Results of base capacity measurements $K_{B8.2}$ (acidity) with three different methods: 1–3: persons 1, 2, 3 using indicators; 4: titration end point using a pH meter; 5: TIC calculation (total inorganic carbon): determination in the laboratory, $n = 3$ except for TIC (data from Lieber 2003).

port is essential to provide the contract awarder and the authorities involved with answers to the questions they had at the start of the project. Usually, such a report is the only summary of everything that was done. Conclusions and recommendations can only be clearly understood when they are presented in an understandable way. The following structure is recommended for a report on mine water conditions in a river catchment (modified and adapted after Brassington 1999):

- executive summary or extended summary
- introduction
- topography and drainage situation of the river catchment
- geological and hydrogeological conditions
- methods
- results (chemistry, heads, flows)
- conclusions and recommendations
- appendix with large format graphs, tables or drawings

Results should be presented in three different ways: in tables, where all measured parameters are listed, in clear graphs that show the relevant changes in important parameters, and summary results after the data are statistically processed (e.g. correlation, principal component analyses, and descriptive statistics).

Tables usually contain all the data measured or obtained within the catchment or in the laboratory. All the data should be given as it was measured, without interpretation. Explanatory tables and data tables should always be separated from each other. Furthermore, it is not appropriate and in accordance with QA/QC protocols to leave out "outlier" data or to interpolate missing data without explaining exactly what and why it was done.

If a large amount of data has been collected within the catchment or the mine, it is advisable to produce graphs that quickly show differences and changes in selected parameters. Several types of graphs are available and widely used for this objective. For a quick classification of water types, Piper (Fig. 70) or Durov diagrams (Fig. 71) give an *impression* of the water composition (Piper 1944, 1953; Lloyd and Heathcote 1985; Klapper and Geller 2001), though such graphs are inappropriate for characterising typical mine waters (ERMITE Consortium et al. 2004). Excellent introductions and explanations of how to interpret Piper or Durov diagrams can be found in Hounslow (1995) or in Lloyd and Heathcote (1985). However, it should be clear that different waters with the same relative concentrations of the main ions plot in the same field and can therefore not be distinguished from each other in diagrams that use relative numbers. For contaminated mine water, a Ficklin diagram (sum of Zn, Cu, Cd, Co, Ni, and Pb concentrations; Ficklin et al.

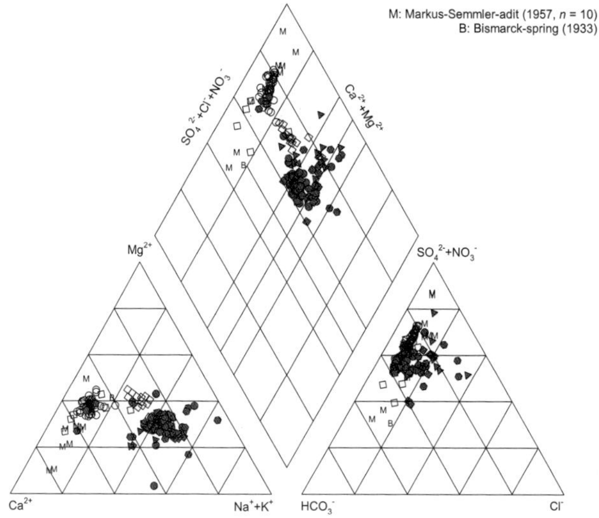

Fig. 70. Example of a Piper diagram for mine waters in the abandoned Niederschlema/Alberoda/Germany uranium mine (from Wolkersdorfer 1996). Coloured symbols are mine waters, open symbols infiltration waters.

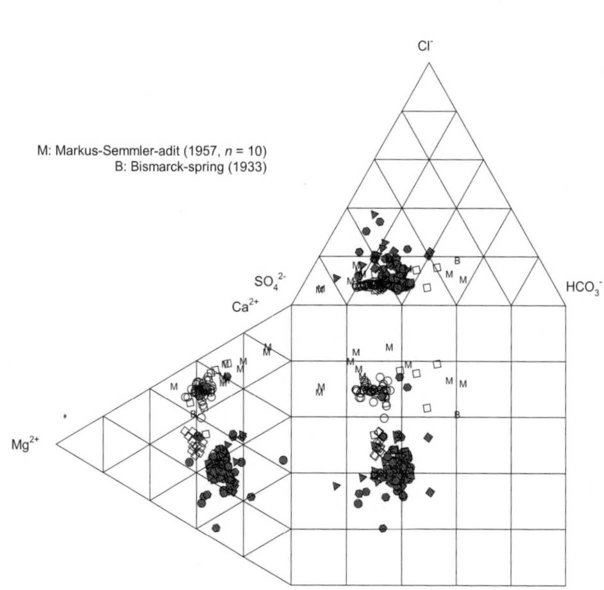

Fig. 71. Example of a Durov diagram for mine waters in the abandoned Niederschlema/Alberoda/Germany uranium mine (from Wolkersdorfer 1996). Coloured symbols are mine waters, open symbols infiltration waters.

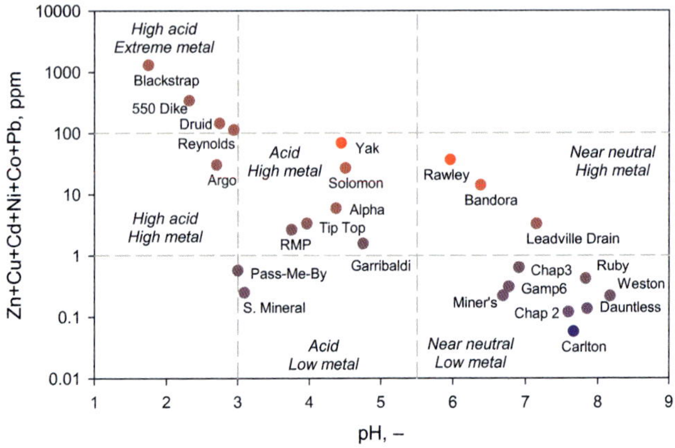

Fig. 72. Ficklin diagram for mine waters of western Colorado (modified after Ficklin et al. 1992).

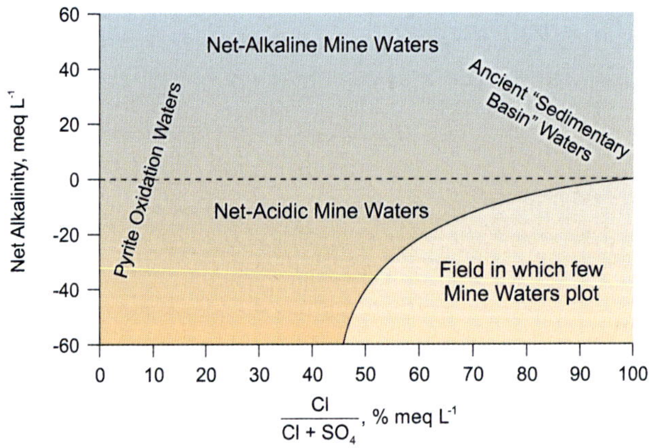

Fig. 73. Mine water classification diagram with net alkalinity against ratio of chloride to sum of chloride and sulphate according to Younger (1995) and Rees et al. (2002).

1992; Fig. 72) can indicate the contaminant potential of the waters being investigated. Another diagram that can be used to classify mine waters is a net alkalinity versus $Cl^-/(Cl^- + SO_4^{2-})$-diagram, introduced by Younger (1995) and improved by Rees et al. (2002; see Fig. 73).

The range of a single parameter is best presented in box-plots that show the mean, median, 5, 10, 90, and 95% quantiles and outliers in a transparent way (Fig. 74). Histogram diagrams can help the reader quickly visualise data clustering (Fig. 75). It is always good scientific practice to indicate error bars

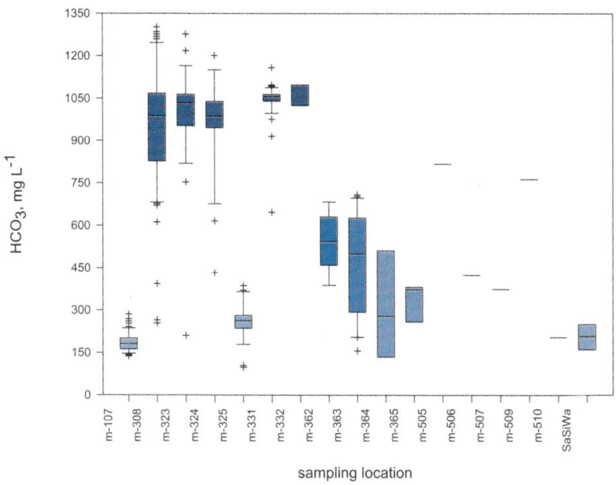

Fig. 74. Example of a box-plot showing 312 data points at 17 sampling locations.

(e.g. Davis 1986, Cumming et al. 2007; Fig. 43). In a report about the total iron concentration within a waste rock dump, a graph was given that, according to the explanation, showed a clear increasing iron concentration with depth and a correlation coefficient r^2 of 0.9 (Fig. 76 left). The original data were listed in a table and in the appendix, the accuracy of the method used was tabulated as well: $Fe_{tot} \pm 0.3$ mg kg^{-1}. Furthermore, the text explained that each sample was analysed from a 0.25 m long core. Plotting the data with the error bars clearly proves that there was no depth correlation at all because the Fe_{tot} values overlap between 1.11 and 1.36 mg kg^{-1}. Therefore, the correct statement should have been: "The Fe_{tot} values of the waste rock pile range from 1.06 to 1.41 mg kg^{-1} and do not show a depth dependence within its error intervals".

If different water types have to be classified, cluster analyses or principal component analyses should be used (e.g. Davis 1986). Especially for cluster analyses it is important to find the relevant parameters which can be identified with a principal component analyses before the clusters are calculated.

With the improvement of GIS-technologies, and the availability of GPS, GLONASS, or GALILEO, monitoring projects should be managed by GIS systems. Stiff diagrams or pie charts are especially helpful on maps produced by these systems for comparing different sampling sites. It is advisable to agree to a common GIS and georeference system before the project starts, as some systems are not fully compatible with each other. The EU is establishing a European geographic reference system that differs significantly from some of the national reference systems.

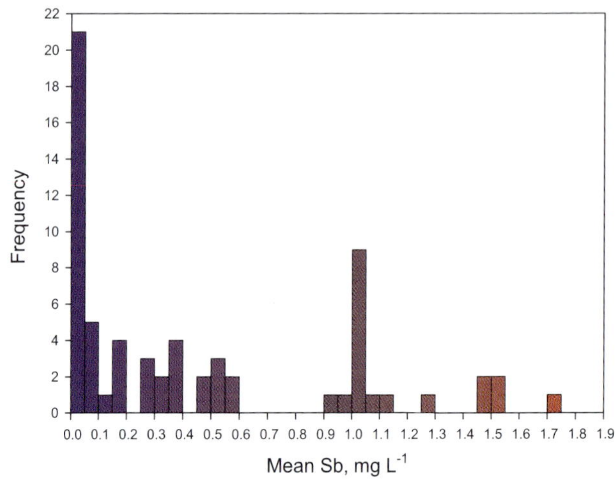

Fig. 75. Example of a histogram, showing the frequency distribution of antimony and the clustering of the data. $n = 66$.

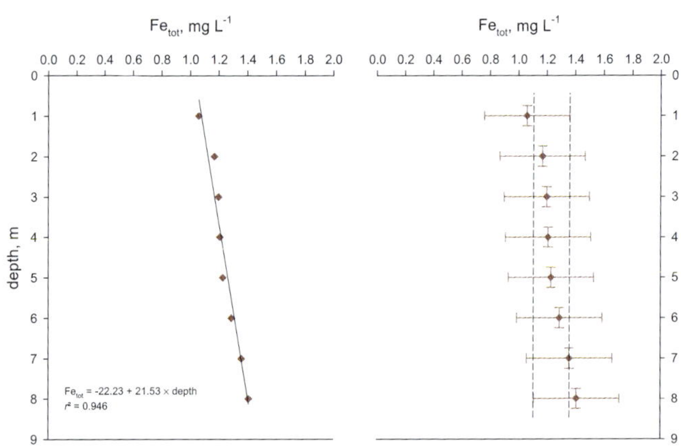

Fig. 76. Faked depth dependencies of Fe_{tot} in a waste rock pile.

Detailed explanations for writing reports, presenting data, or for evaluating hydrological data are given by Lloyd and Heathcote (1985), Shaw (1994), and Hounslow (1995).

9.3.4 Regression and Data Prediction

This section is focused on regression and data prediction rather than statistical interpretations in general. In the broader sense, it falls into the field of data mining or artificial neural networks – without going into the details of those powerful tools, which are commonly used to find relationships between parameters that are not obvious at a first glance (e.g. Savic and Walters 1999; Blockeel et al. 1999; Bourgeois et al. 2003; Hewett 2003; Bessler et al. 2003).

Regression and data prediction is commonly used by many scientists and consultants to estimate or predict missing or future values. A simple example is the estimation of stream or mine water flows by measuring the water height in a given cross section. Usually, 3–10 flow measurements are made at different water levels and then a calibration curve is computed.

Many reports present scatter plots including lines that emphasise relationships between the plotted parameters. Such lines are commonly based on statistical investigations and are given as regression curves or are described as the correlation between the investigated parameters. Unfortunately, many of those curves reflect a lack of statistical understanding; because the software tools are easy to use, they are being used inappropriately. Currently, there is only one comprehensive software package available that fulfils *all* the needs of a proper statistical investigation for curve fitting and equation discovery: TableCurve 2D produced by SYSTAT Software Inc. Richmond, California (Systat Software Inc. 2002; for 3D-Data, TableCurve 3D is available as well). Nonetheless, even if proper tools are used to explain relations between two or more parameters, this does not mean that there is a geochemically-based relationship between the parameters. In cases where analytical solutions or relations are known, those relations must be used for predicting future or dependent values. Often, such analytical solutions are not known and a black box method (a lumped parameter model), a Bayesian, or a Monte Carlo method will be used.

Another fact that has to be considered when computing predictions using PC based software is the floating point precision of the algorithm (not that of the final result). Intel's floating point co-processor has a 19-digit precision, which is used by the common standard graphical tools (e.g. Microsoft Excel). For many predictions, this precision might be good enough, yet, TableCurve uses its own math emulator with a 38-digit precision in addition to the Intel co-processor. One of the most powerful tools of TableCurve is its ability to investigate a prediction using several statistical and graphical tools besides the regression coefficient r^2. Thus, the best prediction can be selected using its graphical and statistical methods.

A thorough regression or data prediction analysis needs to investigate different potential relationships between parameters. Sometimes, these relationships are not obvious from the correlation investigations because, in the case of multiple dependencies of two parameters, the regression coefficient would be small, though the two different relationships could be detected using a regression analysis. As in the case of all statistical analyses or numerical modelling approaches, the interpretation of the results needs a thorough understanding of the governing reactions in the system. Of course, there are cases in the field of data mining where the relation of two or more parameters is not immediately obvious, but if mine water management has to be done precisely and understandably for decision makers, such non-obvious relationships should be encourage more research. If regression analyses or predictions are made into the future, values should always be given in ranges rather than a single value. This means, for example that the iron concentration after 10 years of flooding should not be given as 5 mg L^{-1}, but as either 3–7 mg L^{-1} or 5 ± 2 mg L^{-1}.

9.3.5 Interpretation of Data

Data interpretation is the most important part of consultancy work; it allows conclusions to be drawn and suggests the final recommendations. If results are presented and described without interpretation, a catchment or mine water management study would not have been worth the money and time spent. A good interpretation also needs good data, which means that the data gathering process was conducted using appropriate QA/QC procedures, the data can be verified and back-traced, and that all necessary standards for data acquisition, sampling, and laboratory investigations were followed.

Unfortunately, data interpretation cannot be described here – because it cannot be learned from text books. In general, if a report is clearly structured, the illustrations and tables are presented in an understandable form, and the text is written in an understandable language, it can be concluded that the interpretation of data was given the same priority. This does not mean that all the interpretations are trustworthy, but the likelihood of getting precise results is higher than in cases of "bad" reports.

10 Tracer Techniques for Mines

10.1 Why Conduct a Mine Water Tracer Test?

10.1.1 Objectives of Mine Water Tracer Tests

Usually, tracer tests are conducted in or near mines to find connections from the surface to the mine or vice versa. Typical examples are tracer tests conducted in conjunction with numerical models to assess potential radioactive waste disposal sites (e.g. Lee 1984). All tracer tests in mines can be grouped based on their objectives:

- assessing the risk or cause of mine water inrushes/inundations (Skowronek and Zmij 1977; Goldbrunner et al. 1982; Wittrup et al. 1986; Qiang et al. 1992; Lachmar 1994);
- optimising mining strategy (Adelman et al. 1960; Reznik 1990; Williams and Kirschner 1992; Kirschner and Williams 1993);
- evaluating the feasibility or effects of underground disposal of non-radioactive waste (Fried 1972; Himmelsbach and Wendland 1999);
- evaluating potential underground disposal of radioactive waste (Abelin and Birgersson 1985; Brewitz et al. 1985; Galloway and Erickson 1985; Cacas et al. 1990; Lewis 1990; Birgersson et al. 1992; Hoehn et al. 1998; Sawada et al. 2000; hundreds of studies were conducted in the Stripa mine/Sweden, the Grimsel/Switzerland experimental site, and the Yucca Mountain/USA site);
- subsidence studies (Mather et al. 1969); and
- developing remediation strategies (Aldous and Smart 1987; Doornbos 1989; Aljoe and Hawkins 1993, 1994; Davis 1994a, 1994b; Wolkersdorfer 1996; Wolkersdorfer et al. 1997a; Canty and Everett 1998)

Tracer tests have also been conducted to study "heat mining" in geothermal projects (e.g. Gulati et al. 1978; Horne et al. 1987; Kwakwa 1989; Randall et al. 1990; Aquilina et al. 1998). Because such studies are similar to those conducted in fractured rocks (see Himmelsbach et al. 1992; Käß 1998) and are usually done using boreholes, rather than conventional mines, these studies will not be considered further here.

Many tracer tests are not reported in the literature, or are only available as master's theses (e.g. Anderson 1987; Bretherton 1989; Diaz 1990; Hasche 2001). This may imply that they were unsuccessful or that the results were confidential, but may just reflect the fact that no one was willing to take the time to convert the study into a research paper. So far, only a few published tracer tests have been conducted to trace the hydrodynamic conditions of a flooded underground mine (Aljoe and Hawkins 1994; Wolkersdorfer et al. 1997a; Wolkersdorfer and Hasche 2001a), mainly because no suitable method was available to inject the tracer into the mine water at predetermined depths or without contaminating the mine water above the injection point. Most tests therefore injected the tracer at the surface, allowing the tracer to flow into the mine through fractures (e.g. Lachmar 1994) or used boreholes (e.g. Galloway and Erickson 1985; Cacas et al. 1990).

As will be shown in section 11.6, in-situ remediation and treatment methods usually fail because they require a thorough understanding of the flow regime within the flooded underground mine and where underground dams have to be installed before flooding. Tracer tests in flooded underground mines are therefore imperative for such in-situ options.

10.1.2 Aims of Mine Water Tracer Tests

Tracer tests are commonly used to investigate the hydraulic parameters or interconnections of ground water flow (Käß 1998). Most such techniques are well described and a range of tracers or methods can be selected, depending on the aims of the tracer test and the hydrological situation.

Published results of tracer tests in abandoned underground mines are still uncommon, as already stated by Davis (1994b, 1994a). Since less experience exists in tracing mine water, the expected results of an individual mine water tracer test cannot always be predicted. Summarised, the aims are as follows:

- testing the effectiveness of bulkheads dams,
- investigating hydrodynamic conditions,
- tracing connections between a mine and the surface,
- clarifying reasons for inundations,
- assessing contaminant mass flow, and
- estimating the decrease or increase of contaminants

Historically, the first tracer tests conducted in mines were simply to reveal connections between ground or surface waters and the mine (e.g. Skowronek and Zmij 1977). One of the first tracer tests in a deep flooded underground mine was conducted in 1995 to investigate the more complex hydrodynamic conditions of the flooded mine (Wolkersdorfer 1996).

It would be useful in many mining situations to use a multi-tracer approach. Possible candidates for multi-tracer tests are dyes, salts, and solid tracers. Wolkersdorfer et al. (2002) conducted a multi-tracer test in the Austrian Georgi-Unterbau mine (see section 12.13), using rock salt (NaCl), sodium fluorescein, uranine, and different microspheres in the flooded underground mine. Because the number of tracers that can be used in a single mine water tracer test is limited, my working group prefers fluorescent microspheres. Up to eight different colours can be injected and up to six have been used so far, in a mine water tracer test conducted at the flooded Straßberg/Germany fluorspar mine (see section 12.11).

In the future, tracer tests should become a prerequisite for evaluating remediation strategies for abandoned mine sites. Kimball et al. (1999) gave an example of how to use tracer tests and carry out synoptic sampling of trace metals in surface streams to evaluate the environmental impacts of mine effluents on watersheds. This can also be done in other mine closure cases where the hydrogeological situation is unclear and the choice or implementation of remediation strategies will be affected by the water flow.

10.1.3 Possible Mine Water Tracers

Underground mines generally consist of shafts, adits, raises, and stopes; these are comparable to the features found in karstic terrains. Therefore, a flooded mine can be looked on as a karst aquifer in a conceptual model, and numerical models of flow in karst aquifers (e.g. Liedl and Sauter 2000) might also be used to describe the hydrodynamics of underground mines. Furthermore, the tracer techniques developed for karst aquifers (e.g. Gospodarič and Habič 1976; Käß 1998) should be appropriate for flooded mines, too.

According to the literature, including our own investigations, nine types of tracers have been used in mine water tracing (Table 18), though only successful tracer tests (and not even most of them) are usually reported. Therefore, little can be said about tracers that are unsuitable for tracing mine water, though some researchers have investigated tracer behaviour in the mine water environment. Aldous and Smart (1987) conducted research into the absorbance of seven fluorescent tracers on iron hydroxide and kaolinite (sulpho rhodamin B, sulpho rhodamin G, rhodamin WT, lissamine yellow FF, pyranine, Na-fluorescein, and Tinopal CBS-X). As the behaviour of tracers varies in different environments, no general conclusions are possible. Yet, from their studies, sulfo rhodamin G and pyranine appear to be the best tracers in sediment-laden mine water – though neither has been used so far due to analytical or environmental reasons. For Na-fluorescein, which easily adsorbs to organic materials, we know of successful tests as well as unsuccessful ones.

The chemical composition of mine waters can very often involve high TDS, low pH (e.g. Iron Mountain: pH –3.6; Nordstrom et al. 2000), or high metal concentrations (Banks et al. 1997), which consequently limits the number of tracers that can be used successfully. Even conservative tracers, such as Na-fluorescein, might be unsuitable under certain mine water conditions e.g. low pH, high suspension load, wooden supports, free chloride radicals.

Therefore, laboratory tests using the chosen tracer and typical mine water compositions are essential to obtain positive results. All tracers have to be selected based on their expected or known behaviour in the mine water in question. An important consideration when conducting such tests is the duration of the laboratory test. From our experience, mine water flows with mean velocities of 0.3 to 1.6 m min^{-1} (95% confidence interval of 42 tracer tests investigated; see table 19), which can be used to roughly calculate the expected residence time. The stability of the tracer in the mine water must be guaranteed for the whole duration of the tracer test. Preceding the Niederschlema/Alberoda and Straßberg tracer test, we conducted 45 and 28 days of laboratory tests, respectively. Concurrently, the tracer was sampled and visually analysed to identify alterations (Wolkersdorfer 1996; Wolkersdorfer and Hasche 2001a).

10.2 Tracers and their Use

10.2.1 Introduction

Many people in the mine water field want to use tracers for their studies once they hear what tracers can do. However, very often, the basic knowledge about the properties of tracers and the way in which tracer tests have to be planned or conducted are not clear. The following chapters will merely introduce the broad field of tracer investigations; consequently, the language and the theory are expressed in a simple manner. It is meant for the novice with a basic knowledge of environmental sciences or geosciences and mainly covers tracers that can be used in mines. Therefore, some rarely used tracers, such as the promising DNA-tracers (Sabir et al. 1999), will not be discussed. Yet, all the prerequisites for conducting a tracer test are given and the relevant references are provided. A range of publications about tracers or tracer tests has been cited in earlier chapters and will also be referred to in the following sections. Readers who want to know more details should read one or more of those publications (e.g. Aley and Fletcher 1976; Moser and Rauert 1980; Käß 1998; Field 2003).

10.2.2 What are Tracers?

Tracers or marker substances are used to evaluate characteristic hydrogeological parameters. They can be divided into natural (or environmental) tracers that are “naturally” or unintentionally contained in the water, and artificial tracers. As a rule, the former are already contained in the water, whereas artificial tracers are injected into the water during the field test. Environmental isotopes, environmental chemicals, organisms, or the physical qual-

Table 18. Selection of tracers that have already been used in mine water tracer tests; only selected references are provided.

Tracer	References
NaCl	Mather et al. (1969); Hawkins and Aljoe (1992); Wolkersdorfer (2002b); Wolkersdorfer and Hasche (2004)
Na-fluorescein	Mather et al. (1969); Parsons and Hunter (1972); Nunes et al. (1990); Wirsing (1995)
Rhodamin WT	Canty and Everett (1999)
Club Moss Spores	Wolkersdorfer and Hasche (2001b); Wolkersdorfer (2001); Wolkersdorfer and Hasche (2004)
Microspheres	Wolkersdorfer and Hasche (2001b); Wolkersdorfer (2001); Wolkersdorfer and Hasche (2004)
Sulpho Rhodamin B	Aldous and Smart (1987)
Tritium	Mulenga et al. (1992); Zhou et al. (1993)
CFC chlorofluorohydrocarbon	Parry et al. (2000)
Rare Earth Elements	Worrall (1999); Wolkersdorfer (2002a); Merten et al. (2004)

ities of the water are examples of natural/environmental tracers. Artificial tracers can roughly be divided into water-insoluble (e.g. dyes, phages, microparticles, and spores) and water-soluble tracers (e.g. pigments, salts, and radioactive substances).

The difference between artificial and natural tracers can be clarified by using tritium as an example. Tritium arose from atmospheric nuclear weapon tests and reached ground water naturally in precipitation. Used in this manner, tritium is an *environmental* isotope and is therefore regarded as a *natural* tracer. However, tritium can be injected into the ground water within the scope of a tracer test as a radioactive tracer, and would then be serving as an *artificial* tracer (Geyh and Schleicher 1990).

The increasing environmental awareness of the population and the authorities during the last few decades has led to increasing problems in conducting tracer tests. We have found that it is usually better to use the words “tracer” or “marking substance” instead of the proper tracer names of the tracers, which may sound scary to the uninformed.

Table 19. Distances and mean velocities of worldwide tracer tests in underground mines. Table is given for comparisons only, details concerning geological setting and hydraulic parameters are given in the literature cited. Mean of all 42 tracer tests: 0.3–1.6 m min^{-1} 95% confidence interval of 42 tracer tests; [1] result probably wrong.

Distance, km	veff, m min^{-1}	Reference	Location
0.015	0.0001	Wolkersdorfer unpublished	Rabenstein/Sachsen
0.17	0.0003	Wolkersdorfer unpublished	Rabenstein/Sachsen
0.2	0.001[1]	Aljoe and Hawkins (1993)	Buckhannon/WV
0.044	0.004	Aljoe and Hawkins (1993)	Buckhannon/WV
0.093	0.004	Wolkersdorfer unpublished	Straßberg/Harz Mts
0.093	0.006	Wolkersdorfer and Hasche (2004)	Straßberg/Harz Mts
0.13	0.01	Aljoe and Hawkins (1994)	Buckhannon/WV
0.077	0.01	Canty and Everett (1998)	Red Oak/Oklahoma
0.78	0.01	Wolkersdorfer (1996)	Niederschlema/A.
0.238	0.014	Wolkersdorfer unpublished	Straßberg/Harz Mts
0.048	0.02	Wolkersdorfer et al. (2002)	Brixlegg/Tirol
0.01	0.03	Wolkersdorfer et al. (2002)	Brixlegg/Tirol
1.182	0.07	Wolkersdorfer unpublished	Straßberg/Harz Mts
0.35	0.1	Mather et al. (1969)	Aberfan/South Wales
1.181	0.1	Wolkersdorfer unpublished	Straßberg/Harz Mts
3.539	0.1	Wolkersdorfer unpublished	Straßberg/Harz Mts
0.077	0.12	Canty and Everett (1998)	Red Oak/Oklahoma
0.077	0.14	Canty and Everett (1998)	Red Oak/Oklahoma
0.283	0.15	Wolkersdorfer and Hasche (2004)	Straßberg/Harz
1.773	0.15	Wolkersdorfer and Hasche (2004)	Straßberg/Harz
0.171	0.17	Canty and Everett (1998)	Red Oak/Oklahoma
6.564	0.17	Wolkersdorfer unpublished	Straßberg/Harz Mts
1.7	0.2	Parsons and Hunter (1972)	Ebbw Vale/Mons.
0.229	0.23	Canty and Everett (1998)	Red Oak/Oklahoma
0.4	0.27	Czolbe et al. (1992)	Peißen/Germany
3.6	0.3	Aldous and Smart (1987)	Forest of Dean/Gls.
4.798	0.3	Wolkersdorfer and Hasche (2004)	Straßberg/Harz Mts

Table 19. Continued. Niederschlema/A.: Niederschlema/Alberoda; Ebbw Vale/Mons.: Ebbw Vale/Monmouthshire; Gls.: Gloustershire.

Distance, km	veff, m min^{-1}	Reference	Location
0.15	0.4	Mather et al. (1969)	Aberfan/South Wales
0.172	0.4	Wolkersdorfer et al. (1997a)	Niederschlema/A.
0.216	0.5	Wolkersdorfer et al. (1997a)	Niederschlema/A.
0.22	0.5	Wolkersdorfer et al. (1997a)	Niederschlema/A.
0.2	0.6	Mather et al. (1969)	Aberfan/South Wales
3.18	0.7	Wolkersdorfer and Hasche (2004)	Straßberg/Harz Mts
0.5	1.3	Aldous and Smart (1987)	Forest of Dean/Gls.
2.25	1.5	Wolkersdorfer and Hasche (2004)	Straßberg/Harz Mts
0.776	1.6	Wolkersdorfer et al. (1997a)	Niederschlema/A.
0.04	1.7	Wolkersdorfer et al. (2007)	Freiberg/Sachsen
0.736	1.8	Wolkersdorfer et al. (1997a)	Niederschlema/A.
0.78	2	Wolkersdorfer et al. (1997a)	Niederschlema/A.
2.159	5.7	Wolkersdorfer et al. (1997a)	Niederschlema/A.
2.723	7.9	Wolkersdorfer et al. (1997a)	Niederschlema/A.
0.5	11.1	Aldous and Smart (1987)	Forest of Dean/Gls.

10.2.3 What are Tracers used for?

Tracers are used to characterise flow paths in many different disciplines like geology, hydro(geo)logy, chemistry, physics, water supply and distribution, hydraulic engineering, limnology, or biology. Tracers have been used to pursue underground watercourses since Roman times. Today, tracers are also used in flooded mines, at site investigations of radioactive repositories, and in living beings.

The prime aim of tracer tests is to trace back flow paths, and determine flow velocities, mass transfer, and hydrogeological parameters. By evaluating tracer tests, the following information and characteristic quantities can be determined:

- hydraulic connections of flow pathways,
- ground water flow directions,
- ground water velocity, average interstitial velocity, and mean residence time,

- hydrodynamic dispersion,
- flow and mass transfer rates,
- variability of hydrogeological properties, and
- transport behaviour of dissolved or particulate water constituents as well as their interactions

10.2.4 Requirements for Tracers

A tracer is described as a conservative or ideal tracer in the subsurface medium if it exhibits:

- inert behaviour not prone to chemical reactions with immobile solids,
- unchanging physical qualities, and
- is transported at the same velocity as the flowing water.

In addition, as many of the following criteria as possible should be met:

- no natural occurrence of the tracers in the transport medium,
- good analytical detectability even at great dilution,
- water-soluble or dispersible,
- robust chemically stable,
- no interactions between tracer and medium,
- reasonably priced in terms of acquisition, handling, and analysis,
- physiologically safe (not toxic), and
- no lasting effect on ground water composition

No tracer satisfies all the aforementioned conditions equally well. Rather, the most suitable tracer has to be selected based on the question to be answered. Based on its transport behaviour, tritium would fulfil most of the ideas of a conservative tracer, yet, for reasons of cost effectiveness and analytical procedures, fluorescent dyes are the most common tracers used.

10.2.5 Types of Tracers

"Natural" or "Environmental" Tracers

Stable Isotopes

Environmental isotopes are nuclides of elements that occur in water without having been added or injected on purpose to conduct a tracer test. They are therefore natural tracers. Usually, their concentrations are low and the analysis and the interpretation subsequently turns out to be difficult. In distinction to artificial tracers, they can be used over large areas and long time scales. Besides geohydraulic parameters, environmental isotopes allow the age and temperature evolution of the waters to be traced (e.g. Lippmann et al. 2003). Fractionation of certain stable isotopes also allows the identification of certain biogeochemical processes (e.g. bacterial sulphate reduction; Fauville et al. 2004).

Environmental Chemicals

Environmental chemicals are substances that enter the water cycle accidentally or by slow contamination. They might be the same substances as in the case of artificial tracer tests, but because they reach the water cycle unintentionally they are categorised as "natural" tracers.

Sodium fluorescein or pyranine are two commonly used dyes in cosmetic articles, though a test with several different cleaning detergents showed that other chemicals are now being substituted for sodium fluorescein. Seeping boric acid from house waste disposal sites can be considered a good tracer (Neal et al. 1998). Increased boron concentrations in water analyses usually point to nearby house waste disposal sites or, in certain areas, to naturally elevated boron concentrations e.g. Col and Col (2003). Chlorides are found in the vicinity of streets or airfields, where salt is used for de-icing, or downstream of potash pits. Kolpin et al. (2002) investigated the background concentrations of 95 organic wastewater contaminants in the USA, thereunder caffeine, hormones, and antibiotics. Very good environmental tracers are sulphur hexafluoride (SF_6) or chlorofluorocarbons (CFCs), which are abundant in nearly every water. The same substances can also be used as artificial tracers.

Metals and semi-metals (e.g. Cd, U, As, Cu, and Sb) can be considered as natural tracers downstream of mine water drainage tunnels. One of many examples worldwide is the river Erft in the Eifel/Germany, which receives mine water from the abandoned Mechernich mine (Schalich et al. 1986). Such natural tracers are commonly used in the exploration for ore deposits (Grimes et al. 1995).

One of the most expensive and outstanding "natural" tracer tests happened on 12 August 1901 (Anonymous 1901), when lightning hit the Pernod factory at Pontarlier and about 1 million litres of absinth were flushed down the river Doubs. Two days later, André Berthélot noticed the absinth at the Loue Spring, 50 km north-north-east of Pontarlier.

Organisms

Organisms are rather exotic tracers. Common natural tracers are bacteria, which can be used to trace the environmental influence of excrement. Eels or ducks were also used successfully as tracers, though they should perhaps be listed as "artificial" tracers in the example cited, because the organisms were marked (Käß 1998).

Physical Influences

A physical influence that plays a major role in ground water investigations or the judgement of ground water contaminations is a change in water temperature. New fibre optic measuring methods make it possible to receive a high spatial resolution of temperature differences. Schreck et al. (1998, 1999), and Stoddart et al. (2005) used this method to monitor temperature variations in mine dump sites. Van Berk (1987) was able to show that downstream of a coal mine waste dump, the ground water temperature increased by 3–6 K, consistent with the exothermic pyrite weathering reaction.

Changes in electrical conductivity were used to trace the sources of highly mineralised waters emanating from potassium mine waste rock piles in the German Rhine river valley.

Artificial Tracers

Water-soluble Tracers

Dyes

Dyes are the most frequently used artificial tracers, amongst which fluorescent dyes are today the first choice when carrying out tracer tests. The dyes used are largely non-poisonous apart from the rhodamines and can easily be detected by fluorometers, fluorescence chromatography, or with the support of HPLC (high pressure liquid chromatography). Transportable fluorometers and other sophisticated techniques (e.g. MORES: micro optical remission micro sensors; fibre optics) are now available and allow continuous monitoring of fluorescence intensity. Both the easy analytical verification and relatively low cost have considerably contributed to the distribution of fluorescent dyes as tracers in hydrogeological investigations. They are also used to dye drift particles e.g. spores, microspheres to increase their detectability. Non-fluorescent dyes are of minor significance since they are often sorptive and their analytical verification is less precise than fluorescent dyes.

Salts

The salt most frequently employed for tracer tests is rock salt ($NaCl$), though potassium chloride, lithium chloride, and bromide salts are also used. Though most salts are comparatively cheap and their detectability is reasonably good, large quantities (up to several tons) of salt may be required, and therefore cost might hinder their use in mine water tracer tests. For a tracer test at the abandoned Straßberg/Germany fluorspar mine, 20,000 L of saturated NaCl brine were injected (Fig. 77).

Radioactive and Neutron Activatable Tracers

Radioactive tracers can easily be detected with field or laboratory equipment and though their use requires that radiation protection procedures and regulations be followed, in other ways their handling is no different from that of other tracers used in hydrogeology. However, at present, it is difficult to get permits to conduct tests using radioactive tracers, although they are effective for determining various geo-hydraulic parameters. Neutron-activatable tracers are non-radioactive elements that can be neutron activated after they are sampled in the field (Jester and Raupach 1987). Commonly used elements are ^{80}Br, ^{116}In, ^{165}Dy, ^{152}Eu, ^{140}La, and ^{56}Mn. They can be measured at very low concentrations because the analytical procedure is very sensitive.

Fig. 77. Sampling of saturated NaCl brine used for the Straßberg/Harz/Germany tracer test.

Water Insoluble Tracers

Solid Tracers

A solid substance was used in one of the first tracer tests ever reported: chaff to examine the springs of the river Jordan in the year 10 AD (Käß 1998). However, solid tracers are less suitable for conducting quantitative tracer tests. Should only a qualitative assessment be necessary, solid tracers may still be useful. The semi-quantitative LydiA-method, which uses filter nets and auto-samplers, was developed for mine water tracer tests, and is described elsewhere in this book (Fig. 78).

Phages

Phages are viruses with sizes between 20 and 300 nm. They are inactive when outside of bacteria and consequently, they do not have any active metabolism, which means that their number cannot increase during the tracer test. Their qualities as tracers are very good, but their analysis requires special laboratories. Due to the number of different phages known, they are suitable for tracer tests with multiple injection sites. Like all solid tracers, the

Fig. 78. *Lycopodium* Apparatus (LydiA) for injecting solid tracers into flooded underground mines. Diameter 120 mm.

number of particles found after the sample preparation are counted and used for the interpretation.

Bacteria

Because of hygienic issues concerning drinking water protection zones, bacteria play an important role in certain tracer tests (e.g. 50-day-time-of-travel-line). Numerous bacterial genera have proved suitable for hydrogeological tracer tests. They can be used in ground water, column tests, surface waters, and in coastal waters (e.g. Harvey et al. 1989, 1993; Vega et al. 2003).

Micro Particles: Microspheres

Micro particles are microscopically small polystyrene beads of 0.6–100 μm diameter which are coloured with fluorescent dyes (Fig. 79). In hydrogeological tracer tests, their surface should be neutrally charged to prevent sorption. For special investigations, microspheres with positive and negative surface charges can be purchased. Originally, polystyrene beads were developed for physiological investigations. However, several investigations confirmed that they can also be used as hydrogeological tracers in which the good experiences of the numerous physiological investigations can be used with advantage.

Recently, micro particles containing a rare earth element cocktail were developed. Those tracers can be used at sites where dyes might be destroyed due to certain physico-chemical conditions. Microspheres were used in several mine water tracer tests with generally positive results e.g. Straßberg/Harz

Fig. 79. Selection of microspheres used in the Straßberg/Harz 2003 tracer test.

Mountains, Königshütte/Harz Mountains, Ehrenfriedersdorf/Erzgebirge, Georgi Unterbau/Tyrol, Schlema/Saxony (the latter: Werner Käß, pers. comm.).

Spores

The spore tracer method uses club moss (*Lycopodium clavatum*) spores with diameters about 50 µm and is one of the most important methods used in karst examinations. They can be coloured with different fluorescent and non-fluorescent dyes so that several injection places can be examined in a single test. The evaluation is laborious since the spores found must be counted microscopically. Besides their use in karst areas, spores have also been used successfully in mine water tracer tests to investigate hydraulic connections and parameters as well as effective velocities underground e.g. Niederschlema/Alberoda/Erzgebirge, Straßberg/Harz Mountains.

10.3 Preparation

10.3.1 Hydrogeological examination

The aim and purpose of a tracer test must be defined precisely before a tracer test is carried out. A preliminary hydrogeological investigation is imperative and all available geological and hydrogeological data of the investigation area have to be considered in developing a basic conceptual model. Special attention has to be given to the heterogeneities of the aquifer and the mine layout. Geological maps and existing borehole or pump data must be evaluated for that purpose, because such heterogeneities can have a decisive influence on the selection of injection and observation sites. There are several hydrogeological text books available (e.g. Fetter 1988; Jordan and Weder 1995; Domenico and Schwartz 1998; Brassington 1999; Deming 2002; Hiscock 2005; Hölting and Coldewey 2005). It is beyond the goals of this book to repeat what these authors have already written about good hydrogeological practise.

Occasionally, there is insufficient information to establish a conceptual model of the investigation area or to plan the order of potential injection at observation sites. In those cases, further geological or hydrogeological investigations are essential. If necessary, pumping tests or measuring ground water levels will provide further information (Doherty 1990). In all cases, one or more field inspections should be conducted to get an impression of the examination area and the accessibility of the injection and sampling sites. If necessary, the client or the mine operator may have to modify sampling and injection sites to fulfil the needs of a proper tracer test.

The following data, or at least an idea of their order of magnitude, should be available for the planning of a tracer test:

- ground water contour maps and interfered ground water flow directions,
- ground water flow velocity,
- dilution,
- lithology and hydrogeochemical parameters, and
- water use.

The most suitable tracer has to be chosen based on the results of the laboratory tests, the cost of the tracer, and the analytical requirements for the tracer. In addition, a detailed hydrogeological investigation should clarify the most suitable injection and sampling points, based on a conceptual model of the site. Inadequately prepared tracer tests are the most frequent reason for poor results; preliminary investigations are an essential part of a tracer test, even if, and perhaps especially, when the area of investigation is not well characterised. A lack of knowledge about the hydrogeological situation can lead to poor selection of injection and sampling sites; this topic is addressed below.

10.3.2 Selection of Injection and Sampling Sites

A decisive criterion for whether a tracer test turns out to be successful or not is the correct selection of the injection and sampling sites. At least one of the injection sites must be located so that the tracer will flow through the examination area, satisfying the objectives of the investigation. Although this demand seems trivial at first, unsuccessful tracer tests show that this prerequisite is not always met!

Since tracer tests often bring new and surprising knowledge, it makes sense to choose one or two more sampling sites rather than to fail in finding the tracer injected. As a rule, tracer tests are cost-intensive, and a failed tracer test due to not enough sampling sites is dissatisfying for both contractors and customers! Underground and surface watersheds frequently do not correspond and this fact has to be considered when selecting the sampling sites. When there are uncertainties in the correct location of sampling sites, a simple numerical model that assesses hydrogeological parameters can be helpful.

One of the most unpleasant situations for both contractors and customers occurs when no tracer is found *at all*. Such tests simply cannot be evaluated; it may indicate complete sorption or degradation of the tracers, too low a transport velocity, or no hydraulic connection between the injection and sampling sites. You can at least rule out the last possibility if you use a sufficiently large number of sampling sites! Always include boreholes or shafts directly downstream from the injection site. In addition, do not allow yourself to be so convinced that your working hypothesis or conceptual model is correct that you fail to place a sampling site at locations that do not support that belief. It *is* appropriate to place a sampling site at a location that, according to the pre-considerations, might not be possible at all.

Besides the technical criteria, the relative spatial position of the injection and sampling sites and their accessibility play an important role. Sites that might be necessary from a technical point of view but that can only be reached with a disproportionately large effort should generally be omitted in favor of sites that are slightly less ideal but much more accessible. Finally, you can assume that the client will decide to add more injection and sampling sites the week before the test starts!

10.3.3 Which Tracers shall be used?

Just as important as the selection of the injection and sampling sites is the choice of the correct tracers. This choice is mainly determined by the question to be solved. Besides this, the hydrogeological and geochemical conditions, the required sensitivity, available analysis methods, and restrictions of use (e.g. permit of the authority, water protection areas) play a role in tracer selection. Occasionally, it is required to use several tracers in a multitracer approach. However, compromises have to be made, addressing potenial problems as optimally as possible (e.g. considering costs, transport, permits, and analytical procedures). Käß (1998, 2004) lists the following questions that have to be considered when selecting the optimal tracer:

- sorptivity and ion exchange,
- pore distances,
- chemical and physical water parameters,
- influence of light,
- water use,
- analytical criteria,
- effort of obtaining, transporting, injection, sampling, analytics,
- injection sites,
- sampling sites, and
- long-term behaviour of the tracers.

Although an optimal tracer should behave conservatively, depending on the terms of reference, it might be feasible to use a non-conservative tracer, such as when investigating the transport behaviour of characteristic substances. Table 18 gives a selection of tracers that have already been used successfully in mine water investigations.

10.3.4 How much Tracer must be used?

Numerous equations are specified in the literature for calculating the optimal tracer amount independently of the tracer and the investigation area. In detail, some factors might be contradictory, and therefore, the choice of the correct tracer amount is sometimes tricky. The preliminary values provided by these equations should be regarded as starting points for site-specific considerations.

A comparison of two equations for otherwise identical boundary conditions shall clarify this statement:

$$m = L \times k \times B \tag{68}$$

with m tracer amount, kg
L Distance to the most important sampling point, km
k Coefficient for the tracer substance, if not sodium fluorescein 0.1–20,000
B Factor for general set up of tracer test 0.1–10

and $L = 2.4$ km; $k = 1$; $B = 3$ yields $m \approx 7$ kg for sodium fluorescein (Käß 2004)

$$m = L \times K \tag{69}$$

with m tracer amount, kg
L Distance to the most important sampling point, km
K Coefficient for the aquifer 0.2–2

and $L = 2.4$ km; $K = 1.5$ yields $m \approx 3$ kg for sodium fluorescein (UNESCO 1; Käß 1998)

k, B, and K are tabulated values that can be found in Käß (1998). They will not be further discussed here, because the two equations are only provided to exemplify how tricky the calculation of a correct tracer amount might be.

The values given in Käß (1998) and Field (2003) differ over a range of three orders of magnitude. In general, the empirical values deduced from the experiences of the working group play an important role in calculating the necessary tracer amount. Käß (2004) concluded that expert knowledge is more important than a list of equations and therefore he only refers to equation 68. With respect to this shortage of reliable equations to calculate the necessary tracer amount, Field (2003) published the EHTD (Efficient Hydrologic Tracer-Test Design) computer code, which can be used to calculate the necessary tracer amounts for various hydrogeologically relevant scenarios if one provides values for 12–13 aquifer and tracer parameters. EHTD can also be used to calculate optimised tracer sampling intervals and predict the breakthrough curve. Yet, a mine water scenario is not yet included in the code.

Therefore, in mine water tracer tests, the empirical data can be backed by the following calculation, which gives a good approximation of the solid tracers needed for a tracer test in a quasi-closed system (the formula given in Wolkersdorfer et al. [1997a, 1997b] is faulty, as r_r is given in % instead of the rate):

$$m = \frac{1}{55,000} \times \frac{V_m}{r_r \times q} \times f \qquad (70)$$

with m total mass/volume of spores/microspheres, g
V_m volume of enclosed fluid in the mine, m^3
r_r recovery rate of spores or microspheres, – (not more than 0.02–0.1 might be expected)
q pumping capacity, L min^{-1}
f factor (spores: 1; 15 µm microspheres with a concentration of $1 \cdot 10^6$/mL: 0.25), g L m^{-3} min^{-1}

Equation 70 considers water of volume, V_m, in a mine or another enclosure without a discharge into a receiving watercourse. During a tracer test, the water is pumped out of the mine at a pumping capacity q, which must be relatively small compared to the total water volume. The recovery rate for spores or microspheres is based on experience and usually does not exceed 2–10 %.

Yet, we also used other equations to verify the necessary tracer amount. So far, we have had good experiences with the following procedure: given the detection limit of the tracer is c_{min}, then the discharging mine water must have at least this tracer concentration. The total volume of the water in the mine is V_m and the flow rate at the discharge point(s) is Q with a total potential duration of the tracer test t. Furthermore, we assume a complete homogeneous mixing of the tracer with the mine water. This assumption is very optimistic but it has proven to be a good starting point. Furthermore, the concentration factor, f_c, which depends on the sampling equipment, must be known. If the tracer is directly measured in the discharging water, this factor is 1, if the water is filtered through a filter (e.g. MeFisTo System) then it depends on the ratio of the water pumped through the filter systems q to the discharged water. The expected recovery rate shall be r_r and the sampling interval t_s.

Then the following physically-based equation can be used to estimate the necessary tracer amount:

$$m = \frac{(V_m + Q \times t) \times c_{min}}{r_r \times f_c} \tag{71}$$

$$f_c = \begin{cases} 1 & \text{: direct tracer measurement in discharging water} \\ q \times t_s \times r_r & \text{: aliquot pumping of discharged water} \end{cases} \tag{72}$$

So far, this conservative approximation was successful in all of the tracer tests that we have conducted in fully flooded mines.

10.3.5 Tracer Test Plan

When applying for a permit from the responsible water authorities, a tracer test plan should be made available. This plan should list in detail the injection and sampling sites at which temporal distributions shall be sampled and by whom. At the same time, the test plan has to define why the tracer test is being carried out and should predict, if possible, the anticipated results that will be obtained. Normally, the test plan also contains details on the costs of the complete tracer test, including its evaluation. The success of an application and the tracer test itself depends fundamentally on whether the test plan is complete or not.

During sampling, the local or national recommendations for sampling ground water monitoring wells or surface waters have to be considered. Because national and international standards are changing regularly, even locally, the researcher planning to conduct a tracer test is well advised to get into contact with all of the local authorities that are responsible for permitting the tracer test.

According to Käß (1998) and other relevant guidelines (Schwarz et al. 2002), a tracer test plan should at least contain the following details:

- district and place name,
- description of the injection sites,
- person in charge of the tracer test with phone numbers,
- customer with full address,
- experimental purpose,
- injection time (start of tracer test),
- type of tracer,

- water quantities to flush the injection site and the tracer, and
- a list of the sampling sites and sampling intervals.

Furthermore, the following details are meaningful:

- work schedule for continuous sampling,
- necessary measurements on site,
- work schedule for analytical laboratory, and
- emergency plan.

10.4 Approval Procedure

10.4.1 Legal Regulations

Since the legal regulations for permitting tracer tests differ from country to country, you must contact the local authorities before designing a tracer test. In some countries, tracer tests need a formal, sometimes lengthy approval procedure, while in other countries, a telephone call with the water or mining authorities might be sufficient. A typical procedure for applying a permit for a tracer test is described here for Germany, according to Pascaly (1997).

The responsible authorities act under the principle of § 1 of the Federal Water Budget Law ("Wasserhaushaltsgesetz" WHG), which demands that the waters have to be managed for the welfare of the general public and in accordance with the welfare of individuals. Avoidable impairments of ecological function must be avoided. Additionally, in water conservation areas, the water protection ordinance has to be considered.

According to § 2 (1) WHG, the use of waters requires official permission, and the injection of a substance represents a use in the sense of § 3 (1) № 5 WHG. § 3 (2) regulates measures that can cause harmful and significant changes in the physical, chemical, or biological water parameters, be they permanent or temporary.

The official, revocable permission is granted by the authority in accordance with § 7 (1) WHG. This permission allows using waters (including ground water) for a certain purpose in a specific way predetermined by its manner and degree. In accordance with § 4 (1), the permission can be made conditionally, to prevent or compensate for adverse effects to others.

The responsible authorities are determined by legal state regulations. In Saxony, for example, the State Environment Agencies (Staatliche Umweltfachämter) or the lower water authority are responsible for granting the permission. The authority checks the tracer test plan for consistency with the legal specification that harmful changes and impairments be avoided, and ensures that it complies with the regulations or – if applicable – the existing water protection ordinance. Their concern extends beyond toxicological effects to humans and the environment, including other possible chemical (e.g. solution contents, taste), physical (e.g. colour, temperature), or biological changes to the water.

10.4.2 Necessary Details within the Application

As a basis for its permission, the water authority needs an application with a detailed description of the tracer test plan. The reasons for the test, the tests boundary conditions, and the type of tracers to be used, including their environmental compatibility (Arbeitskreis "Human- und ökotoxikologische Bewertung von Markierungsmitteln in Gewässern" 1997; Behrens et al. 2001), should all be documented. To allow the permitting authorities to understand the procedures of the tracer test and give them a means to assess possible environmental risks, be sure to present the following details: the amount of tracer (including its potential dilution), the volume of flushing water, the amount of water to be removed (for flushing reasons or analyses), the duration of the test, and the potential degradation rates of the tracer. Involving the authority as early as possible will allow you to understand any special concerns that they might have so that these can be addressed in your plan. Furthermore, the responsible special authority is frequently asked to give their recommendation to the permitting authority, documenting the necessity of the tests.

10.5 Execution

10.5.1 Injection Time and Injection Type

It is critical that the samples not be contaminated with even trace amounts of the tracer. This is more challenging than you might think. People who are conducting the injection should not collect the samples or carry out the later analyses in the laboratory. If this is not possible, the individuals should carefully clean themselves and completely change their clothes. Furthermore, if there are several sampling points in the test, the probes used for temperature or electrical conductivity measurements must be carefully cleaned with distilled water after each use. Cleaning them only at the new site risks cross-contamination. Unfortunately, the cover photograph of the Field (2003) publication does not show good practice in tracer injection, because the equipment of the person injecting the tracer is being contaminated with the tracer, and will possibly carry off the Na-fluorescein tracer to the next sampling site.

Independently of the tracer chosen, all containers carrying the tracer have to be flushed out repeatedly to make sure that the tracer is fully injected into the mine- or ground water. In the case of a solid tracer, this also ensures that all of the tracer reaches the water body to be examined. To avoid contamination in the field, it is recommended that fluorescent tracers be pre-diluted in the laboratory.

If necessary, several tens to hundreds of litres of water should be poured into the injection point, especially when the injection site is dry. For tracer injections in the unsaturated zone, one has to ensure that the tracer reaches the ground water. Without additional water, the tracer would remain in the pore space of the unsaturated zone. Of course, this is not an issue if the tracer test is being carried out to examine the unsaturated zone.

When evaluating tracer tests, it is assumed that the injection has the form of a Dirac function:

$$\delta(t) = \begin{cases} 0, & t \neq 0 \\ \infty, & t = 0 \end{cases}$$
$$\int_{t_1}^{t_2} dt\delta(t) = 1 \qquad (73)$$

Therefore, you should completely inject the tracer as quickly as possible. In the past, it was recommended that you inject the tracer slowly, but this is not advisable because it "smudges" the tracer cloud, could feign a non-existing dispersion, and make the results of the tracer test considerably harder to evaluate.

In contrast, Kimball et al. (2002) and Kimball et al. (2007) conducted tracer tests in mine water impacted surface streams in Colorado/USA, Utah/USA, and Tuscony/Itay where they injected soluble salt tracers at a constant rate and calculated the dilution of the tracer. This allowed them to locate mine water inflows, characterise flows below the streambed, and find where stream water was leaking into fractures.

10.5.2 Sampling

Correct sampling is essential when conducting all hydrogeological investigations. Good instructions are given by the Saxonian Agency for Environment and Geology (Sächsisches Landesamt für Umwelt und Geologie 1997) and Ficklin et al. (1999). The following instructions focus on specific aspects of sampling that are applicable to tracer tests. These recommendations are mainly the result of working together with students, mining-engineers, and personnel not normally involved in regular sampling for such tracer tests, and hopefully, will allow you to learn from their mistakes.

It is important that you correctly label all of the sampling containers, record everything, and follow the sampling protocols exactly. All details of the sample have to be written down in the protocol as well as on the sample containers (usually bottles). It is particularly important to denote irregularities (e.g. extreme water levels, insufficient amounts of sample, tracer spilled); mishaps can happen to anybody, and can typically be dealt with, as long as they are properly recorded and accounted for.

Another issue are the containers used for water samples. They must be appropriately selected for each parameter that will be measured in the laboratory. PVC or glass bottles are a good selection for the main ions and the trace elements. TOC samples must be collected in glass bottles and TIC samples must be collected in PVC bottles sealed with aluminium foil. New containers have to be rinsed out three times with the water to be sampled before they are filled. Prior to being reused, all containers have to be washed with hot water and phosphate-free lab-grade soap, rinsed with deionised water thereafter and washed again with dilute acid (e.g. 10% nitric acid). After acid washing, the containers have to be finally rinsed with deionised water.

As indicated above, the sampling has to follow the tracer test plan. In agreement with the customer and the authorities, the test plan can be changed, provided that unforeseen events require this (e.g. appearance of the tracers in till-now unsampled springs, shafts, or drippers; no more tracer analytically detectable; incomplete breakthrough).

A well drawn up sampling plan considers the travel time between the withdrawal points as well as possible difficulties that might develop (e.g. breakdown of equipment, weather conditions). As a rule, sampling intervals are short at the beginning and increase during the tracer test. Without knowledge of the hydrogeological conditions, no rule of thumb can be given. However, it is better to take too many than too few samples. If a numerical model was carried out before the tracer test, the sampling intervals can be deduced from the results of the modelling. Furthermore, the EHTD software code provides tools to calculate the initial sample collection time and the approximate sampling intervals as part of an optimised sampling scheme (Field 2003), based on the predicted tracer mass as well as the aquifer's hydraulic and geometric parameters (Fig. 80).

All staff should be instructed to record the actual sampling times and not simply use the ones in the sampling plan. Many things can cause you to sample earlier or later than projected in the original plan, but a "well-meaning" correction can cause a major problem!

The necessary size of the sampling containers must be discussed with the analytical laboratory. In the case of fluorescent tracers, brown glass bottles are recommended, since fluorescent dyes can be decomposed by light (Käß 1964). For tracer tests with bacteria, phages, spores, or microspheres, a careful selection of the appropriate laboratory is imperative, as those tracers are not easily detectable.

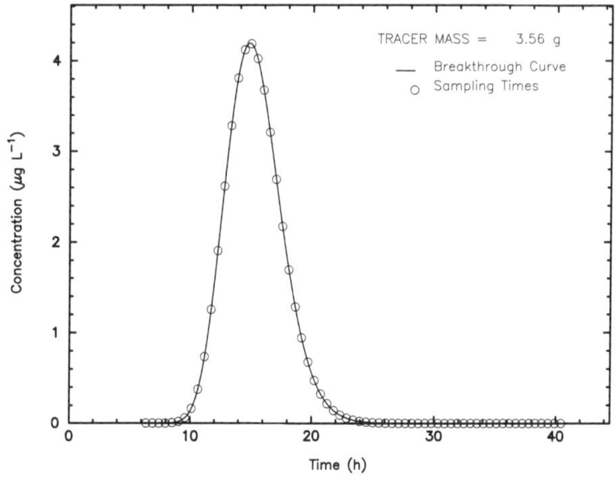

Fig. 80. Sample output of the Efficient Hydrologic Tracer-Test Design (EHTD) software code for a tracer test at the Dyers Spring (data from Field 2003). The circles indicate the optimised sampling intervals.

As in the case of water samples for chemical analysis, tracer samples have to be transported and stored carefully packed and cooled. Since the concentration of tracers can change over time, samples should be brought back to the laboratory and analysed as quickly as possible.

10.5.3 Tracer Analyses

For good results, you should use a high quality laboratory, ideally one that you have used before and trust. The cheapest laboratory will rarely produce the best results! If necessary, some selected laboratories should be compared with each other by using a ring or round-robin test to find out which one has the best reproducibility. Such tests are conducted by sending a sample of known composition or tracer concentration to different laboratories. The results of the different laboratories can be compared with the expected result and the best laboratory selected.

It is important that a laboratory book with numbered pages be used and maintained, and that a precise protocol is followed for every test. Experience shows that reliable laboratories have properly recorded protocols and laboratory books that are in good order! No confidence should be placed in laboratories with "piece of paper economy" and adhesive notes sticking to the sample containers!

You should consult the appropriate technical literature for the analytical methods for the different tracers. A good reference is Käß (1998).

10.5.4 Documentation and Presentation

The results of a tracer test will be evaluated based on the final documentation. Even small tracer tests can produce several hundreds of datasets, each of which must be recorded and presented. The results of a small test can often be analysed using a spreadsheet, while you should use professional, electronic database tools for larger tests. If several institutions are collaborating in a tracer test, a common database structure is recommended. All researchers involved must be urged to follow the database structure provided by the leading team.

Lists or tables are the simplest means to present the sampling place and time, and obviously the tracer amount. Cartesian coordinates (usually in the system UTM WGS84, because such data are provided by modern satellite navigation systems, i.e. GPS, GLONASS, or GALILEO) have to be included when several injection and sampling places are involved. It is generally advantageous to tie the data together using GIS-applications, allowing the reader to immediately see the relationships between the different injection and sampling points. All modern GIS-applications permit coupling of existing databases with graphical and topographical information.

Presenting the data in time-concentration-diagrams ("breakthrough curves") provides a quick understanding and appreciation of the results of a tracer test, as Fig. 81 illustrates. The most important characteristics of the tracer test should also be shown in such a figure.

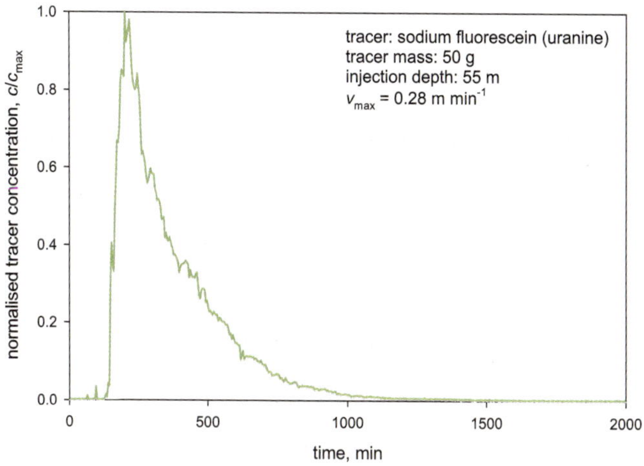

Fig. 81. Normalised breakthrough curve of a characteristic tracer test in the shaft of a flooded underground mine.

10.6 Evaluation and Characteristics of the Tracer Test

10.6.1 Porous Aquifer

Tracer tests in porous aquifers can normally be interpreted easily using existing tools, since there is very extensive experience. Let's start by discussing the governing transport equations for porous aquifers in detail, since the interpretation of nearly all tracer tests (even in mine aquifers) is based on porous aquifer theory. Flow in porous aquifers is determined by Darcy's law (assuming one-dimensional; stationary flow; homogeneous, infinitely extensive aquifer):

$$\frac{Q}{A} = v_{\mathrm{f}} = k_{\mathrm{f}} \times \frac{dh}{dx} \tag{74}$$

with Q flow, $m^3\ s^{-1}$
A sectioned area through which flow passes, m^2
v_f Darcy velocity, $m\ s^{-1}$
k_f hydraulic permeability, $m\ s^{-1}$

$\frac{dh}{dx} = i$ hydraulic gradient, –

and

$$v_{\mathrm{a}} = k_{\mathrm{f}} \times \frac{i}{n_{\mathrm{e}}} \tag{75}$$

with v_a flow velocity, $m\ s^{-1}$
n_e effective porosity, –

Due to the microstructure of the aquifer and its inhomogeneities, a deviation from the ideal flow path is regularly observed. This deviation is expressed by the hydrodynamic dispersion:

$$D_{\mathrm{h}} = D_{\mathrm{m}} + D_{\mathrm{e}} = \alpha \times v_{\mathrm{a}} + D_{\mathrm{e}} \tag{76}$$

with D_h hydrodynamic dispersion, $m^2\ s^{-1}$
D_m hydromechanical dispersion, $m^2\ s^{-1}$
D_e effective diffusion coefficient, $m^2\ s^{-1}$
α Dispersivity (dispersion length), m
v_a flow velocity, $m\ s^{-1}$

Consequently, the mass transport in a homogeneous, three-dimensional aquifer is described by a combination of advection and dispersion:

$$\frac{\partial c}{\partial t} = D_{\mathrm{L}} \frac{\partial^2 c}{\partial x^2} + D_{\mathrm{T}} \frac{\partial^2 c}{\partial y^2} + D_{\mathrm{V}} \frac{\partial^2 c}{\partial z^2} - v_{\mathrm{a}} \frac{\partial c}{\partial x} \tag{77}$$

with c concentration, g L^{-1}
t time, s
x, y, z distance in the x, y, or z direction, m
D_L longitudinal dispersion coefficient, $m^2 s^{-1}$
D_T transversal dispersion coefficient, $m^2 s^{-1}$
D_V vertical dispersion coefficient, $m^2 s^{-1}$
v_a flow velocity, m s^{-1}

As the tracer distribution usually approximates a Gauss normal distribution (Fig. 81) and as the tracer injection is in the form of a Dirac function, the following equation describes the two-dimensional case of the tracer distribution at position x, y and time t:

$$c_{x,y,t} = c_{\max,t} \times \exp\left(-\frac{\left(x - v_a \times t\right)^2}{4 \times D_L \times t} - \frac{y^2}{4 \times D_T \times t} \right) \tag{78}$$

In reality, breakthrough curves seldomly conform exactly to a Gauss normal distribution, as can be seen in the Rabenstein/Germany tracer test example (section 10.5.1). Commercial software allows you to interpret the results of your tracer tests and obtain simulated breakthrough curves, which in turn allows you to deduce the appropriate parameter values by parameter approximation.

10.6.2 Fractured Aquifer

The researcher must know that completely different flow conditions exist in porous and fractured aquifers and that ground water flow behaves entirely differently in the two cases. Although the evaluation of breakthrough curves of those two aquifers is similar, breakthrough curves of fractured aquifers are different from those of porous aquifers. One of the most striking differences can be seen in the recovery rate of the tracers. While more than 50% of the tracers are often recovered from porous aquifers, the values from fractured aquifers are usually less than 50%.

Due to the heterogeneity and the anisotropy of fractured aquifers, preferential flow directions are characteristic of fractured aquifers. Consequently, the assumption that there is a homogeneous, virtually porous system is only partially correct. The fact that breakthrough curves usually do not show a Gauss normal distribution confirms that this assumption only has a low degree of truth.

Fractures represent preferential flow paths for water, pollutants, and tracers within a relatively homogeneous rock matrix. If there is no fracture connection between the injection and sampling sites, no tracer can be recovered at the sampling site. To model the flow in fractured aquifers, different model approaches exist (e.g. artificial or random fracture distribution by fracture generators, e.g. Wendland and Himmelsbach 2002). Whichever approach is used, the researcher must know that it is practically impossible to describe fractured aquifers adequately. This always has to be considered in evaluating the results of tracer tests in fractured aquifers.

In a system with several parallel fractures, variable fracture widths and roughnesses can cause almost all of the water to flow though only a few fractures. Normally, this effect, called "channelling", results in a breakthrough curve with multiple peaks, each of which represents a preferential flow path (Fig. 82).

Another effect arises in fractured aquifers when the matrix porosity has a significant value. In such cases, the tracer from the matrix is slowly transported into the fractures, which usually have a high hydraulic conductivity. When this occurs, the breakthrough curves tail out, which indicates that the tracer from the matrix is continuing to interact with the water in the fractures (dual or double porosity aquifers).

Many tracer tests in fractured rocks show a causal relation of the distance between the injection and sampling sites and the flow velocity: the flow velocity frequently increases with increasing distance. There are similar observations in the case of distance and hydraulic dispersion.

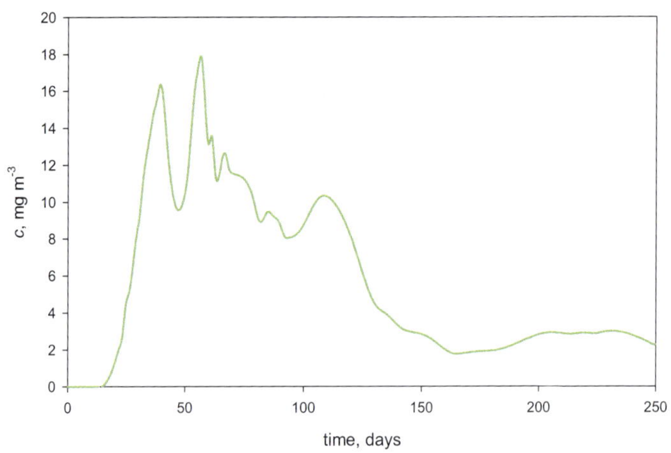

Fig. 82. Breakthrough curve of a tracer test with preferential flow paths (data modified and simplified after Käß 2004).

Often, tracer tests in fractured aquifers turn out to be more difficult than those in karst or porous aquifers since the result of the experiment depends more on a correct choice of injection and sampling sites than in the case of the other two aquifer types. Despite numerous research projects over the last two decades, the number of experiences with tracer tests in fractured aquifers, compared with tracer tests in porous aquifers, is still low. A thorough interpretation of such tracer tests therefore is still difficult, though the single fissure dispersion model (SFDM) generally yields sufficient accuracy (Małoszewski and Zuber 1985; Himmelsbach et al. 1992; Bäumle et al. 2001; Bäumle 2003). At present, research is being conducted in this field in many different countries (e.g. Germany, USA, Slovenia, France, Austria, and United Kingdom).

10.6.3 Karst Aquifer

Tracer tests in karst areas have a long tradition reaching back to Roman times (Käß 1998) and they are a critical tool in understanding the hydrogeological conditions of karst terrains. Nearly all of the potential tracers have been used in karst tracer tests investigations, and were positively tested in several multi-tracer approaches in Slovenia (Gospodarič and Habič 1976; Kranjc 1997). Other, more recent applications are described in the proceedings of the International Symposia on Water Tracing (e.g. Hötzl and Werner 1992).

Karst is named after its typical appearance in the Kraš Region of the Slovene Dinarides (Bates and Jackson 1987). Predominant features of karst areas are dolines, sinkholes, caves, ponors, karren, roofless caves, or poljes. Most of those features were described in the fundamental work of Cvijić (1893). Karst aquifers occur in all water-soluble rock formations, namely limestone, dolostone, gypsum, anhydrite, and even quartzite. The dissolution of the forenamed rocks by the acids in the ground water forms cavities, through which ground water can flow (Šušteršič 1996). In the underground, the diameters of those structures range from several micrometers to several decametres, mostly offering an excellent hydraulic connection for water flowing within that conduit network. As many karst researchers and karst aquifer tracer tests showed, this network of channels gives rise to different flow paths (Brown and Ford 1971); tracer tests proved that water uses different pathways with different effective velocities, reaching maximum velocities of 10–585 m h^{-1} (Gospodarič and Habič 1976; Hölting and Coldewey 2005). Recently, the triple-porosity-model has become more and more prominent in carbonate aquifer research (Worthington 1999) and future tracer tests will have to consider that approach.

As has been seen before, karst aquifers are characterised by high to extremely high flow velocities via discrete flow paths. Modelling of karst aquifers, which is thought to be extremely complicated, has to a special degree been possible in the Tübingen and Göttingen working groups. They developed the computer code CAVE (Clemens et al. 1999; Liedl and Sauter 2000), which has also been used to model water rock interactions in flooded underground mines (Spiessl et al. 2002).

Karst tracer tests are evaluated in the same way as described before, though they often demonstrate multi-peak breakthrough curves and comparatively long tails. These are due to different flow paths through the karst conduits, and are usually interpreted by applying the multi-dispersion-model (MDM; Małoszewski et al. 1992).

10.6.4 Mine Aquifer (Underground Mines)

Underground mines are very similar to karst aquifers (Sammarco 1995; Burbey et al. 2000) and like them, can be viewed as triple-porosity-aquifers with one-, two-, and three-dimensional porosity elements. The one-dimensional elements would be the main adits, shafts, and workings; the two-dimensional elements are the fractures, faults, and bedding planes; and the three-dimensional element is the rock matrix. If a mine is worked in a karst aquifer, the karst conduits would have to be counted as one-dimensional elements as well. Yet, the author knows of no publication that applies the triple-porosity model *sensu strictu* to mine water research.

The triple porosity model clearly shows how important all three porosities are (e.g. Worthington 1999). Comparing the hydraulic conductivities in a flooded underground mine with the nature of the first-flush scenario clearly proves that the most important factor for mine aquifers is still "conduit" porosity. Of course, the long tail of the first flush is due to natural attenuation and the comparatively lower hydraulic conductivities of the minor porosities (fractures, matrix, and panels). Yet, from an environmental point of view, the large volumes are of greater importance in determing the characteristics of the first flush and that of tracer tests.

Due to the similarities of mine aquifers and karst aquifers, the first quantitative tracer tests in deep flooded underground mines used the techniques established for those types of aquifers (Wolkersdorfer 1996). Liebisch and Remus (1966) also concluded that the methods applied in karst tracer tests could be used in mines, and used Na-fluorescein for a tracer test in the Mansfeld/Germany mining area. Despite the similarities, some major differences can be listed (most of which are not valid for prehistoric and medieval mines):

- channels (e.g. adits, shafts, workings) in modern mines are usually not curvilinear and typically meet at right angles,
- diameters usually do not change for a longer distance,

- the water level changes are comparatively small, and
- mine voids are not subject to dissolution.

In detail, mine aquifers comprise a combination of karst aquifers, porous aquifers, and fractured aquifers (see section 7.1). All the larger openings that were used for man or material haulage can be viewed as karst aquifers. Typically, the openings and the effective velocities are large, the storage capacity is small, and the water flow is non-Darcian. Shafts, boreholes, and old abandoned mine workings above the mine water table can be considered epikarstic. Between that zone and the mine water table comes the vadose zone and finally, below the mine water table, the phreatic zone is exposed. As the Straßberg/Harz mine water tracer test shows, those three zones in a mine act very similarly to the real karst features.

A fractured aquifer exists in most mines, especially those near the surface. That fracture network is usually a prominent region of infiltration and has been intensively studied by those investigating the feasibility of underground radioactive waste disposal (e.g. Galloway and Erickson 1985; Abelin and Birgersson 1985; Sawada et al. 2000). Sometimes, this fractured zone is the source of catastrophic mine water inrushes and consequently it has also been studied to prevent inundation (e.g. Singh and Atkins 1985; Singh et al. 1986; Grapes and Connelly 1998; Waterhouse et al. 2003). Though the importance of the fracture network for working or non-flooded mines is obvious, the role of fractures within the flooded part of a mine has not yet been studied in detail. From a hydraulic point of view, their role should not be over-estimated, as most transport occurs in the flooded voids. Yet, the tracer test in the abandoned Königshütte/Harz mining field proved that fractures above the ground water level can be hydraulically well connected during high precipitation periods. An additional fracture network can be found in the immediate vicinity of the mine workings due to the loosening of the rock during mining. Depending on the mining technology, this fracture network can reach several centimetres to several meters into the rock matrix (see section 4.2). Though not important to the overall water flow, it is important to the diffusive transport of the metals into the mine water (see section 4.2).

Finally, every mine, because of its rock matrix, also consists of a porous aquifer. In addition, backfilled areas or the goaf (collapse zones behind longwall mining operations) can be seen as porous aquifers, with porosities commonly in the range of 0.55 to 0.6 (Paul Younger, pers. comm.).

Though mine aquifers have the characteristics of all of the three known aquifer types, mine water tracer tests usually exhibit patterns that are typical for karst aquifers. Therefore, mine water tracer tests should be evaluated using the methods for evaluating karst tracer tests or, in the future, the triple-porosity-model.

10.6.5 Evaluation of Breakthrough Curves

Breakthrough curves are a first means to evaluate the results of a tracer test (Fig. 81). Such curves plot the concentration *c* of each tracer against the duration *t* of the tracer test. From this graph, the following times can be deduced (Fig. 83):

t_0	Start of tracer test
t_{max}	First tracer arrival: maximum distance velocity
t_{dom}	maximum tracer concentration: dominant distance velocity
$t_{med} = t_{50}$	median tracer concentration: median distance velocity
t_{true}	mean tracer concentration: true/mean distance velocity

where the velocity is calculated as follows:

$$v = \frac{x}{t} \tag{79}$$

The longitudinal dispersion coefficient D_L (Fig. 84) is estimated by using the times at which 16%, 50%, and 84% of the tracers are detected (mean value and twice the standard deviation of a Gauss normal distribution). The equation for this is:

$$D_L = \frac{v^2 \times (t_{84} - t_{16})^2}{8 \times t_{50}} \tag{80}$$

As described earlier, the complete dispersivity consists of the three components: lateral, transversal, and vertical dispersivity. In the case of pore or karst aquifers, the predominant dispersivity is the longitudinal one. According to different authors, the transversal dispersivity is between $\frac{1}{10}$th and $\frac{1}{5}$th (Käß 2004) or 1–2 orders of magnitude smaller (Kinzelbach and Rausch 1995) than the longitudinal dispersivity, whereas $D_T \approx 0.1 \times D_L$ has been found to be a suitable mean approximation:

$$D_T \approx (0.01\text{–}0.2) \times D_L \tag{81}$$

By using computer codes or spreadsheet calculations, these equations can easily be used to calculate the characteristic tracer test parameters. You can download the simple MS Windows code, TRACI98 (also known as TRACI95), from the Borntraeger web page (Käß 1998, 2004), or the more

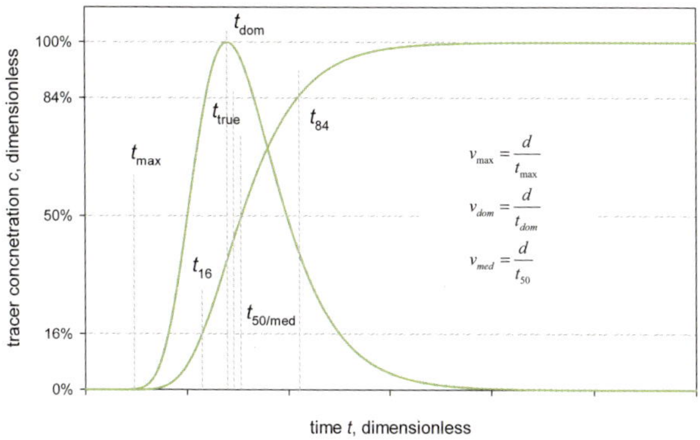

Fig. 83. Velocities that can be derived from a breakthrough curve. t_{16} and t_{84} are needed to calculate the longitudinal dispersion coefficient D_L; t_{true} is exactly in the middle of t_{dom} and t_{med}.

sophisticated programme, QTRACER, from the U.S. Environmental Protection Agency's web page (Field 2002).

10.7 Quality Control and Data Storage

10.7.1 Quality Assurance – Quality Control

Accurate and reliable results and documentation of data acquisition during field and laboratory work are essential when conducting treatment or remediation measures, which, in some cases, might take several decades. Sampling and analyses cost a lot of money and consume a lot of time during a project (Kellner et al. 1998). Very often, the responsible project leader has to rely on external experts and does not personally collect or oversee the analysis of the samples. Furthermore, while laboratories or staff may change during a project, data quality must be ensured at all time. Therefore, quality assurance (QA) and quality control (QC) must be emphasised to exclude variation of the analytical results due to false sampling or inappropriate analytical procedures. This may require a data quality objective (DQO) process (U.S. Environmental Protection Agency 1994), the experience of the project leader, existing QA/QC plans, or decisions made by groups of experts. In the USA, the Environmental Protection Agencies (EPA) employs quality assurance managers, quality assurance officers, or quality assurance representatives, who you can contact for help in setting up QA/QC plans. Furthermore, established procedures are available (U.S. Environmental Protection Agency 1994; EPA QA/R-xx and QA/G-xx series documents).

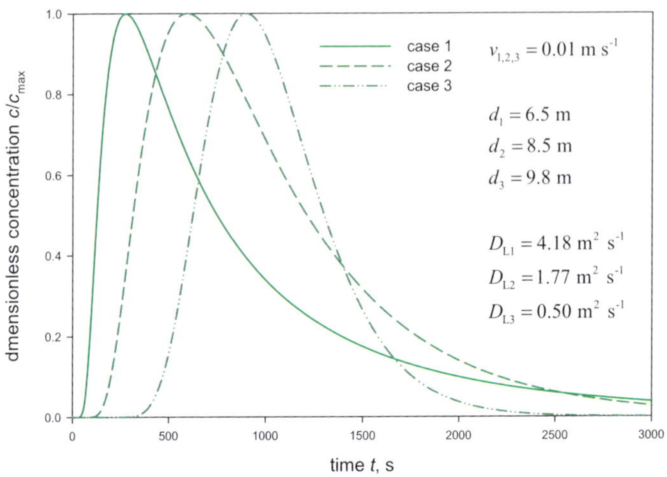

Fig. 84. Graphical representation of different longitudinal dispersion coefficients D_L. v: mean velocity, d: dispersivity.

While QA measures refer to the whole project and ensure that an effective QC is in force, QC guarantees that the quality of the data and the monitoring process are consistent. Most laboratories have their own QA/QC plans (compare Kellner et al. 1998); therefore, there is usually no need to include the laboratories in the QA/QC plans of the project. Nevertheless, the responsible project co-ordinator must record under which QA/QC procedures the relevant laboratories work and, if uncertain about the results of a certain laboratory, standard samples should be send to the laboratory in question (e.g. DIN EN ISO 8403 ["Qualitätsmanagement"], DIN EN ISO 8402, 1995-08, or other relevant national regulations). Calow (1991) listed potential and common laboratory errors, of which only "chemical and physical interference" is not due to human error:

- incorrect identification of samples,
- contamination,
- improper, or inappropriate sample preparation,
- inaccuracy of sample weights, or volumes,
- improper, or inappropriate sample dissolution/treatment,
- chemical and physical interference,
- improper, or inappropriate instrumentation/inaccurate measurement,
- calculation errors, and
- incorrect data handling/reporting

QA/QC measures can be time consuming and expensive. In general, the higher the degree of confidence has to be, the more the QA/QC measures will cost. However, since false data interpretation could result in wrong decisions, which might then result in unnecessary remediation requirements, such expenditures should be viewed as money well spent. Saving money by ignoring principles of quality assurance and control very often results in long discussions about how and where a sample was collected, and, in the worst case, malfunctioning remediation measures. A thorough QA/AC plan will also be the basis for trusting co-operation with the regulators and project reviewers (Downing and Mills 1998). A QA/QC program must:

- document the procedures and methods of sample collection, preparation and analysis,
- provide assurance about reliability of analyses using replicate samples, cross-laboratory checks and relevant reference standards,
- provide assurance about the precision and accuracy from duplicate samples,
- provide assurance about the accuracy from using recognised reference standards,
- provide reliable information regarding the interpretation of data about the behaviour of the mine water during time (longevity of mine water discharge), and
- provide a chain of custody of samples

Field sampling is the first step where mistakes can be made; mistakes made at this stage cannot be corrected or often even identified during any of the later stages. Sampling and monitoring procedures must always be precisely described in your field or laboratory books, which books must have properly bound sheets (adhesive binding should be avoided), numbered pages, and no missing pages (the removal of single sheets is against proper QC methods). Only field or laboratory observations and numbers should be written within the book. If a calculation is conducted, it should clearly be noted as SC ("side calculation") or, in Germany, NR ("Nebenrechnung"). Nothing is allowed to be erased or made unrecognizable; errors should be crossed out with a single line so that the wrong number can still be read. This guarantees that the potential reason for the faulty data can be reconstructed (e.g. 7.45 instead of 7.54; or 27 °C instead of 27 °F). In addition to the measured parameters, "soft" parameters have to be written down, too. These include observations that seem to be abnormal, e.g. animal interference, unusually high or low turbidity, ice covers, or any other observation that might be of interest to co-workers. In the future, electronic field books (e.g. PDAs) connected to GPS/GLONASS/GALILEO-systems will replace traditional data recording (e.g.

Ackman 2003). A QA/QC plan should ensure that this type of data storage is also covered.

Sampling points and samples must be numbered precisely throughout the whole project. It is advisable that all researchers and authorities involved agree to use the same numbering system, even though such systems seem to be complicated (e.g. Mueller et al. 1997). Each sample must be labelled with the name of the project, sampling site, sampling date, responsible research institution, and the person who took the sample. Sample numbers containing a three-letter project code, the date, and the sampling site provide a means to exclude mistakes during sample preparation (e.g. FDR-1202-MP1: Project code-short date-sampling point). Optionally, the laboratory can provide bottles pre-labelled with barcodes. In those cases, the associated barcode number has to be recorded in the field book or PDA.

A well-based QA/QC protocol will also lay out the proper analytical methods and equipment for the catchment study. Project errors often arise from using the wrong equipment or calibration methods. Chemicals or standard solutions used for equipment calibration or during analytical procedures must be fresh; you should regularly check the lifetime of the chemicals. From time to time, blanks should be included in the analytical process to check for contamination or overlapping. Field personnel should measure standard solutions from time to time during the day to ensure that their equipment is still working properly.

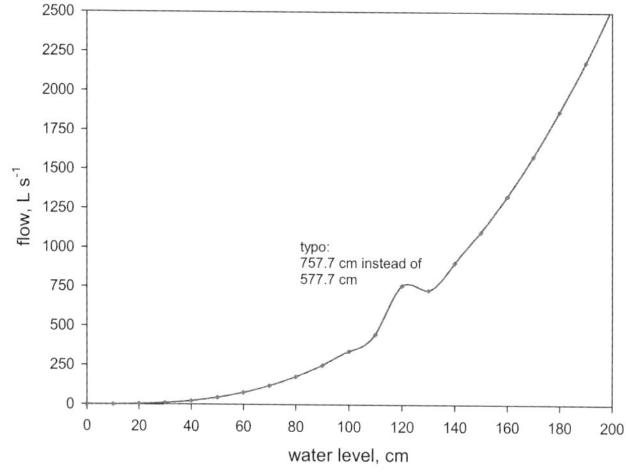

Fig. 85. Example of a scatter plot that would have revealed a typo in a flow/water level calibration table. The table with the wrong data was used for more than six years by a huge German remediation agency without anybody recognising the typo at the 120 cm water level.

A first quality control check is often very easy and can separate "real" outliers, which are mostly typos instead of wrong data. Finding such "simple" errors in data bases of companies that received a lot of money for their work is really annoying, considering how easily such errors could be avoided! One prominent Eastern German remediation agency used a spreadsheet for calculating the flow of a stream and for six years, no-one wondered about extreme flow situations at special water heights in this stream. Those extreme flow situations were due to just one typo in their calibration table! A quick scatter plot would have quickly eliminated this source of errors at its first appearance (Fig. 85).

Quality control thus includes a quick review of current and recent data. When the data has been inconsistent, it is important to ask why. Many errors can be avoided if every person involved in the data gathering and storage chain just followed this simple step. Such simple comparisons do not require a scientific background since they just identify errors or outliers. You should use scatter plots, bar plots, or time series to identify outliers (Fig. 85). Furthermore, at every step, it is important that the data can be traced back through time in a "chain of custody," answering the following questions:

- When, where, and by whom has the sample been taken?
- How, where, and how long has the sample been stored?
- How has the sample been treated and analysed?
- Who handled the sample where and at which time?
- Where have the potential left-overs been stored?

Finally, "it is rather unfortunate that QA/QC reporting is somewhat boring and not readily appreciated by many; however, it is a necessary part of data analysis" (Downing and Mills 1998).

10.7.2 Data Storage and Management

Before a project begins, the project leader has to make sure that a proper data base and data retrieval system is set up and used by all persons, companies, and laboratories working on the project. Many tools are available to assist the project leader in this work, but the data structure and the type of data that shall be stored must be compiled by him, his colleagues, or a working group. For large projects, professional support is available. In Germany, Pangaea, hosted by the Alfred-Wegner-Institute in Bremerhaven, can be used for the management and the storage of geo-projects. They guarantee the persistent storage of the data and they also provide support in setting up the sampling scheme as well as the data structure. In the case of a large and long-lasting project, data storage can be a major task in the project and money should be allocated for this task.

Data storage and retrieval must be given special consideration, because storage systems change quickly and the half-life of an electronic storage technology is in the range of several years, compared to paper storage with half-lives of many decades to centuries. This fact is true for hardware as well as for software. First, widely available computer systems used 9″ disks, which can rarely be read with modern computer systems today, and software produced 10 years ago might have produced data sets that do not produce more than hieroglyphs nowadays. Yet, data storage is not all. Data without a consistent description of the data – if possible in metadata sets – is essential if the data shall be stored for a long time.

Though professional archiving technologies are available that guarantee regular upgrade of the datasets, they are rarely used in mine water research. A discussion of such technologies is far beyond the scope of this book, and as it is very costly, SMEs or authorities are not usually connected to such systems. Therefore, if long life storage is to be achieved, printed reports, tables, and illustrations are the first choice of data storage; from the view of persistency, Hethitian clay tablets and Egyptian hieroglyphs cannot be beaten, yet their storage and handling is rather strenuous and not the means to be used nowadays. Also, ASCII tables should be produced and stored on floppy-discs or CD-ROMs at the end of each project, because such systems are widely used compared to other systems, which guarantees a longer half-life than other electronic storage systems (e.g. streamer tapes, video-cassettes, hard disks). Furthermore, if speaking about proper data storage, a spreadsheet is not a database! It can be used to make quick calculations or draft graphs, but cannot substitute for a real database system – though the reality of consultant work seems to favor it.

Data security (data backup, data encryption) is also important. Many small companies or scientists do not backup their data on a regular basis, because they often think “what shall happen?”. The fact is that all data storage systems can fail to work at any time, no matter what data storage system is used (even Egyptian steles have been reshaped and recycled). Therefore, data backup is an integral part of any mine water and river catchment study and should be included in the QA/QC protocols. Proper data backup has to be conducted daily, with full data backup on a weekly basis. Every month, a copy of a full backup should be securely stored at a place outside the office. All those actions seem to be annoying and time consuming, but from experience, data retrieval after a system crash can take several days and sometimes, data are lost forever, with unforeseeable consequences to the project or the company.

During the writing of a report or after changing a database, it is good practice to add a two or three digit number to the report’s file name (e.g. MINEWATER_01.DOC). This number will then be increased each time the report is stored to the storage system (e.g. MINEWATER_02.DOC). In case a file is corrupted, the previous version can be used instead. At the end of the project, the last file can be renamed (e.g. MINEWATER.DOC) and all the old versions can be deleted (professional desktop publishing systems, such as QuarkXPress™, use a similar system for data security).

Yet, if river catchment projects involve many SMEs, research organizations and authorities, all parties must agree to a common data storage procedure *before* the project starts. Internet or intranet based groupware is commonly used by larger projects. All data should be stored with the corresponding coordinates (keep in mind that GPS/GLONASS/GALILEO data must be recorded with the reference system and the geoid used!), to be usable in GIS-systems. Several countries or authorities provide data storage systems; using such systems allows data exchange between all of the parties involved. Furthermore, since spreadsheets are not databases, data retrieval can be restricted or impossible. In a large project, spreadsheets should only be used for quick reference, and should be organised to provide as much information as possible using as few as possible spreadsheets. A good database system, though its use is more complicated, is always preferred over spreadsheets (although, of course, databases are organised in spreadsheets).

11 Mine Water Treatment and Ground Water Protection

11.1 Introduction

This book covers most of the relevant aspects of mine water management during or after a mine's closure. Because mine water treatment is an essential aspect of mine closure, this chapter will discuss the basic aspects of mine water treatment. However, since 2002, three comprehensive books about mine water treatment have been published (Brown et al. 2002; Younger et al. 2002; Lottermoser 2003) and many research groups have published papers on single treatment options and describe in detail how they can be implemented under special climatic conditions. Therefore, this chapter will only deal with the fundamentals of mine water treatment. After you have read this chapter, you should be able to understand why mine water treatment is necessary, which different mine water treatment options are possible, and how the processes work.

Mine water treatment is an essential task for all mines where polluted mine drainage exists. However, pollution is not an absolute term. Legally non-polluted mine water at one site might be heavily contaminated under the conditions of another mine with more stringent legal requirements. Many countries have limits or regulations that are based on taste and toxicity considerations and background concentrations, but most countries establish site specific discharge limits. Some countries have defined limits that require mine operators to treat their discharge water to values cleaner than the natural background. According to the European Water Framework Directive, future limits must be based on a watershed perspective, but such approaches are not compulsory until 2012 (European Commission 2000; Kroll et al. 2002a; ERMITE Consortium et al. 2004). Though this approach is common to several EU member states, such as the United Kingdom and France, most of the other member states had to significantly modify their water laws (e.g. Germany, Austria; Bongaerts 2002; Timmermann and Langaas 2004).

Areas affected by contaminated mine water are often damaged for many decades, if not centuries. Because the treatment of polluted mine water is usually expensive, active treatment is usually used in heavily populated areas, at working mines, or where governmental money was made available for treatment purposes. The European Water Framework Directive dictates that all European river catchments must have a good ecological and chemical status by the end of the year 2015. It is not clear how that will be reconciled with extreme environments, such as the Spanish Rio Tinto area, which has been heavily polluted by mining activities during the past millennia, and now hosts about 1300 different forms or microorganisms and a macrophyte flora that is unique and well adapted to that extremely acidic environment (Salkield 1987; Ariza 1998; Leblanc et al. 2000).

Conventional chemical or "active" treatment is a standard procedure that – at least in principle – should be known by most mine water researchers. This is not necessarily the case for passive treatment options. However, where feasible, passive treatment technology is generally less expensive. A key reason for conducting the tracer tests described earlier in this book is to provide sufficient understanding of a mine's hydrology, before flooding, so that modifications can be made to lower pollutant discharges. Lowering discharge quantities (and loadings) normally decrease water treatment costs, and often allow passive treatment options to be employed. For all of these reasons, some of the passive treatment options are described in more detail in this book than the more conventional active ones.

If not otherwise specified, the following sections are based on Brown et al. (2002), Younger et al. (2002), and Lottermoser (2003). Most modern active treatment plants use some version of the HDS method (Kostenbader and Haines 1970), though promising investigations are under way to remove sulphate from mine water by a coupled chemical-biological process (e.g. Eloff et al. 2003; Bowell 2004; Maree et al. 2004a, 2004b, 2004c, 2004d). Bowell et al. (1997) have reviewed other sulphate removal options. Several software packages are available to help the researcher or the engineer in deciding which treatment option would be best, including AMDTreat (Office of Surface Mining Reclamation and Enforcement, Pennsylvania Department of Environmental Protection, West Virginia Department of Environmental Protection: McKenzie 2005) and AMDSelect (HERO Research Team: Kruse and Younger 2005).

In the passive treatment field, many different methods are being investigated at sites around the world. Regrettably, there is a tendency to ignore the positive results of other research institutions as can be seen from the literature cited by different authors – language barriers being only one problem. Nearly every year, a new or adapted method is published. This shows that in contrast to conventional treatment, passive treatment is still under development. So far, only passive treatment methods for net alkaline, ferruginous waters can be considered as well settled (Paul Younger, pers. comm.).

Irrespective of which mine water treatment option is chosen, the prescriptive limits set by the authorities must be met at all times. This is usually not a problem for active treatment plants where the discharge and pH are continuously monitored. The amount of chemicals added to the polluted mine water is adjusted, based on the readings. In the case of passive treatment facilities, the situation is different. No control unit is available that would speed up microbial activity where necessary, increase solar input, or reduce the flow velocity to increase retention times. Regulators, therefore, should use slightly different approaches when approving passive treatment plants. Beside a passive treatment concentration limit for the parameters that have to be met, a load limit should be applied as well. Examples for such a procedure are the limits to uranium and radium at the abandoned Niederschlema/Alberoda/Germany uranium mine. In addition to the radium and uranium concentration limits of 800 mBq L^{-1} and 2.4 mg L^{-1}, respecitively, the authorities established mass load-based annual limits of 2 GBq for radium and 20 GBq for uranium (values for 1991–1993; Wismut GmbH 1994).

11.2 Physical Treatment of Inert Solids and Oil

Though most people view mine water treatment as the removal or reduction of chemical constituents, mine water treatment includes physical treatment as well. The first mine water treatment schemes consisted only of physical treatment to remove sediment or coarse material from mine water (Kegel 1950). The characteristic yellow to red precipitates that stain mine water are iron oxihydroxides, commonly referred to as ochre, or yellow boy. Their chemical and mineralogical composition is complex; the term iron oxihydroxides comprises, in reality, a list of different mineral species (Bigham 1994; Schwertmann et al. 1995; Fitzpatrick and Self 1997). The standard methods to remove suspended contaminants are settling ponds, oil/water separators, and lamella clarifiers, often supported by flocculants and coagulants (Skousen et al. 2000). Though flocculation and coagulation can be seen as physical separation processes, the processes leading to flocculation and coagulation are chemically-based surface reactions between the coagulant or flocculant and the metal oxihydroxides (Stumm and Morgan 1996).

Nearly all mine waters contain considerable amounts of suspended matter in different grain sizes: clay, silt, sand, and gravel. Their transportation and sedimentation in the flowing water is a function of the flow velocity, in which the sedimentation is expressed by the Stokes equation and the transport according to the investigations of Hjulström and Sundborg (Hjulström 1935; Sundborg 1956; Selley 2000). Quickly flowing mine water can therefore transport coarser particles than slowly flowing water. According to Hjulström (1935), a 1 mm particle will be transported by water with a velocity of 4 m min^{-1} and a 0.1 mm particle by a velocity of 0.6 m min^{-1}. Therefore, if the mean velocity at the point of discharge is known, the maximum particle size can be estimated and the proper separation technique be chosen (Fig. 86).

It is important that suspended solids, colloids, or floating material be removed before or after the treatment of the chemical pollutants in the mine water. There are standard methods for this that were established in the wastewater treatment field and have been adapted to the special requirements of mine waters (Kegel 1950; Droste 1997).

Settling or sedimentation ponds are the most commonly used approach to remove solid particles from mine water. However, they are only suitable for relatively coarse particles that can simply be separated by gravity. Colloidal particles typically require flocculants or coagulants, which cause them to form larger particles that can then settle in a sedimentation pond. The appropriate size for settling ponds depends on the design flow rate, Q_d, and the average size of the suspended solids. Linsley et al. (1988) and Shaw (1994) give details on how to calculate the design flow rate for treatment facilities. Younger et al. (2002), assuming average sizes of 4 μm, and a retention time of 3–4 hours, suggest the following sizing equation for solid settlement without the addition of flocculants:

$$A_{SP} = f_{SP} \times Q_d \quad (82)$$

with A_{SP} area of settling pond, m^2
f_{SP} sizing factor for settling pond, $1 \cdot 10^5$ s m^{-1}
Q_d design flow rate, m^3 s^{-1}

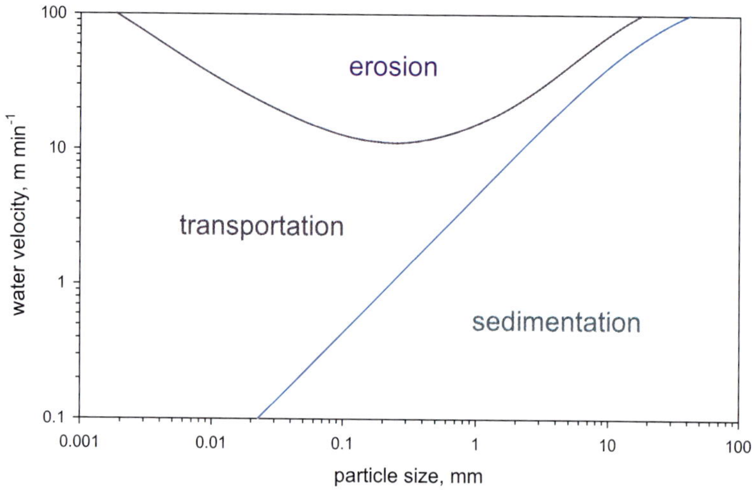

Fig. 86. A Hjulström diagram, relating the areas of sediment erosion, transportation, and sedimentation to the water's flow velocity (Hjulström 1935); the diagram is simplified for the mean particle sizes and velocities expected in the mine water field. None of the lines are absolute, as the transportation/erosion or transportation/sedimentation properties of a single sediment particle depend not just on the velocity of the water but also on the particle's charge, its interaction with other particles, and laminar/turbulent flow conditions. The original transportation/erosion line has an especially broad transition zone.

Kegel (1950) suggests using sedimentation ponds with a length of 15–20 m, a depth of 1.5–2.5 m, and a width that can be calculated from the area calculation in equation 82.

Lamella clarifiers (plate settlers) or tube settlers are more sophisticated than settling ponds. Lamella clarifiers consist of a set of inclined PVC panels, which slope from the bottom to the top of a clarifier (Fig. 87). That way, the flow velocity as well as the Reynold's number are decreased and a turbulent re-suspension of the sludge is prevented. The resultant movement vector of the particles' flow directs them to the lower side of a single lamella from where they slide down into a collecting chamber (Droste 1997). Plate or tube settlers save space, and are either pre-manufactured or assembled according to the special characteristics of the mine water. At the Horden/UK mine water treatment plant, three lamella clarifiers can potentially treat 180 $m^3 h^{-1}$ each. In a normal installation, a 4–5 m^2 lamella clarifier, depending on the particle size, replaces a settlement pond or lagoon of 100–300 m^2 (Siltbuster/Unipure Europe Ltd. Product Information). Kraut and Bondareva (2000) showed that even for rainwater, lamella clarifiers significantly increased treatment effectiveness, and that the addition of flocculants further enhanced treatment performance.

Mine water can also be contaminated with liquid hydrocarbons, namely petrol or oil (e.g. Rätsep and Liblik 1998; Tiwary 2001), which can interfere with the treatment process. At sites where significant amounts of oil are in the mine water, oil/water separators (interceptors) can be used to separate the floating oil from the mine water. If the hydrocarbons, for example phenols,

Fig. 87. Lamella clarifier at the Horden/United Kingdom HDS mine water treatment plant (photograph courtesy Andrea Berger).

are water soluble, they cannot be treated in this way but must be removed by aeration. Experience has shown that the hydrocarbon concentration in the discharged mine water rapidly decreases after mining ceases (Rätsep and Liblik 1998). Therefore, the need for oil/water separators is usually temporary. Oil separators must be installed so that the hydrocarbons cannot explode. Staff cleaning the oil separators must be advised that the residues can be flammable and must be handled with care. Furthermore, the type of separator has to be chosen based on the water quality requirements set up by the relevant authorities and in accordance with the valid directives (Environment Agency 1997).

11.3 Active Treatment

Active treatment plants aim at removing the pollutants in the mine water by physical processes (involving filters or membranes) or by chemical processes. In the case of distillation or solvent extraction, which are not used for large-scale mine water treatment, other physical processes are involved. Active treatment means that human action is required to keep the treatment process running, along with either or both energy or chemicals. Consequently, the term "active treatment" is not normally used if, for example, mine water is pumped from a mine into a constructed wetland.

In most active treatment plants, chemicals are used to adjust the pH or E_H value of the water to reduce the solubility of the metals polluting the water. If the mine water is acidic, which is the most problematic case, reactants are added that neutralize the water and increase the pH to a value that guarantees that the relevant metals precipitate. Some of the commonly used chemicals are caustic lime, sodium hydroxide, and limestone (Skousen et al. 2000). Barium chloride is sometimes added to the mine water to reduce the sulphate content of mine waters (Bosman et al. 1990; Adlem et al. 1994; Maree et al. 2004a) and barium sulphate can be added to eliminate radium (Skeaff and Campbell 1980; Georgescu et al. 1995; Stoica et al. 1998). These chemicals raise the pH of the mine water and simultaneously decrease the solubility of the target pollutants. Thus, precipitates and co-precipitates form, which settle from solution (sometimes after adding flocculants and/or coagulants; Georgaki et al. 2004) in settlement lagoons, lamella clarifiers, or radial flow clarifiers (Fig. 88). Finally, the precipitates are dried and disposed of, either in existing underground mines (e.g. Straßberg/Harz Mountains, "Grube Einheit" Elbingerode/Harz Mountains) or stored in toxic waste deposits, depending on the toxicity of the dried precipitates. Two types of such conventional active treatment processes can be distinguished: low density sludge (LDS) with several percents of solids and high density sludge (HDS) with a 5–10 fold higher sludge density than LDS plants.

Traditionally, mine water treatment is a chemical-physical process. The first chemical water treatment systems were invented by R. A. Henry, a Belgian, and by the German Niersverband in the early 1930s (Kegel 1950). Henry developed a method to treat slurry water from coal-washing plants by adding a flocculant (he used potato starch). He further noticed that alkaline conditions with an optimum at pH 11 and frozen potato starch significantly increased the settling speed of the particles. The first descriptions to use the Henry technique for water treatment in mining operations were published in 1934 (Anonymous 1934; Goette 1934) and the first full scale water treatment plant with the Birtley-Henry treatment process went into operation at the Wallsend Rising Sun pit in Newcastle-upon-Tyne in 1935 (Anonymous 1935). In all cases, the Henry process was used to treat water from a coal cleaning process with 265 mg of quick lime and 6.5 mg of frozen starch per kilogram of slurry water.

Nowadays, such conventional LDS treatment systems, simplified, consist of a reactor where the water reacts with the chemicals, a mixing installation, and a settlement pond. As long as a mine is in operation or somebody is willing to pay the treatment costs, such conventional treatment systems reliably treat the water. Yet, because the reaction kinetics are slow, CaO or other added alkali is sometimes wasted and can be found in the sludge.

In contrast to the LDS process, the HDS process recirculates part of the sludge within the system. Thus, crystallisation "embryos" (or "seeds") are provided at the point where the mine water is mixed with the alkaline material and the recirculated sludge; consequently, the colloids form quicker and reach higher solids/water ratios. The method was first developed by Kostenbader and Haines (1970) to treat $34 \cdot 10^3$ m^3 of mine water at the Bethlem mines of the Bethlem's Ellsworth and Cambria Divisions/USA. Conven-

Fig. 88. Radial flow clarifier of an active mine water treatment plant at As Pontes lignite mine (Spain).

tional LDS treatment did not fulfil the state limits nor was there enough storage capacity for conventional treatment sludge. The aim of their investigation was to improve the settling characteristics of the sludges and the concentration of the solids. They found that by recirculating the sludge, they were able to increase the solid contents from about 1% to 30–40%.

Other large-scale mine water treatment options include electrodialysis, membrane filtering, and membrane filtering combined with distillation. Frequently, such treatment options are used where unusual economic or environmental pressures are involved. For example, Polish coal mine waters are often highly saline with chloride and sulphate concentrations (up to 42 g L^{-1}), and therefore cannot be directly discharged into receiving streams (Gisman and Szczypa 1982; Turek and Gonet 1996). One of the first Polish mine water treatment plants treating saline mine water was constructed at the Dębieńsko coal mine in 1975 (Motyka and Skibinski 1982). Today, this plant produces process water, drinking water, and solid salts for the chemical industry, employing membrane and electrodialyses techniques (Turek et al. 1995; Turek and Gonet 1996).

Over-treatment of mine water is seldom described in the literature and most authorities and legislators are more concerned about low metal and semi-metal concentrations than the potential hazards caused by over-treatment. Over-treatment occurs when an excess amount of alkaline chemicals is used, due to extremely low regulatory limits or to malfunctioning controls. In both cases, this results in a highly alkaline receiving stream, which might pose a new threat to the receiving water course and the catchment area. Because many treatment plants operate in acidic environments, over-saturation is usually of no concern, because the acidic stream is then treated as well. Yet, as in the case of the Straßberg/Harz treatment plant, over-treatment can cause a severe threat to well-buffered circum-neutral fish water streams. Miller et al. (2005) describe a treatment plant in West Virginia/USA, in which the manganese concentration of the water had to be lowered to 1 mg L^{-1} and the pH consequently rose to 8.8–9. Though the manganese concentrations were kept low, the high pH of the water was not suitable for trout. On the other hand, over-treatment can be advantageous when the receiving stream has a low buffer potential and the surrounding area has been affected by acidic mine water (West Virginia Division of Environmental Protection – Office of Mining and Reclamation 2001).

Bogner and Doehler (1984) described the formation of ettringite ($Ca_6Al_2[OH_4|SO_4]_3 \cdot 24\ H_2O$), an end member of the ettringit-metavoltin group, which they attributed to over-treatment of mine water. Drainage water at a coal mine near Kingwood/West Virginia with a pH of 2 was treated by sodium hydroxide (NaOH) and was drained into a 2,500 m^2 settling pond. Within that pond, pH ranged from 4 to 11 because of partly inadequate treatment maintenance and rainfall events. Most of the precipitates in the sediment of the pond were amorphous oxihydroxides and, to a lesser degree, gypsum, quartz, kaolinite, calcite, and dolomite. However, ettringite was found in the northwestern part of the pond where the sediment and water pH as well as dissolved Ca and sediment SO_4-concentrations were found to be at a maximum. Ettringite formation at high pH values (around 11) is a common phenomenon within fly ashes and can bind toxic elements, such as Cd, As, and Cr (Loop 2004). It is also used to reduce sulphate concentrations in mine water, e.g. the SAVMIN and CESR treatment processes (Lorax Environmental 2003). Because the ettringite covers the substrate in the receiving stream and forms an insoluble hard layer, biological activity will be minimised, or even excluded. Therefore, ettringite formation can be seen as an indicator of over-treatment of mine water rich in sulphate or Al, because in most circumstances, pH values under 11 can successfully treat iron rich mine waters. However, where it has been used to lower sulphate concentrations, the treated water can be dosed with acid to lower the pH below pH 8.5 after the sulphate has been removed as ettringite.

11.4 Passive Treatment

11.4.1 Introduction

Several hundreds of passive treatment systems have been constructed during the last two decades (Wolkersdorfer and Younger 2002; PIRAMID Consortium 2003), and a recent evaluation of 83 such systems proved that most of those systems work well (Ziemkiewicz et al. 2003). Their functioning relies on natural chemical and biological reactions, which occur without adding nutrients other than compost or energy (systems where nutrients or energy are repeatedly added after construction are sometimes called enhanced passive or semi-passive treatment systems). The operational and maintenance costs for passive systems are comparatively low, though the investment costs in developed countries are often higher than for active systems, due to the large amounts of land required. Passive prevention or remediation techniques, such as dry or wet covers or reducing gas mixtures, are also sometimes referred to as passive treatment systems (Elander et al. 1998; Skousen et al. 2000; Taylor and Waring 2001; Integrated Geotechnical Engineering Services Specialists in Unsaturated Zone Hydrology 2003). For the PIRAMID consortium (PIRAMID Consortium 2003), passive remediation is:

> an engineering intervention which prevents, diminishes and/or treats polluted waters at source, using only naturally available energy sources (such as topographical gradient, microbial metabolic energy, photosyn-

thesis and chemical energy), and which requires only infrequent (albeit regular) maintenance to operate successfully over its design life.

Passive mine water treatment requires an intensive understanding of mine water chemistry, as will be shown later. As has been argued by Kalin (2004), who takes up the cudgel for a more intensive understanding of bacteriological processes in water treatment, "nature's repair mechanisms may be slow but they're thorough. We must find ways to assist and expedite them".

The first reported scientific investigations of passive water treatment options were conducted by the German researcher Käthe Seidel in the 1960s (Seidel 1966). She investigated 10 water plants with respect to their capability to remove contaminants from water and concluded that only *Schoenoplectus lacustris* would be a potential candidate for further investigations (Table 20). The ability to remove aromatic substances, particularly phenols, was an especially unexpected result at that time. One of the reasons *S. lacustries* influenced the water quality in such a positive way was that the plant supplies the water with oxygen. Because of her investigations, a pilot treatment plant was constructed in Urach/Germany to treate sewage waters from different sources. It was observed that whatever pH the sewage water had (between pH 4.5 and 12), the pH of the system was buffered to a pH value between 6.5 and 8.

Table 20. Metal accumulation of water plants in wetlands. Only the "above ground" parts of the plants were analysed. Scientific names adapted to today's names. Metal concentrations in g m^{-2} (Seidel 1966).

Scientific name	Common name	Fe	Cu	Zn	Mn
Schoenoplectus lacustris	Common Club-rush	2.62	16.13	168.00	4,032.00
Carex elata	Tufted sedge	10.34	15.23	171.36	2,638.4
Iris pseudacorus	Yellow flag	3.22	14.14	124.0	947.36
Typha angustifolia	Small reed mace	1.58	6.77	62.92	1,121.76
Glyceria maxima	Reed manna grass	2.70	11.65	151.84	1,218.88
Phragmites australis	Common reed	4.12	18.82	165.76	743.68
Acorus calamus	Sweet flag	1.55	6.56	60.8	612.8
Sparganium erectum	Bur reed	4.61	7.17	97.28	773.12
Myosotis palustris	Water forget-me-not	3.26	3.90	33.28	640.0
Mentha aquatica	Water mint	5.21	14.28	131.04	640.1

Based on the outstanding work of Seidel (1966), the Othfresen/Germany full-scale passive sewage treatment plant was constructed in 1974 and is still in operation. The treatment method was called the "root-zone-method" (e.g. Kickuth 1977; Bucksteeg 1986; Ebeling 1986; Brix 1987a, 1987b; and section 11.4.4).

Twenty years later, between the 1980s and 1990s, the first publications on passive mine water treatment for coal mine drainage appeared (Kleinmann et al. 1983; Girts and Kleinmann 1986; Kleinmann 1990; Cohen and Staub 1992; Wildeman et al. 1993; Hedin et al. 1994a). These developments were not based on the passive treatment of sewage, but on independent observations of water quality improvements in natural or volunteer wetlands that were receiving mine water from abandoned mine sites. Since then, several hundreds of investigations have been conducted and pilot- and full-scale schemes using various treatment options have been constructed, mainly in the Anglo-Saxon world. A well-described case study of a constructed wetland for mine water treatment was Idaho Springs' Big Five Tunnel, which started pilot-scale treatment in 1987 (Wildeman et al. 1993).

Mine drainage from abandoned surface and underground mines can be passively treated in several ways (Table 21). According to Younger et al. (2002), the different treatment technologies can be categorised into inorganic media passive systems (IMPs), wetland-type passive systems (WTPs), and subsurface-flow bacterial sulphate reducing systems (SFBs). Even enhanced passive systems, such as systems using alcohol to support microbial activity, can be categorised hereunder (e.g. Mascher et al. 2004). A fourth group, not discussed in the above-mentioned publications, is phytoremediation (e.g. Chaney et al. 2000). This term is sometimes used to label all treatment types using plants as a means of pollutant reduction; consequently, some wetland-type passive systems fall under this category as well. Because the term comes from the treatment of soils it is – for good reasons – usually avoided by mine water researchers and is mentioned here, and in section 13.4.8, only for completeness. The limitations and shortfalls of that technology for the decontamination of highly polluted soils and waters have been known for a long time (e.g. Ernst 1996, 2004).

The passive treatment of mining lakes (mine pits that have been allowed to fill with water) has been a goal for many researchers but is still an uncertain task, even after years of research (e.g. Peukert et al. 1982; Kaden 1997; Friese et al. 1998; Herzsprung et al. 1998; Rolland et al. 2001; Werner et al. 2001; Morgenstern et al. 2001; Karakas et al. 2003; Uhlmann et al. 2004; Meier et al. 2004). It is a critical issue in the Eastern German lignite mining areas, where enormous problems with acidic mining lakes exist. In addition, several options that extend beyond passive treatment have been investigated, in Germany and elsewhere, involving in-situ treatment of acid mine lakes.

Table 21. Categories of Passive Mine Water Treatment (adapted after Younger et al. [2002] and other sources).

Inorganic Media Passive Systems	Wetland-Type Passive Systems	Subsurface-Flow Bacterial Sulphate Reducing Systems	Phytoremediation
Anoxic limestone drains	Aerobic wetlands	*In-situ* permeable reactive barriers for ground water	Aerobic wetlands
Oxic limestone drains	Compost wetlands	*In-situ* permeable reactive barriers for mine water	Compost wetlands
Closed system Zn removal reactors	Reducing and alkalinity producing systems (RAPS)		Algal mats
Siderite calcite reactors			deep-rooted hybrid poplars
SCOOFI reactors			bioreactors
Pyrolusite process reactors			rhizofiltration, phytoextraction, phytostabilisation, phytofiltration

These include addition of fly ash, limestone, soda or dolomite powder, organics (to foster biological sulphate reduction), and large-scale water management (Geller et al. 1998; Castro and Moore 2000; Grünewald 2001; Bowell 2002; Koschorreck et al. 2002; Loop et al. 2003). So far, Loop et al. (2003), in which the entire lake was partially filled with alkaline fly ash, is the only full-fledged success. However, as discussed in section 5.5.2 open pit lakes are not the focus of this publication and therefore their potential remediation will not be discussed further.

Even today, many regulatory authorities, researchers, and mining companies believe that the different passive treatment systems are interchangeable. Furthermore, they think that whatever mine water drains out of an adit or shaft can be treated with one of these passive systems. If it was only so simple!

The fact is, the systems are definitely not interchangeable and many mine waters simply cannot be treated with any of the passive treatment technologies that now exist. If an iron-rich mine water of high acidity is treated in an aerobic wetland, the iron might be precipitated to a certain degree, but the pH of the water decreases due to the hydrolysis of the iron ion (eq. 83, Stumm and Morgan 1996):

$$Fe^{3+} + 3\ H_2O \leftrightarrow Fe(OH)_3 + 3\ H^+ \tag{83}$$

See section 4.2 for the full reactions

Though the iron is removed from the mine water, the pH decreases until the system fails to work effectively. At sites where the iron does not precipitate to the expected degree, the money and land expended on an aerobic wetland has been wasted. Such examples are then used by the opponents of passive treatment systems to show that passive treatment is not feasible in the real world of mine water treatment.

One of the key questions is how long will a passive systems work until the system becomes ineffective, because the substrate has to be changed or the precipitates have to be removed? Watzlaf et al. (2004) has provided a good summary of the U.S. studies on this topic. For example, properly designed anoxic limestone drains (ALDs; addressed below in section 13.4.3) should theoretically last 50–100 years. Disregarding those that failed because they were not the appropriate technology to use at a site (e.g. relatively high aluminium concentrations in the influent water), they cite some sites that failed after 9–10 years, while another was still working effectively after 11 years. Similar studies of working RAPS (reducing and alkalinity producing systems) demonstrate that while some systems failed after 5–7 years (due to clogging), at least one was still working 11 years after construction. There have been few such investigations outside of the U.S., where most passive systems treating mine water are younger, but it appears that the compost substrate of the anaerobic Quaking Houses treatment facilty has become less effective after approximately 10 years of work (Paul Younger, pers. Comm.).

Because each mine water is as unique as the mine from which it emanates, designing a passive treatment plant requires careful investigation. Flow charts that explain how to select which passive treatment option to use have been published and explained by Wildeman et al. (1993), Hedin et al. (1994a), McCleary and Kepler (1994), Skousen et al. (1998, 2000), PIRAMID Consortium (2003), and Ziemkiewicz et al. (2003), and are summarised in a simplified form in Fig. 89. They should be used after you have collected water quality and flow data for at least a full hydrological cycle. Neglecting this prerequisite might be one of the reasons that some systems do not work as expected (Johnson and Hallberg 2002). Watzlaf et al. (2004) provides a nice overview of the various passive treatment options and some sizing and design guidelines. Some of the more commonly considered options are discussed below.

11.4.1 Open Limestone Channels

Open limestone channels are rather simple constructions for treating mine water. These can be ditches filled with limestone or larger, open channels or streams in which limestone blocks of reasonable size are positioned. In both cases, the mine water is allowed to flow freely through or over the limestone, thus raising the pH and adding alkalinity to the acidic mine water. To prevent

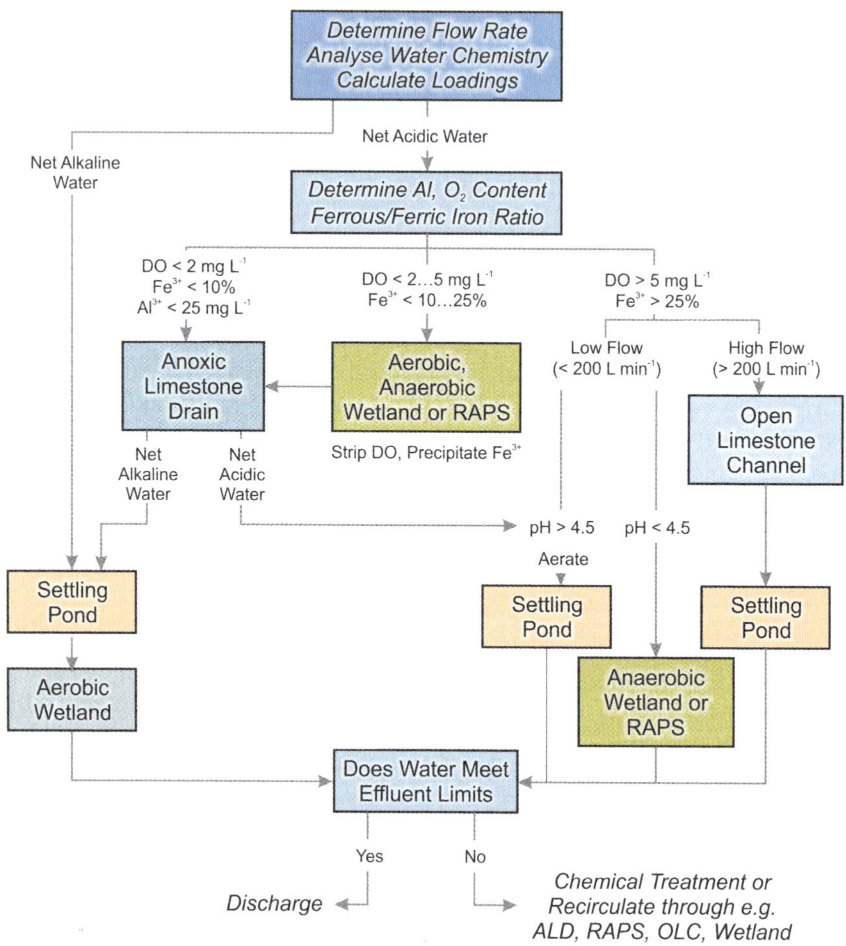

Fig. 89. Flow chart for selecting the most appropriate passive mine water treatment technology depending on the mine water parameters (DO: dissolved oxygen, RAPS: reducing and alkalinity producing system, ALD: anoxic limestone drain, OLC: open limestone channel; modified after Hedin et al. 1994a and Skousen et al. 1998, 2000; a more sophisticated flow chart is given in PIRAMID Consortium 2003).

Table 22. On-site parameters of the Mina de Campanema/Minas Gerais settling pond and open limestone channel. Inflow 1 is not influenced by the mine, while inflow 2 comes from the mine site. The water then flows from the outflow of the pond through the open limestone channel; electr. cond.: electrical conductivity; TDS: total dissoved solids (calculated).

	Inflow 1	Inflow 2	pond	Outflow pond	Outflow limestone channel
pH	5.0	5.5	7.3	6.4	6.2
electr. cond, μS cm^{-1}	8.8	19.5	9.3	55.1	20.3
TDS, mg L^{-1}	5.6	12.4	5.9	35.4	12.8
redox, mV	502	451	432	250	327
temp., °C	25.6	21.9	25.1	24.2	24.4
Fe_{tot}, mg L^{-1}	0.22	–	–	6.53	0.72
Fe^{2+}, mg L^{-1}	0.10	–	–	2.92 (45%)	0.61 (85%)
Fe^{3+}, mg L^{-1}	0.12	–	–	3.61 (55%)	0.11 (15%)
alkalinity, mmol L^{-1}	–	–	–	0.10	0.04
acidity, mmol L^{-1}	–	–	–	0.18	0.10

the metal precipitates from coating the limestone, the water velocity must be high enough to keep the precipitates in suspension and to transport them away from the limestone blocks. Yet, if the flow velocity is too high and the amount of limestone used is too low, the contact time will be too short and the treatment effectiveness will be inadequate. For long-term performance, large quantities of limestone are needed and the treatment efficiency must be monitored on a regularly basis.

However, there are many sites where open limestone channels have not been effective, perhaps because the mine waters being treated had too challenging a chemical composition (e.g. Gernrode/Harz; Simon 2003). At the Mina de Campanema/Minas Gerais/Brazil, a several decametre long open limestone channel was constructed to treat the acidic mine water after flowing through a large settling pond. As can be seen from table 22, iron precipitated during the flow through the channel, but the alkalinity did not increase and the pH decreased from 6.4 to 6.2 due to iron hydrolysis, which produces three moles of protons per mol of iron.

11.4.3 Anoxic Limestone Drain (ALD)

Anoxic limestone drains are constructed and buried channels filled with high quality limestone (Skousen 1991; Hedin et al. 1994b). They exclude contact with the atmosphere, so that the mine water flowing through the channel is not oxidised, which would lead to the precipitation of metal oxides and a coating of the reactive limestone. It is essential that the mine water that flows into the constructed channel contain negligible dissolved oxygen, Fe^{3+}, and Al, as the limestone would be coated in a relatively short time. Such a situation existed at the Lehesten/Thuringia passive treatment plant, which, consequently failed to work after some years (the plant is described in Janneck and Krüger 1999; Rothenhöfer et al. 2000). Within the covered limestone bed, the mine water reacts with the carbonate to generate alkalinity and to consume acidity. Usually, anoxic limestone drains are used to increase the pH and alkalinity of acidic and anoxic mine water before it enters an aerobic constructed wetland. They are also used in combination with RAPS-systems at highly contaminated mine water sites (Barton and Karathanasis 1999).

Their lifetime is based on the amount and quality of the limestone and ranges between 15 and 20 years. Mine waters with high aluminium concentrations and bad maintenance can cause such systems to fail even after less than a year (Robbins et al. 1999; Watzlaf et al. 2000; Ziemkiewicz et al. 2003). If coating of the limestone is expected, such systems can be equipped with a flushing mechanism, which is supposed to enhance the lifetime of an ALD (Vinci and Schmidt 2001); however, questions have been raised on whether such flushing is effective (Watzlaf et al. 2000, 2004).

The first German passive treatment plant for acid, Al-rich mine water began operating in 1997 (in Lehesten/Thuringia). According to Janneck and Krüger (1999), the pH of the treated water ranged between 4 and 5, the Al between 8 and 11 mg L^{-1}, and the Mn between 2 and 4 mg L^{-1}. Fe, acidity, and oxygen contents have been reported during a workshop (*Wasserbehandlung und Rückstandsentsorgung – Konventionelle und innovative Lösungen*, Chemnitz/Germany, 24 to 26 September 1997) but are missing in the official publications. Fe ranged between 0.1 and 10 mg L^{-1}, acidity ($K_{B8.2}$) between 1.5 and 2 mmol L^{-1} (75–100 mg L^{-1} $CaCO_3$). The pilot plant was constructed at a rivulet downstream of an acid-releasing waste rock pile (Janneck and Krüger 1999; Rothenhöfer et al. 2000), and therefore the water flowing into the ALD was saturated with oxygen. After a period of good treatment results, the system failed to treat the water because the limestone in the ALD became coated with Fe- and Al-precipitates and the flushing mechanism failed to work.

11.4.4 Aerobic Constructed Wetlands

As discussed in section 13.4.1, passive treatment of polluted water with constructed wetlands began in 1974, when the first of such systems was installed in Germany (Brix 1987a, 1987b) to treat domestic sewage from the town of Othfresen in Lower Saxony (Kickuth 1977). That system continues to treat domestic waste water for a population equivalent of 5,100 people as well as industrial water from a sausage factory. More than 600 systems of that kind have been installed around the word, among them at British Steel, British Aerospace, Huntsman Corporation, Zurich Airport, Berlin's Schönefeld Airport, Daimler Benz, Cerestar Corporation and an Exhibition at the 2000 World Expo in Hannover, Germany (Abydoz Environmental, Canada, pers. comm.). Walton-Day (1999) outlined the processes involved in such wetlands.

Aerobic wetlands, or reed beds, are constructed (or natural ponds are planted) with typical wetland plants such as common reed (*Phragmites australis* (Cav.) Trin. Ex Steud.; U.S.: giant reed, German: Schilf), reedmace (*Typha latifolia* L.; U.S.: cattail, German: Rohrkolben), common rush (*Juncus effusus* L.; U.S.: soft rush, German: Binse). Though these plants take up metals and contribute significantly to the polishing process (Batty et al. 2002; Batty and Younger 2002, 2003), that is a side benefit. Primarily, these plants reduce the velocity of the water flowing through the system and prevent shortcuts, thus providing time for the microbially catalysed oxidation, precipitation, and co-precipitation reactions to occur (Brix 2003). Consequently, the plants should not be planted in straight lines, as some engineers have a tendency to do (e.g. the Germany Pöhla/Saxony pilot-scale facility).

Aerobic wetlands can be constructed to treat mine water if the mine water is net alkaline ($K_{A4.3} \geq K_{B8.2}$). Mine waters with a pH lower than approximately 5.5 should never be treated with an aerobic wetland as the hydrolysis reactions will lower the pH even further. To increase the wetland efficiency, the CO_2 of the mine water should be stripped off, and the O_2-content increased. This can be done by cascades, riffles, or falls before the mine water enters the constructed wetland (Novak 1994; PIRAMID Consortium 2003).

A slightly different wetland treatment option, involving free-floating macrophyte and algal systems, has been used at some sites in Canada. In those systems, the plants do not grow within the wetland substrate, but float, in anchored rafts, on the water surface. Pilot-scale investigations as well as full-scale treatment plants have proved that those systems can be effectively used to treat polluted mine water (e.g. Kalin et al. 1994; Fyson et al. 1997).

Though aerobic wetlands are now very commonly used for passive treatment around the world and the original technology emanated from Germany, this country has only a few such wetlands being used to treat mine water

Table 23. List of German constructed wetlands. Except Winkler (2001) the authors ignore the pioneering European work of Younger. N.N.: not known.

Site	Site Name, mine type	Date Completed	Mine water type	Passive system type [area, m^2]	Typical influent	Typical effluent
1	Lehesten; Thuringia; shale mining	Start: Dez. 1997; (out of work at the moment), Janneck and Krüger (1999); Rothenhöfer et al. (2000)	Net acidic; Al-rich	ALD, settling pond, aerobic cell [130]	pH ≈ 4.5; 13.9 mg L^{-1} Al; 3.43 mg L^{-1} Mn; 1.24 mg L^{-1} Zn	pH = 7–8; 0.035 mg L^{-1} Al; 0.113 mg L^{-1} Mn; 0.042 mg L^{-1} Zn
2	Pöhla; Erzgebirge; uranium	Start: July 1998; Kießig et al. (2004)	N.N.; drift and underground mine	Aerobic and anaerobic pilot system; build in a former concrete settling pond [300]	pH ≈ 6.9; 0.09–0.15 mg L^{-1} U; 2–3 mg L^{-1} As; 4.2–5.3 Bq L^{-1} Ra; 9–11 mg L^{-1} Fe; 0.6–1.1 mg L^{-1} Mn; 5 mg L^{-1} SO_4^{2-}	0.1 mg L^{-1} U; 0.34 mg L^{-1} As; 1.3 Bq L^{-1} Ra; 0.3 mg L^{-1} Fe; 0.3 mg L^{-1} Mn
3	Helmsdorf; uranium tailing	N.N.	Acidic; drainage water from tailings dam	reconstructed natural wetland [N.N.]	N.N.	N.N.

Table 23. continued.

Site	Site Name, mine type	Date Completed	Mine water type	Passive system type [area, m²]	Typical influent	Typical effluent
4	Straßberg; Harz; flourite	Investigated; Winkler (2001)	Slightly net acidic; underground mine	Aerobic and anaerobic pilot system; settling pond, open channel, compost wetland, 2 aerobic cells [1,700]	pH ≈ 6.5; ≈ 2 mg L^{-1} Fe; ≈ 8 mg L^{-1} Mn; ≈ 8 mg L^{-1} F	N.N.
5	Paitzdorf; Saxony; uranium	Under construction; Kießig and Hermann (2000)	N.N.	Anaerobic pilot system [N.N.]	N.N.	N.N.
6	Schlema; uranium	Start: 2002 (experimental); Kunze and Küchler (2003); Gerth et al. (2005)	circum neutral; underground mine	aerobic and anaerobic wetland [4 cells 72]	U 0.8–1.4 mg L^{-1}	U 0.2–1.5 mg L^{-1}
7	Lausitz (Laasower Fließ); lignite	Experimental; Pietsch and Schötz (2004)	Net alkaline	Aerobic, anaerobic wetlands [small]	pH ≈ 7; ≈ 23 mg L^{-1} Fe; ≈ 1 mg L^{-1} Mn; ≈ 500 mg L^{-1} SO_4^{2-}	pH ≈ 6,4; ≈ 1 mg L^{-1} Fe; ≈ 0.5 mg L^{-1} Mn; ≈ 570 mg L^{-1} SO_4^{2-}
8	Lausitz (Grünewalde RL 78); lignite	Experimental; Pietsch and Schötz (2004)	acidic	Modified natural aerobic wetland [100]	pH ≈ 3; ≈ 70 mg L^{-1} Fe; ≈ 1200 mg L^{-1} SO_4^{2-}	pH ≈ 5; ≈ 10 mg L^{-1} Fe; ≈ 300 mg L^{-1} SO_4^{2-}

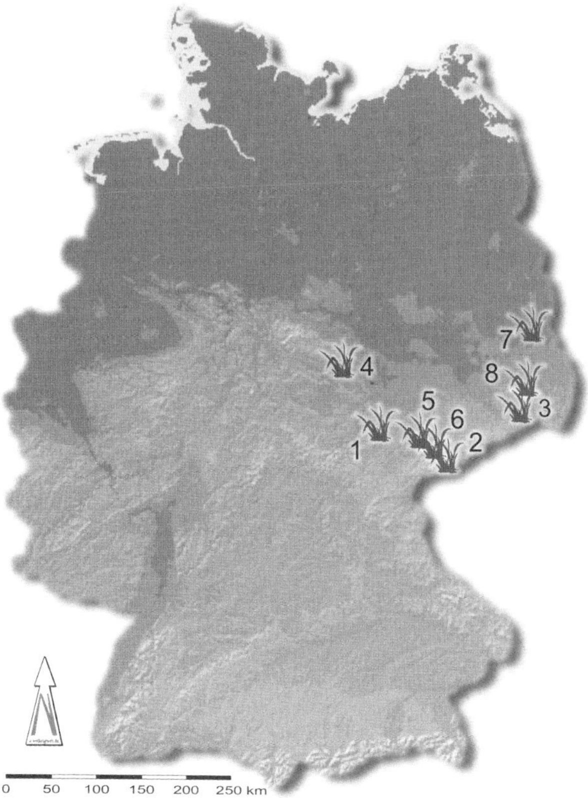

Fig. 90. Sites of constructed or natural wetlands investigated/used for mine water remediation in Germany (mid 2007); details on each site are provided by table 23.

(Table 23; Fig. 90). Aerobic constructed wetlands have proven to be very effective in treating iron-rich mine waters, if the prerequisites summarised in Fig. 89 are met. However, manganese, which very often accompanies high iron concentrations in the mine water, can only be treated in such systems when special conditions apply. To remove manganese in a passive system, the pH of the water must be circum-neutral to basic (Walton-Day 1999). Furthermore, virtually all of the Fe^{2+} must have been oxidized, since Fe^{2+} will reduce any manganese that oxidizes. If both of these conditions exist, an open limestone channel placed at the wetland exit will induce manganese oxidation. The key aspect appears to be substrate surface area in the zone of the air-water interface; once manganese oxidizes and precipitates, the minerals formed help catalyse additional managanese oxidation (Bob Kleinmann, pers. comm.).

It should be noted that the sludge precipitated in aerobic and anaerobic wetlands is usually about 10 times denser than sludges produced at active treatment operations (Dempsey and Jeon 2001).

11.4.5 Compost (Anaerobic) Constructed Wetlands

Compost wetlands (sometimes called anaerobic wetlands or sulphate-reducing bioreactors) comprise a layer of organic material, which should be 0.3–0.6 m thick and have the same type of wetland plants as described for aerobic wetlands. Although compost and aerobic wetlands look very similar to each other, their functioning is fundamentally different. In aerobic wetlands, the water flows above the substrate, while compost wetlands only work when the water flows through the compost layer.

Microorganisms in the substrate and the reducing conditions therein reduce the ferric iron to ferrous iron and reduce the sulphate to sulphide; most metals then precipitate as sulphides. Furthermore, as the protons in the mine water are consumed by the decomposition of the organic material, and H_2S is generated, acidity is removed from the mine water. It is important to note that before significant sulphate reduction occurs, the dissolved oxygen has to be consumed all and the Fe^{3+} must be reduced to Fe^{2+}. This aspect, if ignored can cause an anaerobic system to fail, and why at some sites, aerobic wetlands and sedimentation ponds precede the anaerobic wetlands, to reduce the total iron load as much as possible.

Though the first compost wetlands were constructed with spent mushroom compost (Hedin et al. 1994a) and a lot of research has been conducted using that type of organic material (e.g. Stark et al. 1994, 1996; Manyin et al. 1997; Nairn and Mercer 2000; O'Sullivan et al. 2004), all kinds of organics that release alkalinity into the mine water are suitable for compost wetlands (PIRAMID Consortium 2003), e.g.:

- spent mushroom compost (e.g. Hedin et al. 1994a),
- horse manure and straw (e.g. Younger 1997),
- cow manure and straw (e.g. Cohen and Staub 1992),
- composted municipal waste (e.g. Jarvis and Younger 1999),
- composted conifer bark mulch (e.g. Younger 1998),
- sewage sludge cake (e.g. Laine and Jarvis 2003), and
- paper waste pulp (e.g. Laine and Jarvis 2003)

The first three in this list proved to be better than the latter ones, but whichever carbon source is used, the governing reactions within the substrate are the same.

One important factor of a compost wetland is its permeability. If the permeability of the substrate is too low, the mine water will flow over the substrate rather than through it. If that happens, the microbially catalysed reducing reactions will not occur and the treatment efficiency will decrease. Bolis et al. (1992) found that the hydraulic conductivities of manure in different conditions ranged between 10^{-6} and 10^{-2} cm s^{-1}.

Another important issue is the water depth. If it exceeds approximately 15 cm, little exchange between the water and the substrate occurs, thus reducing the treatment performance (Paul Younger, pers. comm.). Gusek (2002) and Gusek and Wildeman (2002) describe several successful anaerobic wetlands and their advantages, even in cold climates. Furthermore, Gusek (2002) remarks that "a phased design program of laboratory, bench-, and pilot-scale testing has been shown to increase the likelihood of a successful design".

11.4.6 RAPS (Reducing and Alkalinity Producing Systems)

A RAPS (reducing and alkalinity producing systems, originally called SAPS – successive alkalinity producing systems, but also called vertical flow ponds or vertical flow systems) combines the advantages of an anoxic limestone drain (ALD) with those of a compost wetland (Kepler and Mc Cleary 1994; Hedin et al. 1994a). In contrast to classical wetland type systems, the two reactive layers are constructed on top of each other and the water flows vertically through the system. Therefore, they are also referred to as vertical-flow wetlands (Nairn and Mercer 2000). They are designed so that the acidic mine water first flows through the compost part of the RAPS system. In this layer, the dissolved oxygen is consumed and any Fe^{3+} that is present is reduced to Fe^{2+}. Although some sulphate is biologically reduced to sulphide, this is less critical. The key is that the mine water is now anoxic, and can flow through the anoxic layer of limestone and become alkaline without concerns about iron precipitating on the limestone. Mine waters with elevated Al concentrations can still cause problems. Some researchers construct an 'aluminator', which can be used to flush the RAPS and remove, to a certain degree, the Al precipitates but there is some evidence that this is not very effective (Watzlaf et al. 2002, 2004). Currently, about 100 RAPS-systems are installed worldwide and treat acidic mine water of different quality (Edwards et al. 1997; Younger et al. 2002; Brown et al. 2002; Wolkersdorfer and Younger 2002; Rose and Dietz 2002; Rose 2004; Thomas et al. 2004).

Because the chemical reactions within the limestone layer of a RAPS are similar to those of an ALD, the amount of limestone that is needed in a RAPS is calculated in the same way. Due to hydraulic reasons, the thickness of the limestone as well as the compost layer should be around 0.5 m each. Details about designing such systems can be found in the relevant passive treatment literature.

During the installation of the Bowden Close/UK pilot RAPS system (Younger 2002c; Younger et al. 2004), it happened that the limestone and the organic material were mixed with each other (Paul Younger, pers. comm.). Though it was thought that this would negatively affect the performance of the RAPS, the system actually performed better than expected. Mixing of limestone and organic material cannot be excluded when large machinery is used, but this study indicated that there is no certain need to keep the substrates separated. Thomas and Romanek (2002) also showed that such mixing can be advantageous for some types of mine water. In the Gernrode/ Germany experimental RAPS, the limestone drain was only used to protect the drainage pipes from being blocked by precipitates, but above this protective limestone layer, the manure and limestone were intentionally mixed (Simon 2003; Hasche and Wolkersdorfer 2004).

When constructing a RAPS, it is essential to avoid any shortcuts between the inflow of the mine water and the outflow, as this might result in a malfunctioning of the system as well as metal precipitation on the limestone surfaces. Furthermore, there must be sufficient head to force the water to flow through a low hydraulic conductivity layer of compost. If the head is not great enough, as in the Pelenna II (Garth Tonmawr/Wales/UK) system, then the water overflows the RAPS and the treatment efficiency equals zero (Rees and Connelly 2003; Fig. 91).

A recent innovation was the construction of a RAPS in Deerplay/Lancashire/UK where the two reacting components, limestone and organic material, are neither built over each other nor mixed with each other, but constructed side by side. First, the acidic, iron and aluminium-rich mine water flows through organic substrate and then into a limestone bed, which is separated from the organic substrate by an impermeable barrier. To prevent air from entering the system, the organic reactor is isolated from the atmosphere by an impermeable cover (Jarvis and England 2002). An interesting result was that the nitrogen concentration, which comes from the organic matter, decreased relatively quickly after construction. After an initial peak of 10–20 mg L^{-1} NH_3-N, the concentration decreased to 2–4 mg L^{-1} NH_3-N in only two months, and thereafter decreased to about 2 mg L^{-1}, which was roughly twice the concentration of the influent mine water (Fig. 92).

Since nitrate concentrations are always of concern for regulators and authorities when speaking about RAPS or compost wetlands, the reduction of nitrate has been studied experimentally and compared with real case values (Simon 2003; Hasche and Wolkersdorfer 2004). The nitrate concentration mainly depends on the volume of the compost V, an initial nitrate concentration factor f_1, the background concentration c_{t0} in the mine water, and the time t:

$$c_t = \left(f_1 - 80 \times \frac{Q}{V} \right) \times e^{-\frac{t}{b}} + c_{t0} \tag{84}$$

with c_t nitrate concentration at time t, mg L^{-1}
c_{t0} background nitrate concentration of mine water, mg L^{-1}
f_1 initial nitrate concentration factor, mg L^{-1} (usually 20–40 mg L^{-1})
Q measured mine water discharge, L min^{-1}
V bulk volume of compost, L
t time, min

Fig. 91. Mine water overflowing the Pelenna II/Wales system. The hydraulic head is too low to drive the water through the compost (24 March 2000).

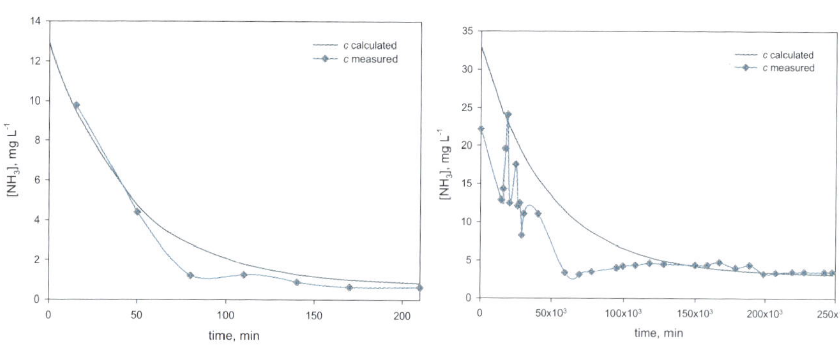

Fig. 92. Comparison of measured and modelled nitrate concentrations in a laboratory experiment (left) and the Deerplay/Lancashire/UK (right) full scale RAPS (Deerplay data from Jarvis and England 2002; modelled data from Simon 2003).

$$b \qquad 100\times\left(\log\frac{Q}{V}-0.2\right)\times\begin{cases}(\log V)^3 & V>1\\ 1 & V\leq 1\end{cases}$$

Only one of those parameters, namely f_1, has to be evaluated in a laboratory test; other parameters can be either measured in the field or determined by the RAPS design. To evaluate this initial nitrate concentration factor, the organic material has to be mixed with the mine water (at a ratio of 3:5) and then the nitrate concentration in the mine water is measured. The factor f_1 is then calculated by multiplying this concentration by 0.7 and rounding the result to the nearest mg L^{-1}. However, to use a real-world example, f_1 was determined to be 33 in the TU Bergakademie Feiberg/Germany laboratory test, while at the Deerplay/Lancashire/UK site, f_1 must have been around 22. The difference was caused by the relative maturation of the organic material. For the laboratory tests, relatively fresh horse manure had to be used, as the supplier did not have composted manure available. It can be assumed that more mature manure was used at the Deerplay RAPS, which caused the initial nitrate concentration factor to be lower than in the laboratory experiments. Consequently, it is possible to show regulators that with time, the nitrate concentrations in the mine water will come down to nearly natural background values.

11.4.7 Reactive Barriers

Reactive "barriers" are subsurface engineering constructions placed in the flow path of polluted water. The material within the barrier, which must be permeable, reacts with the contaminants and decreases their concentration. Different materials, such as zero-valent iron, activated carbon, zeolites, iron oxides/oxihydroxides, phosphates, and clay minerals have been used so far. Reactive barriers have been constructed downstream of tailings and waste rock dumps but several systems haved failed, so their application is not widespread. Yet, several investigations throughout the world give hope that such systems could be useful (e.g. Benner et al. 1997; Younger 2000c; Schneider et al. 2002; Bolzicco et al. 2003; Younger and Elliott 2003; Younger et al. 2003; Jenk et al. 2003; Simon et al. 2003; Gibert et al. 2004; Roehl 2004; Munro et al. 2004).

So far, no reactive barriers have been constructed downstream of a flooded underground mine, though the construction of vertical subsurface limestone drains should be a promising option for large-scale ground water treatment. "Reactive barriers" within a flooded underground mine will be described in section 11.6, along with other in-situ and on-site treatment options.

11.4.8 Phytoremediation

As mentioned earlier, phytoremediation is a commonly used term for all types of treatment technologies that include plants and the uptake of the pollutants into the plant (Cunningham et al. 1997; Adams et al. 2000). Enhancement techniques for the rhizosphere are becoming more and more prominent and enhanced biodegradation technologies will probably result once plant–rhizome interaction is better understood. Generally, the term phytoremediation refers to a way to remediate contaminated soils in which metal tolerant plants are used to take up pollutants (e.g. PHYTO-DEC 5th EU framework project EVK1-1999-00024, EVK1-2002-00552); some promising organisms show extreme accumulation rates (Bryan and Langston 1992). Phytoremediation is also used to stabilise soils and minimise soil erosion (e.g. Cai et al. 1990; Baker et al. 1991; Chaney et al. 2000). During the last decade, attempts have been made to extend phytoremediation to the treatment of polluted mine waters, yet, the technology is still as far away from large-scale implementations as it was 10 years ago, when Wilfried Ernst wrote: "there is still a long way to go from the potential small-scale to a realistic large-scale approach" (Ernst 1996). Especially for mine waters, where large amounts of metals have to be removed from the water, phytoremediation *sensu strictu* cannot be viewed as successful technology – yet, because the methodology has large potentials for remote areas, research should continue to find optimised plant-root-bacteria communities that foster the uptake of potentially poisonous elements from mine water. Merten et al. (2004) reported some promising results for elements in mine water that are toxic at low concentrations. Furthermore, it is an excellent method to increase the biodiversity of abandoned mine sites (Batty 2005).

Three types of phytoremediation are relevant at mine sites:

- *phytoextraction:* (semi-)metal tolerant plants are grown and harvested with the intention to remove the accumulated elements from the soil or water (e.g. Chaney et al. 2000; Karathanasis and Johnson 2003)
- *rhizofiltration or phytofiltration*: (semi-)metals are adsorbed on the surfaces of roots or the whole plant. To remove the contaminants from the environment, the plants are harvested and deposited off (e.g. Dushenkov et al. 1997)
- *phytostabilisation*: (semi-)metal tolerant plants are grown on polluted sites to stabilise the pollutants in the soil and to minimise soil erosion at the contaminated site. Usually, the plants are not harvested (e.g. Stoltz and Greger 2002; Bleeker et al. 2003; Chan et al. 2003; Wong 2003).

The disadvantage of these methods is that the clean-up of contaminated sites might take hundreds if not thousands of years and that annual harvesting as well as the disposal of the contaminated plants or, after their burning, ashes, is imperative.

Because phytostabilisation reduces the amount of erosion on tailings, waste rock dumps, or decommissioned mine sites, this methodology is very promising for the fixation of pollutants within the soil. Trees, depending on their age and type, can transpire between 30–1,500 L of water per day; thus, they take in a large amount of pollutants and can lower the ground water table (Adams et al. 2000). Consequently, the dissolution of metals from the upper soil horizons can be minimised and the transfer chain from the soil to the ground water can be stopped, or at least hindered. Thus, the pollutant load into the receiving water course and the subsequent watershed can be significantly reduced. If it is possible to enhance the metal tolerance of the plants, by manipulating the microorganisms living in symbiosis with the plants, acid draining from the dumps into the ground water might be positively controlled. Such research has been conducted by Quinn et al. (2001), who planted 420 poplars with roots reaching into the contaminated aquifer at a depth of 8–9 m, and another 389 willows and poplars for soil and shallow ground water remediation.

11.5 Natural Attenuation

Natural attenuation of mine water has occurred for as long as people have mined raw materials. It is not a new invention or technology, just a name for a process that occurs even without human interference. Some of the smallest extremophiles up to large macrophytes that provide nutrients or surfaces for microorganisms, as well as filter mechanisms for suspended solids or obstacles in the mine water flow, play an important role in natural attenuation. Ironically, most of the sedimentary raw material deposits and as far as we know many other metal deposits are a result of "natural attenuation" (Pohl 2005). Possibly the first mine water related paper using the expression "natural attenuation" appeared in 1994 (Kwong and Van Stempvoort 1994). They describe the natural reduction of Cu, Zn, and sulphate in acid rock drainage downstream of an open pit in central Vancouver Island/Canada by a natural wetland system. This was clearly not the first work on "natural attenuation" – they were just the first to use that name for that type of remediation in a mining impacted environment. For example, Barnes and Clarke (1964) described the process when they stated that: "It is well known that the acid streams recover in the course of travel as the iron precipitates and the acid is diluted and neutralized by contact with alkaline minerals in the streambed and with diluting waters". As discussed earlier, it was the observation of natural attenuation of mine water contaminants in wetlands that led to the passive treatment of mine water.

From observation, it is known that nature is able to take care of itself and that natural attenuation has a large potential to treat mine water (e.g. Wolkersdorfer et al. 2003; Höglund et al. 2004). If the processes involved are understood, it is possible to help nature to take care of itself or to avoid any interference that might prevent natural attenuation. At the abandoned Straßberg/Harz/Germany fluorspar mine, part of the drainage water flows through a natural wetland, and the mine water chemistry is improved while it flows through the system (Neef 2004). The natural wetland did not exist before the mining at this site started in the 18th century. It is clearly the result of mine water seeping through leaky settling ponds, which were constructed to treat the pumped mine water. In the course of time, an outstanding flora with wild orchids and wetland plants became established; several decades ago, the mining impacted site was declared a nature reserve called "Eisenwäsche", which – ironically – means "iron wash" (Fig. 93). Due to this fact, it was not possible to divert part of the mine water from this wetland as the authorities feared an irreversible impairment of the flora through the treatment of the iron-enriched mine water.

However, where natural attenuation is being relied on nowadays, it typically involves monitoring, and incorporates the potential to shift to enhanced natural attenuation or to one of the above-mentioned treatment processes if the monitoring indicates a severe negative impact to the anthroposphere.

Natural attenuation comprises mainly the following processes (Höglund et al. 2004):

Fig. 93. Natural attenuation occurring at the Straßberg/Harz mine site and the *Eisenwäsche* ("Iron wash") natural wetland. Flocculants of iron hydroxide and a colourful film of iron-hydroxide/bacteria aggregates can be seen in and on the water.

- abiotic
- physical
- biotic

Some typical abiotic processes are the chemical precipitation of iron or aluminium minerals, such as jarosite, schwertmannite, ferrihydrite, siderite, and pyrite. Physical processes might be the filtration of colloids or their settling in a quiescent stream, while biotic processes might be the fixation of contaminants on roots or branches, or by microbiological activities.

An example of natural attenuation is the Buttonwood Tunnel discharge in the Eastern Pennsylvanian Wyoming Basin/USA. This drainage adit was regularly monitored between 1968 and 1980 (Ladwig et al. 1984). The mean sulphate concentration decreased from 2,130 mg L^{-1} in 1968 to 982 mg L^{-1} in 1980, and the minimum and maximum values also decreased continuously (Fig. 94).

Because natural attenuation happens without human interference, there would seem to be no need to discuss natural attenuation concerning mine water issues. Yet, cases of natural attenuation all over the world clearly show that this "treatment" option can be hampered by human activities or in the worst case consultants who want to remediate a polluted mine site, and destroy the natural attenuation process. For example, at the German Gern-

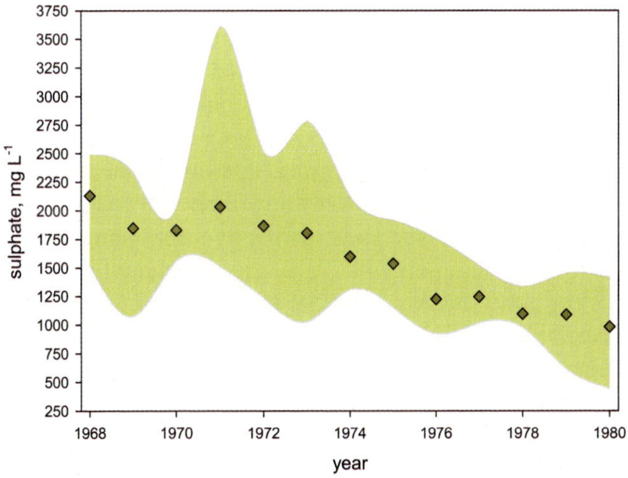

Fig. 94. Natural attenuation at the Buttonwood Tunnel, one of the main discharge adits of the Wyoming Basin in Eastern Pennsylvania/USA. The mean sulphate concentrations (diamonds) as well as the maximum and minimum values (shaded area) decreased continuously between 1968 and 1980 (modified from Ladwig et al. 1984).

rode/Harz Mountains fluorspar mine, polluted mine water discharges from an abandoned adit. A consultant straightened the drainage gully and thus prevented the water from flowing over an already established, natural cascade into a settling pond. The mine water is now directly diverted into a mountain stream without taking advantage of the already established natural attenuation potential of the cascade and the natural settling pond downstream of the cascade. In the meantime (mid 2007) the situation has even worsened.

11.6 In-Situ and At-Source Treatment Methods

Mine water treatment generally involves the treatment of polluted mine water discharges from a mine. Yet, mine operators should try to use "in-situ" or "on-site" techniques to minimise the amount of polluted water. In most instances, it is cheaper to discharge water that does not require treatment, even if that requires diverting, pumping, or otherwise intercepting it. It has already been shown that the prevention of convective loops in flooded underground mines lowers the maximum concentration of pollutants and that by using transparent mine closure techniques, polluted mine water can be either kept in the mine or treated in-situ (e.g. Wolkersdorfer 1996).

In-situ and at-source treatment means that the source of contamination is treated essentially at its origin: within the flooded mine or the acid-producing rock (Pearce and Ries 1982). It is commonly used to preventing or abating the generation of acid rock drainage in overburden material at a surface mine, but has not found wide application in flooded underground mines. Theoretically, nearly every acidic mine water could be converted to alkaline water if only enough alkalinity would be added to the drainage system. Yet, experience demonstrates that the situation is not that simple and further investigations are necessary.

Alkaline material that has so far been used to inhibit the generation of acid mine drainage includes limestone, dolostone, lime kiln dust, coal combustion residues such as fly ash, caustic soda, and active treatment sludges (Waddel et al. 1986; Larew and Skousen 1992; U.S. Environmental Protection Agency et al. 2001). Limestone and dolostone are most commonly used because they are widely available und relatively inexpensive (Prediction Workgroup of the Acid Drainage Technology Initiative 2000). If enough alkaline material is added, the neutralization will be sustainable; as a general rule, positive results have been obtained at remediation projects where the amount of alkalinity added exceeded the calculated requirement, based on acid-base accounting (Perry et al. 1997; Smith and Brady 1998). The calculations result in a net neutralization potential (NNP); enough alkalinity has to be added to generate a positive NNP. In addition, the most suitable place for the alkaline addition has to be found, preferably based on the results of mine water tracer tests, to guarantee good mixing between the alkaline material and the acidic mine water. While the acidity load is positively influenced, this may not be the case for the metal loads and the sulphate concentration. Detailed discussions of different in-situ methods and alkaline addition can be found in The Pennsylvania Department of Environmental Protection (1998), Skousen et al. (1998), Prediction Workgroup of the Acid Drainage Technology Initiative (2000), and U.S. Environmental Protection Agency et al. (2001). Note that the "Alkaline Addition" sections in U.S. Environmental Protection Agency et al. (2001) are nearly identical with Smith and Brady (1998).

Fully flooding a mine, to prevent or limit additional pyrite oxidation, is an in-situ method that is quite effective. Adding alkalinity to underground mines, especially to partially or fully flooded mines, is more complex. For example, how much of the overburden should be included in an acid-base accounting, and how can you add the alkalinity at the right location, or even know where the right location is? Sometimes, it could surpass the storage capacity of the underground space for the deposition of alkaline material. Yet, it should be considered if a controlled backfill with alkaline material before flooding can effectively prevent the first flush or reduce its extent. A cost-benefit analysis of the various options might be a feasible way to find an environmentally sound solution. In addition to alkaline addition, other options that have been investigated in the laboratory or in the field include:

- surface inactivation by encapsulation or coating (Evangelou 1995; Fytas and Bousquet 2002),
- reducing the hydraulic permeability of pollutant bearing rocks (Ziegenbalg 1999; Waring and Taylor 1999; Ziegenbalg 2005),
- injection of chemically reducing gas mixtures (Taylor and Waring 2001; Houston et al. 2005), and
- reactive material ("reactive barriers").

Encapsulation and coating can be conducted with a range of substances, such as Fe-phosphate, silica, sodium silicates, or "simple" but effective clay caps (Brandberg et al. 1993; MEND [Mine Environment Neutral Drainage Program] 1994; Evangelou 1995; Roy and Worral 1999; Fytas and Bousquet 2002; Carlsson 2004). The "oxidation proof silicate surface coating on iron sulfides" method, though still unproven, is patented (U.S. Patent 5,494,703), so it cannot be freely used!

Alkaline material can also be deposited at potentially acid producing areas within the abandoned mine before flooding. For example, Hause and Willison (1986) reported how anticipated water treatment costs were greatly decreased at Beth Energy's underground Mine № 105W near Buckhannon in West Virginia/USA by rockdusting areas of the mine that would not be flooded with 700 tons of ground apatite, and by adding NaOH and lime slurry to the mine pool as it formed. In total, 120 t of lime and 568 m^3 of NaOH were injected between September 1984 and June 1985. Compared to conventional treatment costs, estimated at about \$4.7 million (U.S.), the in-situ method had total costs of only \$1.3 million (U.S.). The in-situ treatment increased the discharge pH from 3.9 to 7.0. Iron concentrations decreased from 133 mg L^{-1} to ± 1 mg L^{-1}, sulphate decreased from 2,500 to 1,000 mg L^{-1}, calcium decreased from 385 to 120 mg L^{-1}, alkalinity rose from zero to 350 mg L^{-1} $CaCO_3$, and acidity decreased from 470 to 20 mg L^{-1} $CaCO_3$ (Hause and Willison 1986). Even in 1989, the pH and acidity of the mine water remained at an acceptable level, although iron concentrations had increased to 15 mg L^{-1} (Aljoe and Hawkins 1993).

However, deciding to add alkalinity after the mine has been abandoned is much more challenging. It is difficult to decide where to remotely add the alkaline material, while ensuring adequate mixing is virtually impossible; there have been many failures. For example, the Keystone State Park site in Pennsylvania/USA was hydrogeologically investigated before alkaline injection. Keystone State Park hosts the abandoned Salem № 2 coal mine, which closed in 1954. A tracer test was conducted but no reliable data could be obtained from the first test, as no tracer arrived at the sampling points (Aljoe and Hawkins 1993). A second test was conducted at a later stage (Aljoe and Hawkins 1994; the mine's name is not reported in this paper, but from the drawings it is obvious that this mine is Salem № 2 as well), which had positive results. In 1990, 1,890 kg of NaOH (as a 25% solution) were injected into two boreholes of the Salem № 2 mine, which were thought to be enough to treat all of the water in the mine. Yet, two months after alkaline injection, only a minor change in the sodium concentration occurred. Calculations of the recovery rate showed that only about 5–6% of the injected sodium mass passed through the drainage portal; the rest stayed in the mine or seeped into the groundwater. The acidity, total iron, manganese, and aluminium levels in the discharge indicated no neutralization occurred after the alkaline addition.

Even more hydrogeological and biological investigations were conducted at the Red Oak coal mine near Red Oak in Oklahoma/USA before the injection of alkaline fluidised bed ash (Canty and Everett 1998, 1999). An extensive tracer test, historical, and GIS-based site investigations were conducted to define the potential flow paths and the hydrochemical development of the acidic mine water within the flooded mine. From old mine maps, a mine volume of $1.6 \cdot 10^5$ m^3 was calculated with the mine pool having a volume of $1.1 \cdot 10^5$ m^3. During the hydrogeological monitoring phase, the mean pH was 4.4, the electrical conductivity was 1.2 mS cm^{-1}, and acidity averaged 475 m L^{-1} $CaCO_3$. Finally, 418 t of fluidised bed ash was injected into the abandoned mine voids through wells using high pressure pumps (1.7–

$2.1 \cdot 10^3$ kPa) at an average rate of 84 m^3 h^{-1} (110 yd^3 h^{-1} in the original report were wrongly given as 28 m^3 h^{-1}). After alkaline injection, the pH and alkalinity increased, and the concentration of Al decreased significantly, while most of the other parameters did not show significant changes. The authors concluded that only some of the goals were reached but that the desired water quality in the receiving stream could have been attained with a greater alkaline addition. Based on their geochemical calculations, an additional 2,000 t of alkaline fluidised bed ash would have been necessary.

At the abandoned Broken Aro coal mine near Coshocton Ohio/USA, fixated flue gas desulphurization sludge was used to create a permeable mine seal. The sludge consisted of 35% calcium oxide, 21% sulphur dioxide, 10% silica and other oxides, and thus contained a lot of potential alkalinity. The mine water quality significantly improved after the installation of the plug, and the contaminant load to the receiving water course was reduced up to 98% (Rudisell et al. 2001). However, the site has to be reclaimed again because the measures taken were not enough to guarantee the ecological quality of the Flint Run watershed. In principle, such mine seals can be considered as reactive barriers. Reactive mine seals were also investigated by Penrose (1974). He investigated different limestone types and sizes for the optimization of a permeable plug mine seal and concluded that the acid discharge can be reduced if such "reactive" mine seals are used.

It is technically feasible to pump the mine water out of a mine, add alkalinity to the water stream, and pump the alkaline water back into the mine. Because low-density sludge from mine water treatment operations contains large amounts of alkalinity, this sludge can also be pumped back into the flooded mine workings. Experimental and modelled results clearly showed an increase in pH and, after an initial mobilization, a demobilization of iron and aluminium (Hansen 1998; Klinger et al. 2000a). The analytical data of the Straßberg/Germany and *Grube Einheit* Elbingerode/Germany mines clearly prove that the re-injection of such sludge does not pose negative environmental consequences to the receiving water courses (as long as the water in the mine is not too acidic); in fact, the alkalinity of the mine water increased (Kubiak 2007). It is possible that stratification in the mine water bodies of the two mines favoured this effect, but this aspect has not yet been investigated. A similar approach is being conducted at the Spreetal Nordost/Germany open pit lake, which receives the LDS sludges from the *Schwarze Pumpe* active treatment plant (discussed later in this section).

Another mine where in-situ alkalinity injection was attempted was the Sherman Helms № 2 mine in West Virginia/USA (Burnett et al. 1995). This mine had a 115 m^3 mine pool with an average flow of 4 L min^{-1}. About 400 t of fluidised bed combustion by-products were pumped pneumatically into four boreholes to improve the mine water quality and to prevent surface subsidence. Yet, the water quality was not affected by the fly ash during that first investigation, as the fly ash did not get into contact with the mine pool, possibly due to a misinterpretation of the hydrodynamics of the mine water. Later, more effective injection techniques were established, and the acidity dropped by 83% (NRCCE – National Research Center for Coal and Energy – News: 2 February 2005).

A more sophisticated in-situ project was conducted at the Tide Mine in Pennsylvania/USA. The project consisted of three phases with different in-situ treatment options involved (Houston et al. 2005):

- addition of sodium hydroxide, a bromide tracer, and an organic carbon source into a recirculation scheme,
- carbon dioxide gas injection into the mine pool to suppress the generation of acid mine drainage
- passive alcohol addition to the mine pool to sustain sulphate reduction

In fact, using the terms in-situ and passive to describe the demonstration study is inaccurate, and the injection of carbon dioxide cannot suppress the generation of acid mine drainage. The chemicals were injected into the mine water on-site and recirculated into the mine pool, which, along with the addition of alcohol into the mine water, are definitely active processes. Furthermore, unless the carbon dioxide displaces the oxygen far above the mine water table, in the unflooded disulphide-bearing strata, it cannot stop the formation of acid mine drainage at its point of generation. Although the initial results of the study seemed to be promising, it appeared that the treatment was not sustainable, as can be seen from the last measurements they made.

Remining is another in-situ technique that can be used to remediate water quality from underground mines. Remining, or daylighting, as it is also called, converts the old underground mine into a surface mine; it improves the water quality when the remaining coal (pillars or coal that could not be economically mined at some time in past) is the major source of acid generation, and where the alkalinity in the overburden exceed the acid generating potential. Additional alkalinity can be added to the overburden, the mine backfill, or the daylighted mine environment (Hawkins 1995). By using acid-base accounting tests, the necessary amount of alkaline material can be calculated (Skousen et al. 2002). Remining, if done properly, can significantly reduce the formation of acid mine drainage.

Other in-situ treatment options have been attempted in the Königstein/Germany uranium mine and other mine sites in Saxony, including the injection of chemicals into an uranium-bearing sandstone to reduce the hydraulic conductivity of the rock matrix, and the use of large tanks filled with different types of reactive materials, such as resins, wood chips, and zero-valent iron (such as scrap iron). In the first case, $Ba(OH)_2 \cdot 8H_2O$, Na_2SO_3, water glass, and a precipitation inhibitor was injected into the rock matrix at the Königstein mine to prevent the formation of acid mine drainage in some selected parts of the mine (Jenk et al. 2005). Although large-scale experiments were conducted, some leached blocks were partially passivated (selected references are: Klinger et al. 2000b, 2002; Ziegenbalg 2005), none of those promising in-situ treatment options were used as a final solution at the abandoned Königstein uranium mine, presumably because large amounts of mine water would still have to be treated. Schneider et al. (2001) conducted tank experiments at two abandoned uranium mines and had promising results. They used glass columns at those sites, which were filled with zero valent iron, iron and lignite, lignite, and sewage sludge. Yet, the chemical reactions involved the release of hydrogen gas, which would have to be removed to avoid underground explosions.

An extraordinarally simple in-situ/on-site treatment option was suggested by Merkel (2002). Based on geochemical calculations, he concluded that mixing the mine water with unpolluted water could inexpensively treat the large amount of acid mine water expected from an abandoned mine. This was a twist on the "dilution is the solution to pollution" rule that used to be the norm before environmental regulations went into effect. It seems as if some people listened carefully to Merkel and that the mine operator found the option potentially attractive (Zeißler et al. 2005). Yet, this measure does not reduce pollutant loads and is therefore not advised.

Red mud, a residue from aluminium production, and fly ash were considered as inexpensive alkaline materials that could be used for in-situ treatment in the abandoned Freiberg/Germany silver-lead-zinc mines. It was suggested that parts of the Reiche Zeche shaft (Fig. 15) could be filled with the material to create a small reactive barrier within the shaft. Because a large amount of the polluted water flows through that shaft, it was expected that the mine water could be treated in-situ (Zoumis et al. 1999, 2000; Zoumis and Calmano 2000). Yet, what they disregarded is that installing a relatively impermeable reactive barrier would reduce the hydraulic conductivity of the shaft. Consequently, the mine water would no longer drain through the Reiche Zeche shaft but through one of the other shafts. Zoumis (2003) finally concluded that the barrier was impractical for large-scale application because either the break through time would be too short or the permeability of the barrier would be too low.

Finally, let's briefly address the in-situ remediation of acidic mine lakes, which commonly exist where large open pit mines once operated. Several "in-lake" techniques have been applied to acidic mining lakes in the Lausitz/Germany lignite mining area, with various amounts of success (Benthaus and Nitsche 2005; Scholz 2005). Recent investigations have used CO_2, coal combustion ash, and LDS slime to add alkalinity (Loop et al. 2003;

Merkel 2005). Yet, the addition of alkalinity without simultaneous reduction of sulphide results in less success than the coupled acidity-sulphate reduction associated with the addition of organics to stimulate bacterial sulphate reduction (Castro and Moore 2000; Koschorreck et al. 2002; Meier et al. 2004).

To finish on a more positive note, the abandoned Spreetal Nordost/Germany mining lake is a flooded open pit lignite mine that annually receives $2 \cdot 10^6$ m^3 of alkaline residues (sludge) from a mine water treatment plant at the *Schwarze Pumpe* industrial site. Because the low-density sludge of this treatment plant contains large amounts of unreacted lime (CaO), the lake pH ranges between 4 and 6. The buffering capacity in the lake is not yet high enough to keep this pH level stable if the sludge addition were to stop (Wolfgang Rabe, pers. comm.), but a research project is underway at the TU Bergakademie Freiberg to investigate possible ways to achieve a sustainable pH in this abandoned mining lake (Merkel et al. 2005a; Merkel 2005).

11.7 Use of Mine Water

This section describes using treated as well as untreated mine water during and after mining, and partially overlaps with section 5.2, which described the use of post-mining areas in a more general way. A common post-mining use of abandoned mines, especially underground mines, are visitor mines. Their number goes in the thousands and because no special considerations – unless safety procedures – are necessary, they will not be considered here – which does not mean that they are economically unimportant as shown by the Schwaz/Austria visitor mine with more than 100,000 visitors annually (Palme and Ingenhaeff-Berenkamp 1993; Palme et al. 2002).

Use of mine water mainly depends on its chemical and physical quality. If the quality criteria meet drinking water standards, mine water can be used as drinking water (Burbey et al. 2000; Teaf et al. 2006), and several hundred smaller and larger communities depend on such water supplies. Regular monitoring is of course necessary to make sure that any changes in mine water are recognised at an early stage. One of the larger cities that uses mines for its drinking water reservoirs is Freiburg/Breisgau/Germany, though there the mine water is only used to supplement the normal drinking water supplies, when necessary (Wirsing 1995). In fact, several hundred mine water adits are used for drinking water in German towns and cities, including Bayreuth, Selbitz, Kassel, Bensheim (used as an emergency back up supply), Obernkirchen, and Plettenberg (Thomas Krassmann, pers. comm.). Stengel-Rutkowski (1993, 2002) lists at least 49 various abandoned mines in the Rhenish Schiefergebirge, with discharges between 1 and more than 80 L s^{-1}, that are used for drinking water purposes. Stengel-Rutkowski (2002) also discusses potential problems with drinking water supplies from underground mines in the Taunus region: high arsenic (up to 1.4 mg L^{-1}) and iron contents (up to 18 mg L^{-1}) in a polymetallic underground mine, and too low inflows in relatively dry periods. None of those problems occurred to that degree in the Lahn-Dill area. In the German Harz Mountains, the Teichtalstollen adit near Königshütte is used as a secondary drinking water supply, but a tracer test revealed a good hydraulic connection between a nearby abandoned day shaft (Ahrendfeldschacht) and the drinking water gallery (Wolkersdorfer and Hasche 2003; Ohly 2005). Because the shaft is situated in farm land, which is used for sheep grazing, the drinking water may be affected by bacteria or other pollutants.

Luckner and Hummel (1982) discussed quantitative and qualitative measures that could be taken to guarantee the drinking water supply in the Eastern German lignite mining areas. However, mine water is used as a primary or secondary source of drinking water at many locations around the world. In the U.S., Varnell et al. (2004) reported that 70 communities in the state of West Virginia use mine water, in most cases without any treatment other than chlorination (Hobba 1987; Pack 1992). At the other extreme, at the St. Aidan's opencast coal site, in Leeds/United Kingdom, polluted mine water was treated by reverse osmosis to supply drinking water (Wilson and Brown 1997). It was the largest reverse osmosis plant of that type, treating about 20,000 m^3 of mine water daily, and was planned to work for only one year (Paul Younger, pers. comm.). Experiments were also conducted in Poland, for a reverse osmosis mine water treatment plant with a daily treatment capacity of 2,500 m^3, at the Pokój coal mine (Magdziorz and Sewerynski 2000).

Underground mine waters have also commonly been used in spas, especially for people with diseases of the respiratory tract. One of the most prominent European spas associated with underground mining is Bad Gastein in Austria. Its 19 thermal springs, known since the 14th century, were described by Paracelsus, and the list of well-known persons searching for health there includes kings, politicians, and Pope Benedict XVI. Documented gold mining in the area dates back to early medieval times (Ludwig and Gruber 1987) and the most recent mining activities were conducted in the 1940s. The latest development was the *Gasteiner Heilstollen* (formerly *Paselstollen*), a radon health gallery, in Böckstein near Bad Gastein. During the Second World War, the deposits of the Hohe Tauern were exploited and it was found that in one part of the *Paselstollen* adit, elevated rock temperatures occurred (the air temperatures range between 37 and 41 °C). The mine workers reported positive health effects while they worked in the mine, so after the War, the *Paselstollen* was investigated. It was found that it could be used for recreational purposes and has been used since the 1950s. Today, the *Gasteiner Heilstollen* can be used by up to 600 people. Like Bad Schlema, in Germany, radon is the reason that people come to the Gasteiner Heilstollen, but unlike

Schlema, the water from the mine and the famous thermal springs are also used (Hayek and Wierer 1951; Exner 1952; Ludwig and Gruber 1987; Böttcher et al. 1991; Zötl 1995; Sohrabi 1998; Zötl 2001; Franz Xaver Rieser, pers. comm.).

The Franziskibad in Bad Häring/Tyrol uses the sulphate-rich mine water of an abandoned hard coal mine which already was worked in the 19th century (Angerer and Lechleitner 1893). The current spa water comes from an exploration borehole that was sunk in 1951 to explore the coal and copper reserves (Lechner 2003). Other German spas that use mine water are Bad Suderode/Harz Mountains, Feengrotten Saalfeld/Thuringia, and Alexisbad/Harz Mountains. Another spa, used since the 17th century but closed in 1905, was the Silbertaler Bädli in Vorarlberg/Austria, which used iron-copper sulphate water flowing out of an abandoned medieval copper mine (Vogt 2001). Mine water from an adit at the *Bad in der Au* near Brixlegg/Tyrol was once used to supply a spa, but today the Ca-SO_4-water is bottled and sold as mineral water.

Mine water is often used to drive turbines for the mine itself (Döring 1993) or for the surrounding communities. For example, at the Max-Braun-Adit, Biberwier/Tyrol, a very small engine (less than 50 kW) that dates back to the end of the 19th century is still in use (Wolkersdorfer 1991). This was one of the earliest turbines ever used to produce electricity in an underground mine. Two larger German installations, one in Clausthal and another in the Freiberg mining area were abandoned in the late second half of the 20th century (Döring 1993; Galinsky et al. 2001).

There are many examples of treated or untreated mine water being used for recreational purposes, and only one outstanding case shall be mentioned here. The Passagem de Mariana Gold mine, near Ouro Prêto/Minas Gerais/Brazil, was once a very productive gold mine but it was sold after the death of the last owner; since then, it has been used as a visitor mine for divers and as an underground pool. As in many gold mines of Minas Gerais, the arsenic content is very high and reaches 1.7 mg L^{-1} (Williams 2001), which apparently does not deter people from swimming in the water (Fig. 95).

Another use of mine water is for power plant cooling and process water. As long as the mine water is of sufficient quality (or can be satisfactorily treated) so that it does not interfere with the technical operation of the cooling system, it can be used for that purpose. Veil et al. (2003) investigated mine water in Pennsylvania and West Virginia/USA, where more than 120,000 abandoned mines and consequently a large number of flooded mine pools are available. In this area, six power plants are already using mine water for cooling and process purposes. Among them is the Tremont 31-MW block, which has used the treated Lyken mine discharges since 1986. The GenPower LLC Maidsville 600-MW coalfired power plant (known as the

Fig. 95. Pupils swimming in the arsenic contaminated mine water pool of the Passagem de Mariana/Minas Gerais/Brazil Gold mine.

Longview Power Plant) is currently being developed; commercial operation is targetted for mid 2007. GenPower is going to use 1,600 m^3 h^{-1} of treated mine water from the Shannopin mine as a source of cooling water. The water will be treated by a public/private partnership of Pennsylvania Department of Community and Economic Development, the Pennsylvania Infrastructure Investment Authority, AMD Reclamation Inc., and Dana Mining.

Treated open pit water is used as cooling and process water at the German Jänschwalde, Boxberg, and *Schwarze Pumpe* coalfired power plants. In 1999 $21 \cdot 10^6$ m^3 of treated mine water were used for that purpose (Arnold and Baumgarten 1996; Lausitzer Braunkohle Aktiengesellschaft 2000). At the South African Tutuka Power Station, Eskom mine water (treated by a spiral reverse osmosis plant) is used as cooling water (Buhrmann et al. 1999). Like the Shannopin example, this allowed the company to simultaneously address their cooling water requirements and the environmental problems caused by a mine water discharge.

In some areas with a shortage of irrigation water, treated and untreated mine water is being used for irrigation (e.g. Du Plessis 1983; Voznjuk and Gorshkov 1983; Annandale 1998; Annandale et al. 1999, 2001; Jovanovic et al. 2001; van Zyl et al. 2001; Annandale et al. 2002; Jiries et al. 2004). Most of these studies have been conducted in South Africa due to the abundance of mine water and the arid climate there. A key question being asked is what kind of long term effects will using mine water for aquicultural crops have on the soil, and ground or surface water? Soil balance models and weather data generators were used to model the long term impacts of mine water on the soil functioning. Jovanovic et al. (2001) modelled precipitation of gypsum in the soil profile after an 11 year irrigation cycle, while van Zyl et al. (2001) discussed the desalination procedures possibly needed for irrigation water of that type. In another investigation, Annandale et al. (1999) modelled a 30-year irrigation cycle and confirmed that gypsum would precipitate in the soil but stated that the soil is an effective sink for the salts in the mine water and that there was a low potential of remobilisation. Given this, they conclude that using mine water for irrigation "should be an effective and economical means of utilising large quantities of gypsiferous water without causing irreparable damage to soil resources". Though this is correct for the chemical effects of the precipitates, it is not clear what long-term effect the blockage of the soil pores might have on the hydraulic conductivity of the soil.

Mine water has also been used for heating and cooling purposes. In the Henderson Molybdenum Mine (Empire/Colorado/USA), one of the world's largest molybdenum mines, the ventilation air was heated with heat exchangers from mine water (Jensen 1983). The Freiberg/Saxony Hospital uses mine air from a 2.7 km long abandoned mine gallery in their ventilation system. The temperature of the mine air is constant throughout the year and has been proven to be microbiologically clean and dust free. Though mine water is not directly connected to the Freiberg ventilation system, the energy needed to evaporate the mine water in the gallery is one reason for the cool temperature of the mine air (Weißenborn and Benthin 2003).

Another use of mine water is the recovery of metals from highly enriched acidic waters by a process called cementation. The earliest reported use of this technique was in China, as early as 1086 A.D. It has also been commonly used in Europe since medieval Times, and in the USA since the 19th century (Lung 1986). Lai (1980) reports on extracting copper from synthetic mine water through a liquid membrane. Wright et al. (1995) also describe a successful on-site experiment with unsupported liquid emulsion membranes for the extraction of copper in an Arizona/USA copper mine. In all such experimental set-ups, an extraction rate of more than 90% could be achieved. The Mynydd Parys mine site in Anglesey/Wales (compare section 5.7.4) is another mine where metals were extracted from the mine water (Boult 1996). A similar experiment with liquid emulsion membranes was conducted in Chile and resulted in similar recovery rates (Valenzuela et al. 1999). This experimental situation was modelled by Valenzuela et al. (2002) with a good fit between the experimental results and the numerical model. Rare earth elements were extracted through a combined extraction/electrostatic pseudo liquid membrane process by Yang et al. (1999). All those studies were conducted in the laboratory or on an experimental field scale. However, at the Hungarian Mecsekérc uranium (Pécs/Hungary) mine water treatment plant, mine water polluted with radionuclides is treated using a membrane process, and uranium is recovered from the treated mine water. In a second step, uranium oxide is produced and sold to the world market (Benkovics et al. 1997; Gombkötö et al. 2002; Csövári et al. 2004).

11.8 Ground Water Protection

Ochreous precipitates emanating from mine portals provide obvious evidence of polluted mine water at hundred of thousands of sites around the world. Yet, the real challenge for the future is the protection of ground water and aquifers from polluted mine water. Despite many site-specific studies that have shown local impacts, it is not clear how much of a problem this might pose to aquifers and even drinking water reservoirs.

We know that nearly every mining operation affects the natural ground water system by either lowering the ground water table, polluting the ground water with chemical compounds or suspended solids, changing the ground water flow path, or by just changing the natural flow rates. The treatment of aquifers polluted with acidity, alkalinity, or metals is not as simple as the treatment of water discharging from shafts or mine portals, so the protection of an aquifer has to be given special consideration. However, due to the low flow velocities of ground water in slightly permeable rocks, natural attenuation plays an important role in remediating aquifers affected by polluted mine water. Karst aquifers may have much higher permeabilities, but also have very high neutralization capability. However, in highly permeable fissured aquifers, the situation might be problematic, as the natural attenuation processes are too slow or not effective enough.

Because the potential pollution of ground water by mining activities is obvious, many procedures exist to predict that contamination. Both, geochemical modelling and reactive transport codes are used to predict the degree of impact on the aquifers; most of the models used are sophisticated and produce reasonably good results (see chapter 8). It is therefore interesting to see that potential ways to protect the ground water from mining induced pollution are limited and rarely used. Even Teaf et al. (2006) restrict themselves to listing potential dangers and potential treatment options; they do not give case studies where ground water was treated or protected from underground mine drainage.

In-situ treatment options have already been described in section 11.6. A frequently used option is a permeable reactive barrier (PRB). The first PRB installed at a mining location was probably the one constructed at the Nickel Rim mine site near Sudbury, Ontario/Canada (Benner et al. 1997). In Europe, at least four PRBs have been installed at mining locations: at the Aznalcóllar/Spain site (Bolzicco et al. 2003; Gibert et al. 2004), the Pécs/ Hungary uranium mine (Gombkötö et al. 2002; Simon et al. 2003; Csövári et al. 2004), the Shilbottle/United Kingdom mine spoil site (Younger 2002a;

Younger and Elliott 2003), and between the open pit lakes Skado and Sedlitz/Lausitz/Germany (Schöpke et al. 2006). An experimental set-up with good treatment results was installed at the Königstein/Germany uranium mine (Klinger et al. 2000b, 2002; Jenk et al. 2003). The long-term reliability of the systems still has to be validated and many researchers have described problems with the decreasing permeability of the system after it became operational, especially in the Lausitz case, where researchers finally concluded that the set-up could be used to construct sealing walls (Schöpke et al. 2006). Further investigations must ensure that those systems should work at least as long as the peak of the first flush.

Classic pump-and-treat measures, commonly used in ground water remediation, are very uncommonly used to treat ground water polluted by mine water, though it is the standard procedure at many flooded abandoned mines. To prevent the pollution of the Basal Permian Sands aquifer, a pump-and-treat operation was installed near Horden/UK (Coulton et al. 2004).

One of the best known conflicts between safe mine closure and the interest of protecting an aquifer used for drinking water purposes involves the flooding of the German Königstein uranium mine, where the flooding might impact an aquifer that is used to supply the cities of Pirna and Dresden (Biehler and Falck 1999; Merkel 2002; Schreyer et al. 2002). Some potential remediation measures for the Königstein mine are described in section 11.6. Another, large scale, ground water pollution area is the German Lausitz lignite mining area with pH-values as low as 3 (Nixdorf 1999; Grünewald 2001; Rolland et al. 2001; Nixdorf and Deneke 2004; Fyson and Nixdorf 2005). Based on the assumption that all the ground water in the area is to some degree polluted, a volume of $2 \cdot 10^{10}$ m^3 of ground water is affected. After more than 15 years of research and an unknown number of remediation techniques, the ground water situation has not significantly improved.

One key procedure to protect ground water from polluted mine water is the sealing of waste disposal or tailings sites (see section 5.4.2). Covering such facilities results in a significant reduction of pollutant loads and therefore can be seen as an imperative measure to protect ground water reservoirs (e.g. Sheibach et al. 1982; Brandberg et al. 1993; MEND [Mine Environment Neutral Drainage Program] 1994; Cabral et al. 1997; Elander et al. 1998; Waugh et al. 2001; Carlsson 2004). Modern tailings dams, such as the Río Narcea El Valle and Carlés gold mines and the Aguablanca nickel mine (Asturias/Spain), are underlain by a geomembrane to prevent seepage water polluting the ground water. Such geomembranes, which can also be used for tailings impoundments, typically concist of HDPE or PVC. Another option to prevent or even minimise the impacts of tailings or waste rock material is to add alkaline material (Wisotzky 2001).

This is a good place to discuss the negative impacts of abandoned ISL (in-situ leaching or solution mining) mines on ground water quality. Solution mining for uranium, copper, or sulphur significantly changes the aquifer because the acid, basic, or hot solution as well as the pumping alters the host rocks and mobilises metals and arsenic, which can pollute nearby aquifers (U.S. Environmental Protection Agency 2000). Yet, there is usually not much open space to be flooded and therefore ISL mines are not really addressed by this book. Two exceptions would be the Königstein/Germany uranium mine and the White Pine/USA copper mine. Both mines started as conventional underground operations and were converted to in-situ leaching mines thereafter, and have already been discussed in earlier chapters or in the case studies hereafter (e.g. Lewis et al. 1997; Jakubick et al. 2002). A well investigated ISL site is the Stráž/Czech Republic uranium deposit that was mined from 1967–1996 and is intensively discussed in the literature about this mine site (e.g. Frydrych et al. 2000; Novák 2001; Smetana et al. 2002). Because of the experiences gained during the abandonment of these three ISL mines, modern mines of that type are planned and operated in a more environmentally safe manner (Kalka et al. 2005).

In conclusion, the protection of ground water from polluted seepage emanating from spoil heaps, waste rock piles, or tailings dams is a common procedure that is used wherever aquifers have to be protected from negative influences. However, the protection of ground water from flooded mine sites is not yet a common procedure. In general, the engineering methods are not fully developed yet and the costs of preventing polluted mine water from impacting the ground water are not economically feasible.

12 Flooded Underground Mines: Case Studies

12.1 Introduction

In the following chapter, 13 case studies for flooded underground mines are described in detail (Fig. 96). The first six examples focus on sites with stratified mine water, where no tracer test has been conducted so far, whereas another six examples are flooded underground mines where extensive tracer tests have been carried out to evaluate the hydrodynamic regime of the mine. The last section summarises salt mine examples. All known and published investigations are described and discussed in this chapter. How the knowledge gained about the hydrodynamics of mine water and potential stratification might be used as a remediation measure is addressed at the end of section 12.9.4.

Down-hole probes for physico-chemical measurements in flooded mine shafts or wells connected to the mines usually measure pressure, temperature, electrical conductivity (or resistivity; due to space restrictions some of the following figures are labelled with 'conductivity' instead of 'electrical conductivity'), pH, redox potential (E_H) and oxygen. While the measurement of pressure, temperature, or electrical conductivity, even at great depths, is less problematic, the electro-chemical probes for pH, redox potential, and oxygen are restricted in their application (Schindler et al. 1995), though the modern LDO oxygen sensors can also be used under higher pressure conditions (Körtzinger et al. 2005). This results from the mechanical arrangement of the probes, which are constructed of glass parts and membranes being exposed to the depth or salinity dependent pressures. For the Hope/Germany investigations, mis-measurements due to pressure-effects of the probes were

Fig. 96. Worldwide location of the flooded underground mines discussed in this chapter.

excluded by using a specially constructed down-hole probe for pressure-free measurement of all the relevant parameters within the water (GSF – Gesellschaft für Strahlen- und Umweltforschung 1986). Unfortunately, this probe has a diameter that usually excludes the probe from being used in small wells or shafts with damaged pit frames, but in-situ probes with smaller diameter are now available (Schindler et al. 1995). Due to the above-described limitations of the probes, some measured parameters, especially pH or the redox potential, must be interpreted with precaution. Another effect, which was observed in the Niederschlema/Alberoda uranium mine, is the overlap of two potentials: the self-potential of the rocks and the redox potential of the mine water. In some cases, the differences between the redox potential measured in the shaft and at the surface was as high as 200 mV, so that calibration of the probe failed. Usually, the self-potential method is used for geophysical investigations of ore deposits (Wöbking 1982; Giese et al. 1994; Corwin 1997).

Fluctuations of the electrode signals, due to random signals or noise, are common in the situations described here. Such random signals or noise is typical for signal detection and information transmission and can be due to mechanical or electrical noise (Carlson et al. 2002). If they are ergodic, they usually do not cause problems for the measurement of physical parameters. Yet, such fluctuations can also be real, as in turbulent flow situations, or a malfunctioning electrode (Fig. 97). In some cases, both effects might be in the same range, but usually, extreme signal fluctuations are due to electrode-dependent drift or noise – whatever the cause. Systematic errors during down-hole measurements usually shift the measurement to higher or lower values, while random errors cause peaks in the curve.

Although these effects must be kept in mind, they will not be further discussed here. Readers interested in them and what kind of filter algorithms deduced from communication systems can be used are referred to the literature (e.g. Carlson et al. 2002). In the following chapters, several examples of stratified and unstratified flooded mine shafts will be described using a similar style for each case study. It will not be possible to give all the geological, hydrogeological and mining relevant data for each of the cases reported, either because they are not available or due to a lack of space. However, the relevant literature for each mine is given in the respective introduction or can be found in the literature cited.

Stratification of the water body occurs in many mines. Most published papers describe the stratification effect, without trying to explain the physical causes. A very similar situation can be found for publications on mine water tracer tests. Though described in several publications, only one publication so far summarised the results of different mine water tracer tests and draws some first conclusions out of those tests (Wolkersdorfer 2002b).

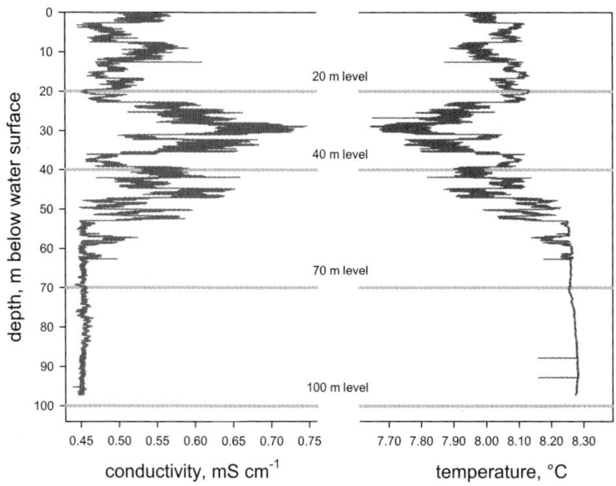

Fig. 97. Temperature and electrical conductivity depth profiles with a malfunctioning downhole probe. Several reasons are possible and they cannot always be deduced from the measurement. The probe came back to normal operation at a depth of about 63 m.

12.2 Northern Anthracite Fields, Pennsylvania, USA

12.2.1 Description of Location

Mining in the Pennsylvanian Anthracite Fields/USA started in 1807 and ceased in the mid-1970s. During that time, about $5 \cdot 10^9$ t of anthracite were mined in four areas, of which the Northern Anthracite Field covers 456 km^2. An extensive dewatering scheme was installed during operation to dewater the mines, which were as deep as 450 m below sea level. Even then, the mine water was highly acidic (Felegy et al. 1948), causing damage to the pumping machinery. The Henry colliery, for example, pumped some 28 $m^3\ min^{-1}$ in 1950 and discharged the untreated water to surface streams. After the mines were abandoned, the pumping stopped and the flooded area now covers approximately 175 km^2. More than 95% of the open voids were flooded in 1972; the Wanamie colliery was the only working colliery left (Stuart and Simpson 1961; Erickson et al. 1982; Ladwig et al. 1984).

When the water table rose, the mine water became enriched in contaminants and created a strongly acidic mine drainage that discharged through several point- and non-point-discharges. In addition, the rising mine water table caused flooding of homes and surface subsidence (Erickson et al. 1982). Four drainage adits ("outfall discharges") were developed to limit the extent of the flooding, and discharge a maximal annual mean flow of $63 \cdot 10^6$ m^3 of highly contaminated mine water from the Wyoming Basin. Due to the long mining history in the area, with many interconnections between the mines, goaf, or collapsed voids, the flow system is highly intricate, with a general flow direction towards the drainage adits. Though the mine water quality steadily improved, the contaminant levels were still extremely high in the 1980s. The pH ranged between 5 and 6, sulphate between 1 and 2 g L^{-1}, and the acidity between 100 and 400 mg L^{-1} $CaCO_3$, with a strong trend to change to alkaline conditions (93% were already alkaline in 1984). The first report of stratification in flooded underground mines was published from the Pennsylvanian Anthracite fields (Stuart and Simpson 1961; Fig. 98).

Geologically and structurally, the Northern Anthracite field is a crescent-shaped syncline striking in a northeastern direction. It comprises many discontinuous folds, flexures and faults, of which a major saddle structure divides the basin into the southwestern Wyoming and the northeastern Lackawanna Basins. The host rocks of the anthracite are Upper Carboniferous (Pennsylvanium) conglomerates, sandstones, and shales interbedded with 18 workable coal seams and a total thickness of 610 m. They are underlain by Devonian sandstones as well as conglomerates and partly overlain by Pleistocene till (Erickson et al. 1982; Ladwig et al. 1984).

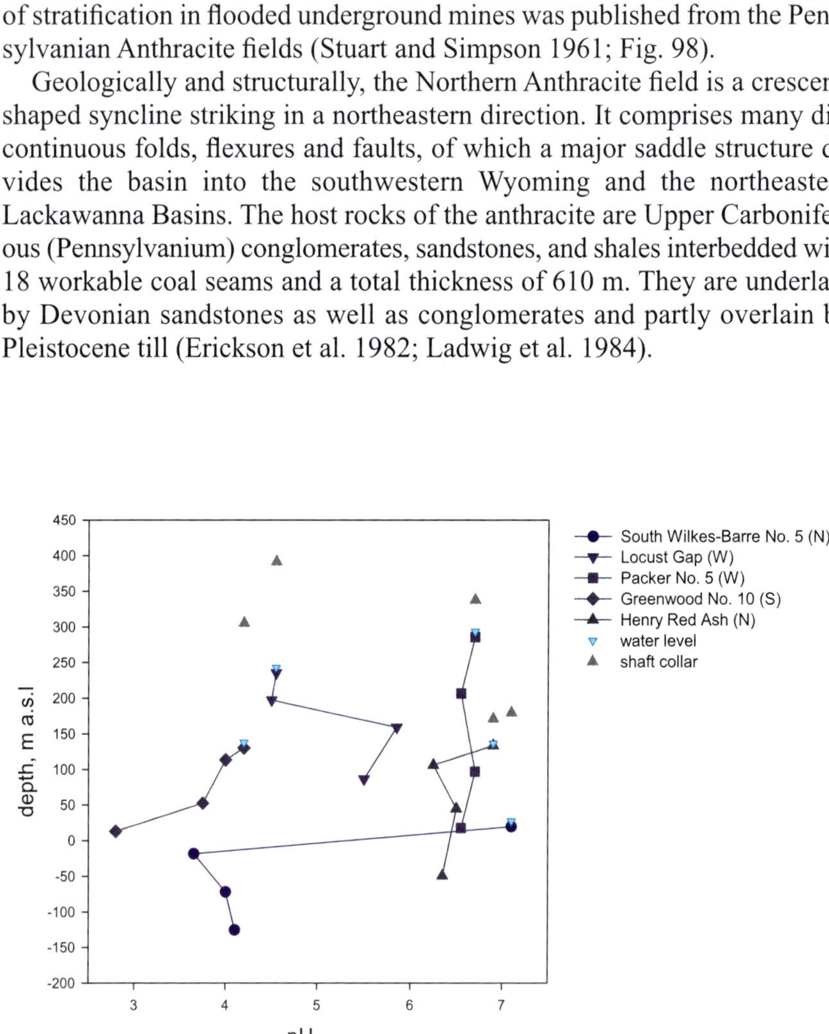

Fig. 98. First known, depth-dependent pH measurements in the Pennsylvanian Northern Anthracite field (modified after Stuart and Simpson 1961).

12.2.2 Appearance of Stratification

Between 1961 and 1984, twenty shafts, eight of them more than once, were studied by Stuart and Simpson (1961), Barnes et al. (1964), Erickson et al. (1982), and Ladwig et al. (1984; Table 24). Stuart and Simpson (1961) measured pH in flooded shafts, realising that in nearly all of them, water with higher pH overlaid mine water with low pH, indicating a clear stratification with two major mine water bodies in the pools (Fig. 98). Comparing their results with those of Erickson et al. (1982) and Ladwig et al. (1984), measured 20 years later, indicates that the stratification was stable between the 1960s and 1980s while the mine water quality in the flooded mine workings notably improved (Fig. 99).

Five of the nine shafts investigated by Ladwig et al. (1984) showed a clear stratification, indicated by abrupt changes in the parameters measured (Fig. 100). It can also be seen that the stratification appears at a depth of 20–100 m a.s.l. (meters above sea level) even though the distance between the shafts is 15 km (from NW1 to SEU1, Fig. 99 to Fig. 101). The electrical conductivity tends to increase in an east–west direction, from about 400 µS cm^{-1} to 1,000 µS cm^{-1} in the shallow layer and 1,000–1,400 µS cm^{-1} in the deeper one. Based on their similar physico-chemical behaviour, the shallow layers seem to be more isolated from colliery to colliery, while the deep ones are better connected. According to Ladwig et al. (1984), "the deep zone reflects

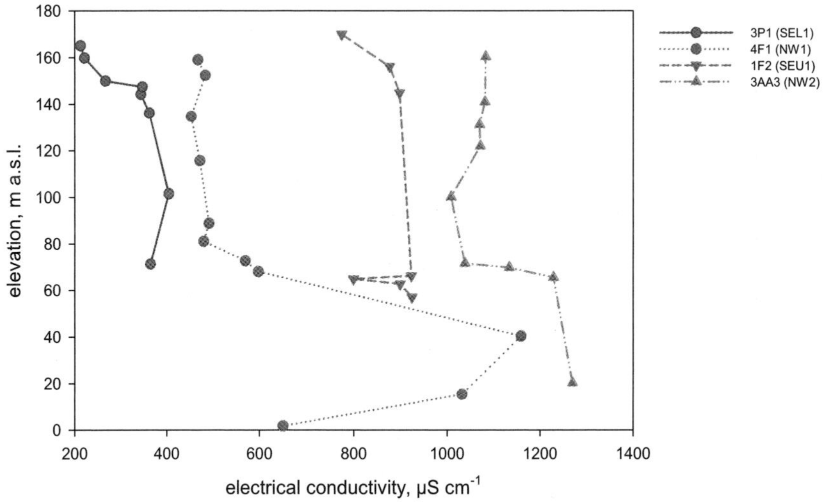

Fig. 99. Depth–conductivity plots of four shafts in the Northern Anthracite/USA mining field (modified after Erickson et al. 1982).

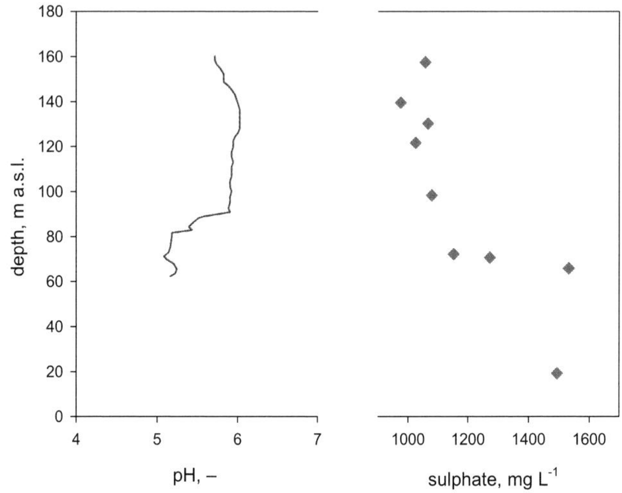

Fig. 100. Depth pH and sulphate plots of the Gaylord Colliery NW2 shaft on 26 June 1981 (modified after Ladwig et al. 1984).

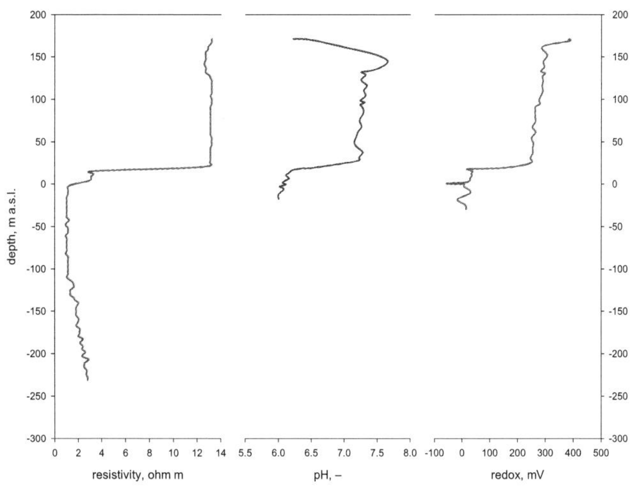

Fig. 101. Depth resistivity, pH, and redox plots of the Truesdale-Bliss Colliery SEU2 shaft on 17 to 18 November 1981 (modified after Ladwig et al. 1984).

the poor water quality in a more stagnant subsystem". If that was true, the variations between the measurements in a single shaft should be very small, those between the different shafts be significant large, and there should be no fluctuations during a down-hole measurement. In fact, the physico-chemical

Table 24. Shafts where mine water stratification were studied in the Pennsylvanian Anthracite fields. Each line corresponds to a single shaft, each of which was named differently by the authors. κ: electrical conductivity; res: resistivity; T: temperature. References: [1]: Stuart and Simpson (1961); [2]: Erickson et al. (1982); [3]: Ladwig et al. (1984).

Mining field	Shaft	Colliery	Parameters
North-west[1, 2, 3]	NW1, 4F1, Red Ash	Henry Fig. 99	κ, pH, E_H
North-west[2, 3]	NW2, 3AA3	Gaylord Fig. 99, Fig. 100	κ, pH
North-west[2, 3]	NW3, 5R4	№ 6	κ, pH, E_H
North-west[2, 3]	NW4, 3HH3	Kingston	κ, pH, E_H, res
North-west[2]	6G2	Nottingham	κ
North-west[1]	Red Ash	Exeter	pH
North-west[1]	№ 1	Schooley	pH
South-east Upper[1]	№ 2	№ 7	pH
South-east Upper[2, 3]	SEU1, 1F2	Bliss Fig. 99	κ, res, T, E_H
South-east Upper[3]	SEU2	Truesdale Fig. 101	pH, E_H, res
South-east lower[2, 3]	SEL1, 3P1	Hollenback Fig. 99	κ, pH, E_H, res
South-east lower[2, 3]	SEL2, 4H2	Peach Orchard	κ, pH, E_H
South-east lower[2, 3]	SEL3, 2F1	Loomis	κ, pH, E_H, res
South-east lower[1]	Clear-Spring	Clear-Spring	pH
South-east lower[1]	№ 5	South Wilkes-Barre	pH
Eastern Middle[1]	Hazelton	Hazelton	pH
Western Middle[1]	Locust Gap	Locust Gap	pH
Western Middle[1]	Reliance	Reliance	pH
Western Middle[1]	№ 5	Packer № 5	pH
Southern[1]	№ 10	Greenwood	pH

measurements and some of the chemical analyses imply that turbulent flow likely occurs in the deeper layers.

Erickson et al. (1982) concluded: "All the parameters monitored in the preliminary field studies indicate that most sharp changes in water quality occur near the elevation of mined-out seams intercepted by the shafts. Consequently, such voids must be significant flow conduits for mine pool circulation. Smoother depth profiles would be expected if the mine pool complex was well mixed or if flow was slow enough to allow chemical equilibrium".

12.3 Picher Mining District, Oklahoma, USA

12.3.1 Description of Location

Located in northeastern Oklahoma/USA, the Picher Mining District was one of the world's largest lead-zinc mining fields, covering an area of 186 km^2 (the studied area covers about 40 km^2) and originally containing ore reserves of $1.22 \cdot 10^8$ m^3. Mining of galena and sphalerite started between 1891 and 1901 and ceased in 1976 with a peak period between 1915 and 1920, when 230 mines were under operation. During the Second World War, 63 major pumping stations pumped a maximum of 60 m^3 min^{-1} of mine water out of the mines; this decreased to 24 m^3 min^{-1} after the War. The deepest mines in the Picher Mining District reached a depth of about 200 m below the surface; the volume of the flooded voids totalled up to $4 \cdot 10^7$ m^3. Due to the room and pillar mining technique, pyrite- and marcasite-bearing waste material was deposited in the mined-out areas, giving rise to highly contaminated acid mine water with pH values of 4 to 5. Furthermore, the mine water is enriched in zinc and iron, reaching up to 150 mg L^{-1} of each. In 1979, the water levels reached the surface, finally discharging into Tar Creek (Sheibach et al. 1982; Spruill 1983, 1987; Parkhurst 1987, 1988; Suman et al. 2000).

The oldest rocks in the Picher Mining District are Ordovician dolostones with interbedded sandstone layers, which are in places overlain by Devonian shales. Above the shales and dolostones, lies the ore-bearing lower Carboniferous (Mississippium) Boone Formation with a thickness of 110 to 120 m. This formation consists of greyish, karstified limestones and cherts and is a good drinking water reservoir outside of the mining area. Pennsylvanian age strata, mainly shales and sandstones, lie above the Mississippian rocks. Bedrock strata in the area are faulted and folded and dip slightly to the northwest with angles of around 0.2°. Sphalerite and galena were the economic minerals mined in the area. Above that, chalcopyrite, enargite, pyrite, and marcasite were found in uneconomic quantities. Common gangue minerals were jasperoid, dolomite, calcite, quartz, or barite. Besides that, secondary minerals such as smithonite, calamine, anglesite, cerussite, malachite, epsomite, or gypsum were found (McKnight and Fischer 1970; Tar Creek Superfund Task Force Water Quality Subcommittee 2000a). Genetically, the mineralisation in the Picher Mining District belongs to the Mississippi-Valley-Type (MVT) deposits.

In January 2000, the Tar Creek Superfund Task Force was appointed in the Picher Mining District to "[develop] a comprehensive remediation plan for the ill-fated area of north-eastern Ottawa Country" (Griffin et al. 2000). During its eight months of work, the Task Force, consisting of eight subcommittees, developed a remediation plan for the whole area. In addition, the Mine Shaft Subcommittee identified more than 1,300 shafts in north-eastern Oklahoma and assumed that there were up to 2,600 abandoned shafts in the whole Tri-State mining district. They recommended an extensive 5-year shaft and hole closure programme to plug all known shafts in the area (Suman et al. 2000).

12.3.2 Appearance of Stratification

In 1978, Playton and Davis (1978) and Playton et al. (1978) published the first results of their measurements in abandoned mine shafts of the Picher Mining District. Two years later, they published a comprehensive report on their findings (Playton et al. 1980). Seven mine shafts were sampled by physico-chemical parameters at different depth intervals, taking up to 8 water samples. Most of the sampled shafts showed a clear stratification identified easily by the specific electrical conductivity and the pH, and given the extent of the mine field, it can be concluded that it is the largest stratified mine water body known so far. Two water layers were observed: an upper one characterised by lower electrical conductivities (0.7–1.9 mS cm^{-1}) and higher, circum-neutral pH, and a lower one with higher electrical conductivities (3–4 mS cm^{-1}) and acid pH. In most shafts, the boundary between the two mine water bodies could be observed at a depth between 185 and 195 m a.s.l.; shafts above that elevation showed no clear stratification. Playton et al. (1980) realised that no clear spatial variation of the physico-chemical data could be observed. Furthermore, they discussed the fact that cooler water overlay warmer water and concluded that "an unstable thermal-stratification condition apparently exists" which is masked by the higher density of the higher mineralised water. Yet, they forgot that the stability of a stratified water body is a result of water density, which is a function of the water temperature *and* its mineralisation. Applying the UNESCO polynomial (Fofonoff and Millard Jr. 1983), the density of the upper layer at 14 °C and 1 mS cm^{-1} is around 999.7 kg m^{-3} and the density of the lower layer at 15 °C and 4 mS cm^{-1} is around 1001.2 kg m^{-3} (Fig. 41). No explanation was given for the stratification of the mine water body.

Sheibach et al. (1982) also reported that less mineralised mine water was overlying a mine water body with a higher mineralization and lower pH values, though the separation of the two water bodies was at a depth of about 50 m (Fig. 102). In the three shafts where they took samples (Kenoyer, Consol 2, and Lawyer shafts), the total dissolved solids ranged roughly between 1 and 3 g L^{-1} in the upper and 3.5 and 5 g L^{-1} in the lower mine water pool. In the same shafts, the pH was circum-neutral in the upper and acidic (pH 4–5) in the lower parts. Without further investigations, they concluded that the stratification might have been a result of acid-producing waste rock left in the deeper parts of the mines.

More than 20 years later, DeHay (2003) and DeHay et al. (2004) reinvestigated the Picher Mining District. Though the electrical conductivity had decreased from a range of 3.5–4.5 mS cm^{-1} to 2.5 mS cm^{-1}, and the pH had increased from 5.5–6 to over 6, the stratification patterns in the mining field were similar to those of the late 1970s and early 1980s. The boundary between the lower and the upper mine water bodies was approximately at 185–202 m a.s.l. and was indicated by a significant jump in the electrical conductivities and in the concentrations of the water constituents. Age determinations in the shafts with stratification showed that the lower mine water body had a significant higher mean residence time than the upper one (1980s in the lower, 1990s to early 21st century in the upper layer, with some exceptions from this general rule). Though the water was flowing through the mine pool, the stratifications stayed very stable during the past 20 years. Not enough data are presented by the authors for a more detailed description.

12.3.3 Conclusions

Based on the data presented by the several investigators between the 1980s and 2004, the stratification in the Picher Mining District remained very stable for nearly 25 years with physico-chemical parameters of the upper and

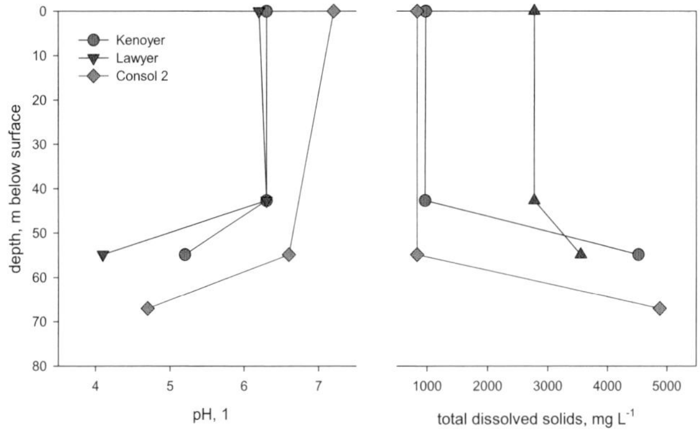

Fig. 102. Depth dependent pH and TDS measurements in three shafts of the Picher mining field (modified after Sheibach et al. 1982).

lower mine water body being very similar in all the investigated cases. This shows, on the one hand, that the hydrodynamic connection between the upper and lower mine water bodies is extremely poor and no convective flow between those two is taking part. On the other hand, the similar physico-chemical parameters in each of the water bodies leads to the conclusion that they are hydraulically well connected, as can also be deduced from the mean residence times of the mine water. Furthermore, the measurements in the shafts show that the flow is dominated by convective flow and not by diffusive flow, because the electrical conductivities and temperatures in the different depths are very similar in each shaft. One exception is the Royal shaft, which has, from the top to the bottom of the upper layer, an increasing electrical conductivity from 1.2 to 8.5 mS cm^{-1}. Diffusion-controlled flow appears to dominate in this shaft and the age determination of the water in fact differs from the other shafts investigated: it is about 35 years old.

No tracer test has yet been reported for the Picher Mining District. It would be interesting to see if the water age determinations would be validated by a tracer test or not. Furthermore, such a test could be used to characterise the type of stratification in the District.

12.4 North-Rhine/Westphalian Coal and Ore Mines, Germany

12.4.1 Description of Locations

Coldewey et al. (1999) investigated 19 shafts and wells in the German Ruhr (Niederrhein-Westfalen: Lower-Rhine/Westphalia) and the Aachen coal field as well as the Sauerland ore mining areas (Table 25). Besides the physico-chemical parameters measured with down-hole probes (temperature, electrical conductivity, pH, E_H, and pressure), they analysed (in the laboratory) water samples collected from discrete depths in the shafts. Furthermore, they conducted a laboratory experiment to investigate and understand some of the stratification patterns in the flooded shafts. The following descriptions are only based on the work of Coldewey et al. (1999) if they are explicitly cited. Otherwise, what is presented here is a personal interpretation of their raw data, which in parts deviates from their findings.

All three mining areas differ geologically, although all of them belong to the same geological system: the rheno-hercynian zone of the middle European Variscan orogeny. While the comparably small Aachen coal mining area is part of the Ardenne mountain chain, the Sauerland, easterly from it, is part of the Rhenish Schiefergebirge, and the Ruhr area, north of the Rhenish Schiefergebirge, is in the molasse trough of the Rhenish Schiefergebirge. Devonian and Carboniferous rocks dominate and due to the excellent surface outcrops and underground profiles, many locations became the *locus typicus*

Table 25. Shafts and wells investigated by Coldewey et al. (1999). κ: electrical conductivity; T: temperature.

Mining Area	Shafts, wells	Parameters
Aachen	Eduard-Schacht, Emil Mayrisch 1, von-Goerschen-Schacht, Willem II	κ, T
Ruhr	Shafts: Friedlicher Nachbar, Friedrich, Hermann 1, Hermann 2, Radbod 7, Wehofen 1; Wells: B4B, Landwehr 1, Landwehr 2, Schacht 7, Stuckenbusch 1, Stuckenbusch 2, Millinger M2	κ, T, pH, E_H
Sauerland	Sicilia Schacht	κ, T

for their stratigraphic units (e.g. Siegenium, Emsium, Eifelium, Westfalium, or Namurium). All investigated Aachen shafts are within the Wurm syncline, the easternmost part of the Namur syncline, filled with Carboniferous sandy, silty, and clayey deposits of Namurian C to Westfalian B (± lower Pennsylvanium) age. This syncline runs in a southwest–northeast direction, dips with less than 20° in a southerly directions, and, because of faulting, the syncline is divided into several blocks that intersect the coal seams. Of minor importance are the overlying deposits of the Buntsandstein and Cretaceous ages. During the Tertiary and the Quaternary, the Carboniferous rocks were overlain by 200–300 m thick sediments. East of the Ardenne mountains, and also east of the river Rhine is the Rhenish Schiefergebirge, with several thousand smaller and larger lead, zinc, and pyrite ore deposits and a large number of small, but long west-southwest—east-northeast striking synclines and anticlines as well as faults. The oldest rocks are of Silurian age and the youngest belong to the Westfalium C. In the close vicinity of the Meggen pyrite-zinc mine, which hosts the Sicilia shaft, the rocks are of middle Devonian age and consist of sandy to clayey sediments, overlain by upper Devonian schists and limestones; the overall thickness reaches 1,000–1,600 m. By far, the largest German coal mining district is the Ruhr area north of the Rhenish Schiefergebirge, which hosts about 100 coal seams of different productivities and thicknesses. A characteristic of the Ruhr area are up to 100 km long west-south-west—east-north-east striking synclines and anticlines, which are intensively internally folded and faulted. Those tectonic structures caused severe problems for the miners, as they had to extract the coal from steep seams; following the coal seems was sometimes difficult for them (Walter et al. 1992; Schönenberg and Neugebauer 1997).

Of the three mining localities investigated, the Aachen area and the Meggen mine are abandoned, whereas the Ruhr coal mining area is still in operation. Both Aachen and Meggen were abandoned in 1992 and have flooded since then (Heide and Hasse 1997, 1998; Heitfeld et al. 2002), whilst the Ruhr coal mining area has only been flooded in the parts that are not directly connected to the still working mines in the northern part of the Ruhr coal mining area.

12.4.2 Appearance of Stratification

In total, 24 continuous and discontinuous measurements were conducted in mine shafts and 23 measurements were made in wells. At a first glance, it seems as if no general conclusions can be drawn because the results of the measurements vary from shaft to shaft and from well to well. Yet, there are some characteristic similarities in the depth-dependent parameters:

- zones with constant physico-chemical parameters
- zones with gradually changing parameters
- sudden changes in the parameters
- transition zones.

The electrical conductivity and temperature, usually, but not always, increased with depth, and sudden changes of parameters usually occurred at the shaft's onsetting stations (Fig. 103), which was seen nicely at the Meggen Sicilia shaft (Fig. 104). These sudden changes were very constant with respect to depth, so that even measurements in different years were relatively consistent. The geothermal gradient did not seem to have an important influence on the temperature-depth profiles of the shafts, but did in the small diameter wells. The ground water in the Ruhr coal mining area is usually highly saline; the electrical conductivity in the shafts and wells can range from several milli-Siemens up to 189 mS cm^{-1}. Because the deeper mine zones are commonly overlain by ground water of lower conductivities, the density differences are very large and the stratification is therefore constant for a long time. Though the transition zones are commonly small between different water bodies, there can also be transition zones, with thicknesses of up to 20 m, which seem to be more typical for wells than for shafts (the Landwehr 1 shaft shows such a 20 m transition zone). This occurs because the flow in a shaft is better established than in a well. Some shafts show a stepwise change in the salinity or temperature (e.g. Hermann 1, Sicilia, Fig. 104), which could, at least for the Hermann 1 shaft, be due to double-diffusive convection (see section 7.2.1 for explanations), though there is insuffcient data in the report to verify this hypothesis. In the Sicilia shaft, the steps are clearly linked to the

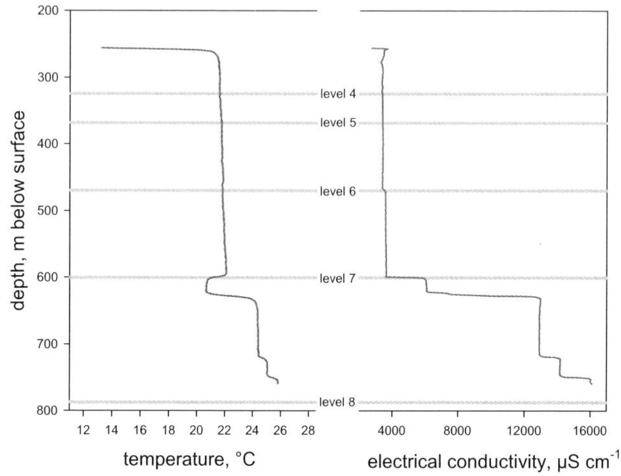

Fig. 103. Temperature and electrical conductivity measurements in the *Friedliche Nachbar* shaft (modified after Coldewey et al. 1999). Horizontal lines indicate the levels in the mine.

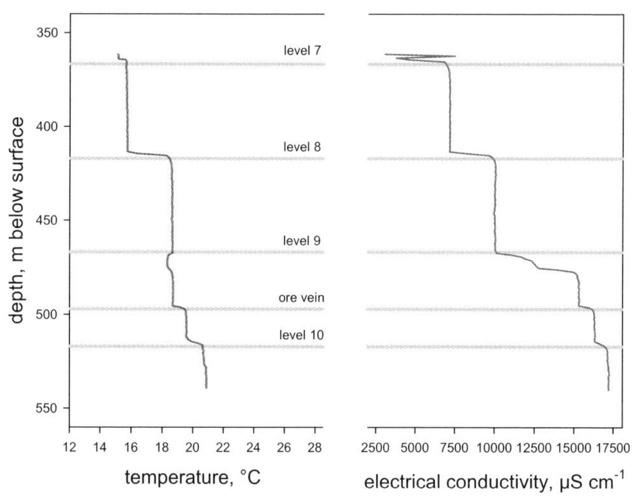

Fig. 104. Temperature and electrical conductivity measurements in the Sicilia shaft of the abandoned Meggen/Germany mine (modified after Coldewey et al. 1999). Horizontal lines indicate the levels in the mine.

on-setting station and might therefore reflect different flow cells or the inflow of water with different physico-chemical parameters.

The 23 measurements in the seven wells are of limited use in interpreting stratification patterns. Water in wells is not affected by free convective flow because of the small diameter to length ratio, in which the Rayleigh number (Ra) might be smaller than the critical Rayleigh number (Ra_{kr}). This is true for both cased and uncased wells. Therefore, the temperature and electrical conductivity patterns observed in the B4B, Landwehr 1, Landwehr 2, Schacht 7, Stuckenbusch 1, Stuckenbusch 2, and Millinger M2 wells could be due, at least in part, to the hydrodynamics occurring in stagnant water in the small diameter pipes (see also Ronneburg uranium mine). However, at least some of the wells clearly show the characteristic depth profiles for diffusion-dominated transport.

12.4.3 Laboratory Experiment

Coldewey et al. (1999) built a "laboratory mine," with a 0.5 m diameter and 0.8 m height, to investigate and understand some of the stratification patterns in the flooded shafts. Because no similarity calculations were conducted and no Rayleigh numbers were used to downscale the real world situation, the laboratory experiment is of limited use in explaining the stratification or hydrodynamic situation of the flooded underground mine. The temperature and salinity distribution during the 14 months of the test clearly proves that the flow within the system was – as could be expected from the experimental set-up – controlled by diffusive flow. Their interpretation of an upward/downward flow in the shafts is – due to the scaling of the system – physically not comprehensible and their temperature and salinity logs clearly show that the difference between the two shafts was negligible.

Overall, the results of the experimental mine show more similarities with the measurements from the wells than with those of the shafts. Therefore, there is no urgent need to discuss more of the results of this laboratory mine.

12.4.4 Conclusions

The final temperature–salinity profiles of a shaft or a well are clearly unpredictable. It is highly likely that steps or transition zones will occur at onsetting stations, but which onsetting station will exhibit such steps seems to be unpredictable. Interestingly, once stratification occurs, it appears to be rather constant, even over time scales of several years or even decades. Stratification in shafts is characterised by small transition zones between different mine water bodies, where wells have usually larger transition zones. Another interesting feature is the fact that – though the depth of the different mine water bodies seems to be rather constant – this is not true for the electrical conductivities within the different layers. There seems to be a common trend that electrical conductivities decrease with time, such as in the Willem II and

Friedrich and in the lower parts of the Sicilia as well as the Glückaufsegen 3 shafts. These differences prove that the water is not in a steady state condition.

12.5 Ronneburg Uranium Mine, Thuringia, Germany

12.5.1 Description of Location

Ronneburg is located in Thuringia/Germany, 8 km southeast of Gera and 50 km west of Chemnitz. Uranium mining by the former Soviet-German company, SDAG Wismut, started there in 1951 and stopped in 1991. During that time, several interconnected underground mines with a total volume of $68 \cdot 10^6$ m³ were under operation, of which $44 \cdot 10^6$ m³ are backfilled; a now backfilled surface mine with a volume of $160 \cdot 10^6$ m³ of open space covered a total area of 60 km². There were 22 mine levels at depths of 30 to 940 m below the surface, interconnected by 40 shafts, totalling to a length of 3,000 km. In 1998, flooding of the underground mines started, following the installation of about 120 underground dams to separate areas with high pollution potentials from those with low ones. Only the deep and isolated mine parts, adding up to $0.8 \cdot 10^6$ m³, were already flooded between 1989 and 1997. Final seepage of the mine drainage water is expected to occur at several topographic low areas, especially in the Gessenbach valley (Unland et al. 2002). To control the flooding process and to prepare for possible unexpected flooding situations, a 600 mm monitoring and pumping well and 20 monitoring wells were constructed, as the shafts were backfilled. While the mine water in the southern fields has to be treated to remove uranium, radium, iron, arsenic, and potentially toxic metals, the mine water in the northern field will probably not need treatment. The end of flooding was projected to occur between 2003 and 2005, and it is anticipated that treatment will be required for another 20 years (Gatzweiler et al. 2002; Paul et al. 2002). As of March 2007 no water has discharged yet.

All of Ronneburg's uranium deposits (Schmirchau, Reust, Paitzdorf, Beerwalde, Drosen, Dittrichshütte, Lichtenberg, and Stolzenberg) are located within the northwest—southeast-striking Nejdeck-Crimmitschau-fault-zone, which also hosts the important Joachimsthal/Bohemia and Aue/Saxony uranium deposits (Zentralinstitut für Physik der Erde 1989; Lange and Freyhoff 1991; Büder and Schuppan 1992; Meinel 1993). Typically, the strata-controlled deposits form lenses and stock works within Palaeozoic pyritic black shales, limestones, and diabases (Dahlkamp 1993: Type 16). Besides the Silurian Graptolithenschiefer, Ordovician shales and Silurian carbonates ("Ockerkalk") are abundant and the rocks contain 0.5–7% sulphides, 0.5–9.5% organic carbon, and 40–60 ppm uranium. A total of 111 kt of UO_2 were produced (Szurowski 1968; Lange and Freyhoff 1991; Dahlkamp 1993).

An extensive effort to model flooding and post-flooding scenarios was previously undertaken at the Ronneburg mine site. Gatzweiler et al. (2002) attributed the insufficiency of the modelling results to the "complex mine structure, size and nature of the influenced ground water regime and changing infiltration water quality due to ongoing remediation work at the surface". As discussed earlier, the reasons the modelling failed was mainly that such systems are chaotic systems (see section 7.1.2); they cannot be numerically modelled far into the future because the accuracy of such models is quickly lost!

12.5.2 Appearance and Interpretation of Stratification

Stratification was observed in the uncased monitoring wells since the beginning of depth-dependent temperature and electrical conductivity measurements at the Ronneburg mine site. Paul et al. (2002) gave details of three measurements within the e-1260 observation well, located in the Ronneburg northwest mine field (March 2000, December 2000, March 2001). Unlike the other case studies, the shafts and most of the mine voids had been backfilled. Therefore, monitoring is being conducted in specially drilled wells of comparatively small diameters. In these wells, the water will not be affected by free convective flow because of the small diameter to length ratio (see section 12.4), so it is possible that the observed effects are due, at least in part, to the hydrodynamics that occur in small diameter pipes. The figure published in Paul et al. (2002) is typical for standing water in wells and the conclusions drawn here and by the authors, therefore, must be interpreted with caution (Fig. 105).

Nevertheless, it becomes clear that there is a stratification in the Ronneburg north-west uranium mine with highly contaminated mine water (electrical conductivity above 4 mS cm^{-1}) in the lower levels and less contaminated mine water (1–3 mS cm^{-1}) in the upper parts of the well. Furthermore, there is no evidence for an overall convective flow, as the temperature profile mirrors the geothermal gradient very well. This absence of mixing would explain the differences in the chemical composition of the mine water in the various mine fields and also proves that dams and backfilling delay the development of large scale mixing. However, as the experimental investigations of Robillard et al. (1986) or the field observations of Le Carlier de Veslud et al. (1998) showed, porous media does not exclude long-term thermal convection.

Some dynamic processes can be observed between levels 300 and 345 (the levels in Fig. 105 are taken from the original publication. Differences between the numbers given here and the figure are because mine levels usually are not horizontal, so that the levels meet every shaft at different depths).

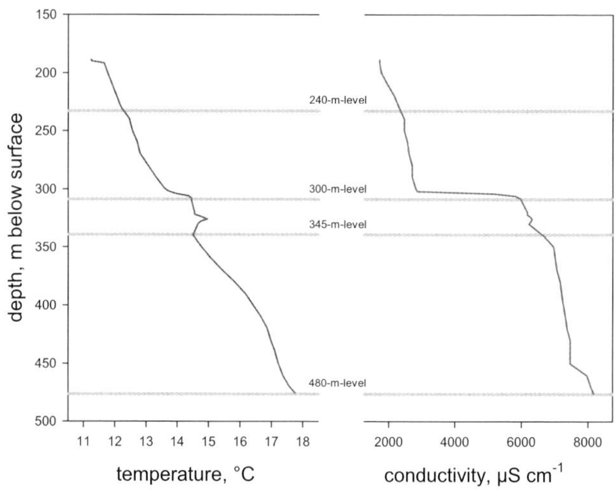

Fig. 105. Temperature and electrical conductivity plots of the Ronneburg/Germany north-west well e-1260 (modified from Paul et al. 2002).

Mine water with a 1 K higher temperature seems to enter the well at the level 300 and interacts with the water at the level 345. Certainly, there is a short-cut between both levels, causing a small ca. 45 m long convective flow cell, as shown by the temperature log.

Summarised, the measurements in the Ronneburg northwest mine field show that there are two separate mine water bodies. No convective flow occurs in either and the temperature is due to the regional geothermal gradient. It appears that the construction of dams delays mass transport because they obstruct the establishment of a regional convective flow regime.

12.6 Frazer's Grove Mines, County Durham, United Kingdom

12.6.1 Description of Location

Frazer's Grove fluorspar mines are situated in the United Kingdom's North Pennines, 40 km southwest of Newcastle-upon-Tyne, near the river Rookhope Burn in the Weardale District. Large-scale mining in this area started in the 18th century and stopped in 1999. Frazer's Hush underground mine was opened in the early 1970s and closed as the last working mine in the North Pennines. Similar to the Straßberg/Germany mine, the area around the Frazer's Hush mine was first worked for lead and zinc, and later for barium and fluoride. All four mines, Frazer's Hush, Rake Level, Firestone and Grove rake workings, reached down to about 200 m below the surface and worked on 5–9 interconnected main levels (Johnson and Younger 2000, 2003).

During the operation of the mine, a daily water make averaging $1.9 \cdot 10^3$ m^3 was pumped, which mainly originated from the Carboniferous Great Limestone. Grove Rake's № 2 Shaft had a depth of 160 m and is connected to the 30, 40, 50, 60, 80, and 90 fathom (fm) levels (55, 73, 91, 120, 146, 165 m). Though the 40 fm level is not fully connected to the № 2 shaft (Paul Younger, pers. comm.), the temperature-depth plot indicates that "something" happens at that level. Therefore, the 40 fm level is drawn in the figures 106 and 107. Unfortunately, the water level in Johnson and Younger (2002) figure 3 is not correctly drawn, causing some problems in understanding the hydraulic situation.

Geologically, the Frazer's Grove fluorspar mines are located in the Northern Pennine Orefield, which covers an area of approximately 1,400 km^2. The mine itself is located in the Eastern Fluorine Zone. An east—west trending syncline, the Stainmore syncline, divides the Ore field into a southern (Askrigg block) and northern part (Alston block), the latter one hosting the Frazer's Grove mines. Host rocks there are Lower Carboniferous sandstones, limestones, and shales of the Great Limestone between the overlying Millstone Grit and the underlying Weardale Granite. The mine worked the Greencleugh Vein, which consisted of the following minerals: galena, fluorite, quartz, sphalerite, siderite, calcite, pyrite, marcasite, bismuthinite, and pyrrhotite. Genetically, the mine is similar to the Mississippi-Valley-Type deposits of the USA Tri-State district, such as the Picher Mining District described in section 12.3 (Taylor 1978; Dunham 1990 [1948]; Fisher and Greenbank 2000; Johnson and Younger 2002).

12.6.2 Appearance of Stratification

Between 1998 and 2002, Johnson and Younger (2002) intensively investigated Frazer's Grove mines, concentrating on the mine water hydrogeochemistry. They found the mine water to be slightly acidic with electrical conductivities ranging between 500 and 3,200 µS cm^{-1} with Zn concentrations up to 111 mg L^{-1}, iron concentrations up to 68 mg L^{-1}, and sulphate as high as 2,000 mg L^{-1}. Furthermore, in 1999, they noticed that the № 2 shaft had stratified, with a pH of 3.8 at the lowest levels and 6.2 at the surface of the mine water. While the water quality has gradually improved, the stratification was preserved, although the character changed.

In June 2000, electrical conductivity, temperature, and velocity were measured in the № 2 shaft with continuously-measuring down-hole probes (Fig. 106). Three water bodies could be observed by then, being separated from each other at the 50- and 60-fm levels. At the top of the mine water body, the

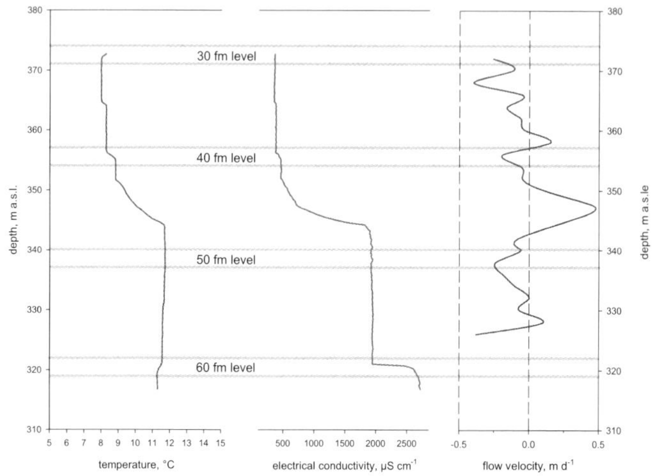

Fig. 106. June 2002 temperature, electrical conductivity, and flow velocity measurements in the Frazer's Grove № 2 shaft. All elevations of the four levels are taken from a graph, and might vary within several meters. The height of the individual levels has been estimated from a photograph taken by Wilkinson (published in the Internet at MinDat.org) and the photographs in Younger (2003) to be approximately 2–3 m (modified and supplemented from Johnson and Younger 2002).

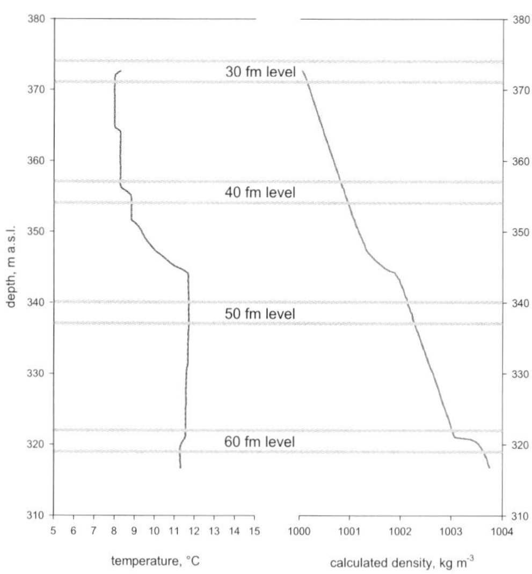

Fig. 107. June 2002 temperature measurements and calculated density in the Frazer's Grove № 2 shaft (explanations as in Fig. 106; modified and supplemented from Johnson and Younger 2002; density calculated according to Fofonoff and Millard Jr. 1983).

water quality was relatively good (electrical conductivity ≈ 500 µS cm^{-1}) and in the deepest accessible parts, the water quality was relatively bad (electrical conductivity ≈ 2,600 µS cm^{-1}) with a slight temperature inversion occurring there. The mine water's density and the temperature are shown in Fig. 107, which clarifies that the density of the mine water steadily increased from the mine water surface to the deepest measuring point reachable with the down-hole probe.

12.6.3 Results and Interpretation

Based on the velocity profile, Johnson and Younger (2000) interpret the results as follows: water is moving downward from the water surface to 350 m a.s.l. and there water enters the shaft from the Great Limestone, thus causing the water to flow upward. Yet, as Younger (2003) noted, the shaft is fully brick-lined and so it is unclear how water should enter the shaft from the aquifer through the bricks. Hydraulically, this interpretation is highly problematic, because the interpretation does not explain where the downward-moving water and the water entering the shaft at 350 m a.s.l. are flowing towards or how the water is entering the brick-lined shaft from this aquifer. One of the reasons for their misinterpretation might be that they forgot to draw the 40 fm level in their publication (though it is correct in table 2 of Younger 2003). Furthermore, if water would enter the shaft from the Great limestone (to be more precisely, at the 40 fm level), a forced flow would occur; after a small mixing zone between the warmer water swelling upward and the cold water from the aquifer, there should be a relatively homogenous temperature, electrical conductivity, and velocity profile. Considering the hydrodynamic relationships explained above, the situation can be explained easily:

- First, the velocity and the temperature profiles prove that there is a turbulent convective flow. No sign for a forced flow is given, as it is the case of the № 530 shaft of the Straßberg mine. Therefore, there are several convective cells and in each of the cells, water flows up and down, depending on the time of measurement and the current occurrence of the convective cell.
- Second, most of the water discharging from the № 2 shaft seems to come from the 30-fm level as there is no clear indication that water is up welling from the 50-fm or 40-fm levels to the 30-fm level. If that were the case, the linear range between ca. 344 and 352 m a.s.l. could not be stable. On the other hand, the logs clearly show that there is convective flow between the 60-fm and 50-fm levels as the temperature and conductivity stay nearly stable.

12.7 Frances Colliery, Scotland, United Kingdom

12.7.1 Description of Location

In 1995, pumping stopped in the Scottish Frances and Michael coal mines (East Fife Coal Field). It was predicted that after the mine water table recovered to a certain level, it would be necessary to start pumping again, as the mine water would be polluted and contamination of the area could only be prevented by treating the mine water (Nuttall et al. 2002; Nuttall and Younger 2004). As mine water pollution ranks fifth in the causes of Scottish water pollution (Anonymous 1999a), the treatment of such types of water are given special consideration; therefore, an intensive pumping test was carried out at the Francis Colliery site.

The Fife coal field, together with the Ayrshire, Sanquhar, Douglas, Central, and Midlothian coal fields (Wood et al. 1999), is part of the Midland Valley tectonic structure, an intra mountainous Old Red Sandstone basin filled with Ordovician to Devonian rocks. After the Old Red development came to an end, Carboniferous sediments were deposited, among them the thick Westphalian "Coal Measures" (Schönenberg and Neugebauer 1997). They were folded at the end of the Palaeozoic and within the 500 m thick sequence at the Frances Colliery, about 20 coal seams of workable thicknesses with up to 1.5 m width are reported (Nuttall et al. 2002; Nuttall and Younger 2004). Some of those sequences, namely the Dysart Main Seam, contained up to 3% of pyrite, and so were a potential source of acid drainage generation. Though initial pH values in the Frances Colliery were nearly neutral, they decreased after several days of pumping and reached values as low as were reported from the active mining period (Nuttall and Younger 2004).

12.7.2 Appearance and Interpretation of Stratification

Stratification in the Frances mine was not known before the 1995 pumping test. Because the quality of the pumped water decreased rapidly during the six weeks pumping test, it was assumed that this was the result of a stratified mine water body. At the start of the pumping test, the electrical conductivity was 5.6 mS cm^{-1}; it increased to 25.6 mS cm^{-1} after day one and reached a value of 26.5 mS cm^{-1} at the end of the test (Nuttall et al. 2002; Nuttall and Younger 2004). At the same time, the iron concentration increased from 7 mg L^{-1} to 597 mg L^{-1}, thus exceeding the pre-flooding iron concentrations of 7–41 mg L^{-1}. Several months after cessation of the pumping test in 1995 and at two later occasions, in 1997 and 1998, another set of vertical samplings were conducted in the Frances Shaft. It could clearly be shown that the mine water was stratified, with the two layers separated at a depth of 260–

270 m below ground level with electrical conductivities of about 3 mS cm^{-1} in the upper portion and 20 mS cm^{-1} in the lower portion of the mine.

For the duration of the pumping test, Nuttall et al. (2002) calculated Reynolds numbers of Re = 190–350 and concluded that the flow in the four interconnected adits, consequently, was turbulent. Whilst this might be true for the duration of the pumping test, there is no evidence that during non-pumping times the flow in the adits must be turbulent as well.

From the available data, it can be concluded that a major inflow from an adit of less contaminated water exists between 260 and 270 m below the surface. This adit is hydraulically well connected with other adits at higher levels and, within the part of the shaft above the stratification horizon, laminar or turbulent convective mixing is occurring; otherwise, the physico-chemical parameters of the water column above the stratification horizon would be different. It can clearly be seen that the lowest portion of the mine is not taking part in the convective flow of the higher portion of the mine and that the mine water quality in those deeper parts is therefore rather poor.

12.8 Stahlberg Iron Mine, Siegerland, Germany

12.8.1 Description of Location

The Stahlberg Iron Mine is located near Müsen (part of Hilchenbach), 14 km northeast of Siegen in the German Rothaar Mountains of the Variscan Rhenish Schiefergebirge. Another 16 km north of Müsen is the well known Meggen iron mine with the Sicilia shaft, another mine where mine water stratification with turbulent convective flow can be observed (Coldewey et al. 1999; Hasse 2000; see section 12.4).

Lusznat (1985) and Döring (1999) gave a comprehensive description of the mining history and geological situation of the Stahlberg iron mine, which is dewatered by the *Kronprinz-Friedrich-Wilhelm-Erbstollen* drainage adit (1826–1875, 144-m-level). Today, the only accessible shaft is the № II shaft in the recreation area Stahlberg, with a depth of 670 m (-243.7 m a.s.l., 660-m level). Using it, 14 sub-levels are accessible, 12 of which are flooded. At the 304-m level, the № II shaft is connected to the 312 m deep Stahlberg shaft, which is also drained by the *Kronprinz-Friedrich-Wilhelm-Erbstollen* adit. Below the 304-m level, the № II shaft was used to explore the *Neues Mittel* ore body, which was mined in an overhead stoping technique.

Mining dates back to the year 1313 (open pit mining) and the mine was worked continuously until 1930. The host rock of the two main siderite ore bodies *Altes Mittel* und *Neues Mittel* (Fe: ≈ 37%; ≈ Mn: 8.12%) are the Lower Devonian Siegen and Müsen series, consisting mainly of shales and schists. Where the siderite died away, quartz was the gangue material, accompanied by pyrite, copper pyrite, fahlore (0.98% Ag) and galena (0.06–0.07% Ag). During its more than 600 years of mining, a total of $1.7 \cdot 10^6$ t of iron ores were produced (Lusznat 1985).

12.8.2 Appearance and Interpretation of Stratification

So far, only one measurement has been conducted in the № II shaft of the abandoned Stahlberg Iron Mine. On 12 November 2000, a 500 m cable dipper with a resolution of 0.1 K was lowered to a depth of 500 m below the surface and temperature was measured at every 10 m. Two water samples were collected, at 160 and 210 m below the surface (Fig. 108). Due to a malfunction of the on-site instruments, no pH, redox, or electrical conductivity readings were made. At the water surface, which was 143 m below the top of the

Table 26. Selected parameters of the Stahlberg № II shaft mine water. Depth in m below surface, Temp in °C; trace elements and main ions in mg L^{-1}.

Sample	MÜS-1211-160	MÜS-1211-210
Depth	160	210
Temp	12.8	13.1
Na	4.4	4.4
K	2.2	2.2
Ca	45	45
Mg	46	46
Li	< 0.05	< 0.05
Cl	9.6	9.5
NO_3	< 1	< 1
SO_4	64	61
Zn	1.12	0.94
Mn	0.27	0.29
Sr	0.13	0.14
Ba	0.05	0.06
Ni	0.03	0.02
Sb	0.01	0.01
Cu	0.01	0.01
As	0.005	0.005
Pb	0.004	0.004

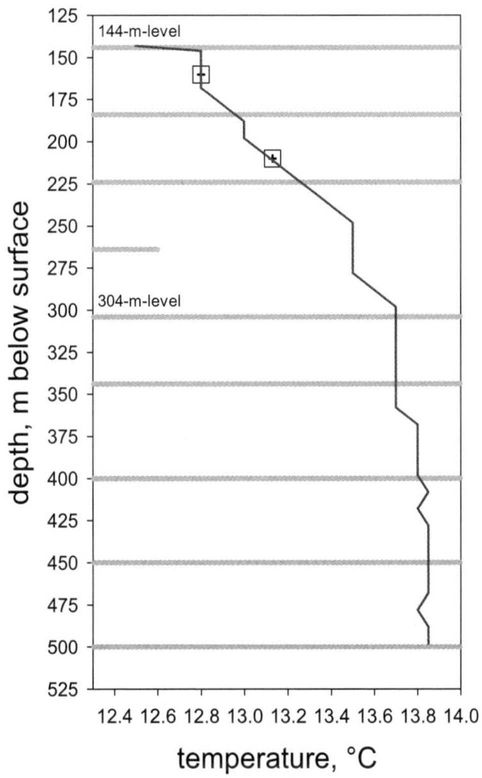

Fig. 108. Temperature–depth profile in the № II shaft of the abandoned Stahlberg/Germany mine (measurement on 12 November 2000). The gray horizontal lines indicate the depth of the different mine levels and the two squares the depth of the water samples. Resolution of the temperature probe: 0.1 K.

shaft, the water temperature was 12.5 °C; it was 13.8 °C at a depth of 500 m below the surface, thus showing a temperature difference of 1.3 K. Between the 410 and 500 m depths, the temperature measurements fluctuated between 13.8 and 13.9 °C, indicating turbulent flow conditions. The metal concentrations of the mine water were reasonably low, and overall, the chemical composition of the mine water did not differ significantly between the two sampling points (Table 26).

Disregarding the small temperature changes in the depth profile, the mine water body appears to have two parts: a lower part with a relatively stable temperature below the 304-m level and the uppper one with decreasing temperatures between the 304 and the 144 m level drainage adit. Considering the theoretical considerations and the experience from better examined shafts, turbulent convective flow seems to be taking place below the 304-m level, while the upper part is dominated by a stagnant, diffusive flow situation, thus indicating a connection between the Stahlberg and № II shafts.

12.9 Niederschlema/Alberoda, Saxony, Germany

12.9.1 Description of Location

The Niederschlema/Alberoda uranium deposit is located 28 km southwest of Chemnitz/Saxony in a densely populated area of Germany. The mine has a depth of nearly 2,000 m and consists of 50 main levels with more than 50 shafts connecting the different levels of the mine. The length of the mine works sums up to 4,150 km (Büder and Schuppan 1992) and the total volume to $36 \cdot 10^6$ m³. For 45 years, the mine has been operated by the SDAG Wismut, a Soviet-East German Company. Due to economic, environmental and political problems of uranium mining and milling, the Government of East Germany decided in 1990 to close one of the largest gangue deposits for uranium in the world. Over its lifetime, the mine produced about 81,000 t of uranium ore (Böttcher et al. 1991; Mühlstedt 1992; Bundesminister für Wirtschaft 1993; Barthel 1993).

Geologically, the mine is situated within the Lößnitz-Zwönitz Syncline, a southwest—northeast striking fold zone north-west of the central Erzgebirge (Ore Mountains). The host rocks are of Ordovician, Silurian, and Devonian age, which are contact-metamorphosed in the vicinity of the post-kinematic granites of Aue, Kirchberg, Oberschlema, Auerhammer, Lauter, and Eibenstock. In conjunction with the granite intrusions, aplites, and kersantites intruded during the Stefanium and Autunium. Important for the emplacement of the mineralisation is the Gera–Aue–Joachimsthal thrust zone, which also hosts the famous Joachimsthal/Bohemia and Ronneburg/Thuringia uranium mineralisations. Typical rock types of the Niederschlema/Alberoda deposit are skarn, amphibolites, graphite-schists, meta-diabases, kersantite, quartzite, limestone, granite, marble and different types of schists. During mining, nearly 200 mineralised gangues were exploited (Janischewskij and Konstantinow 1962; Zentralinstitut für Physik der Erde 1989; Lange et al. 1991; Büder and Schuppan 1992; Stemprok 1992; Meinel 1993). About 140 minerals are important in the uranium mineralization and the hydrogeochemical development of the mine water (Schröder and Lipp 1990a, 1990b). They formed during one or more of the mineralization cycles of the Erzgebirge. The most important uranium and arsenic minerals were uraninite (UO_2) and loellingite ($FeAs_2$).

The controlled flooding of the mine began in 1991 and was projected to be completed in 2006. As the mine water circulates through the open mine works, it becomes enriched in various elements, some of which are toxic. Therefore, the remediation plans for the Niederschlema/Alberoda mine include active and passive treatment of the discharged water as soon as the water table reaches the Markus-Semmler drainage adit (Gatzweiler and Mager 1993; Hagen and Lange 1995; Gerth et al. 2005).

Before the inundation process began, little was known about the hydrogeochemical, thermal, and hydrodynamic processes that would take place. Numerous experts were consulted and a Ph.D. thesis was undertaken (Wolkersdorfer 1996), but still the hydrodynamic conditions during the flooding and within the water body have only been outlined in general (Merkel and Helling 1995).

Meyer et al. (2002) stated that "according to current knowledge, convective exchange will diminish and continue in mid- and long-term" without giving any references to their "current knowledge". As discussed above, "convective exchange" cannot "diminish" or "increase"! Convective exchange either exists or does not exist.

In the deepest parts of the mine, rock temperatures of 70 °C were measured. It is not known if the water there currently has the same temperature, but in January 1991, a water temperature of 52 °C was observed at level -1800 (Günther Fröhlich, pers. comm.) and between June 1992 and December 1994, a maximum water temperature of 41 °C with an average of 36 °C was measured (Merkel and Helling 1995). Convective flow will therefore exist as long as the energy source exists and as long as no measures are undertaken to interfere with the convective cells to prevent convective exchange and mass transfer. Wolkersdorfer (1995) has predicted that a temperature of 33–35 °C will exist at the end of the flooding process.

12.9.2 Investigations conducted

Between 1992 and 1993, the author conducted discontinuous and continuous in-situ temperature and electrical conductivity measurements (Wolkersdorfer 1996). Discontinuous temperature measurements were conducted with a 500 m downhole dipper at intervals between 1 cm and 10 m, depending on the degree of stratification known from each previous measurement. Continuous multi-parameter measurements were made with electronic probes from the Gommern/Germany geophysical apparatus company at predefined depths to investigate the course of the water rise and the change of the physico-chemical parameters during the rise of the water level. In addition, since the flooding process started, the mine operator has conducted intensive chemical and physico-chemical monitoring of the water according to the mine abandonment plan, though the data since 1993 are not available for discussion in this publication. A continuous downhole measurement in the № 371 shaft has been presented by Meyer and Schramm (2004) and is used to interpret today's hydrodynamic situation within the mine.

At the end of 1995, Wolkersdorfer (2001) conducted a multi-tracer test in the mine. This test evaluated the hypothesis that the water within the mine is well mixed because of convective flow within the well-connected parts of the mine. Four differently coloured bulks of *Lycopodium clavatum* spores were injected at different, pre-defined depths in 4 shafts.

12.9.3 Appearance of Stratification

During the active flooding period, the water in the mine showed a clear density stratification with usually higher temperatures in the deeper parts of the mine and lower ones in the higher parts (with some exceptions). Most of the temperature logs of the in-situ shaft measurements showed relatively constant temperatures and electrical conductivities in the lower parts of the mine while the temperature and the electrical conductivity in the upper parts decreased at lower water depths. A continuous in-situ measurement reported by Meyer and Schramm (2004) showed that this situation remained stable in 2004. Yet, only three levels are connected to the measuring location shaft № 371: the -540, -990, and -1080 levels. As Fig. 109 shows, the flooded portion below the -540 level (-227 m a.s.l.) shows a relatively constant temperature profile, whilst above that elevation, the temperature steadily increases with lower water depths. This is characteristic of convective mixing in the lower portion and diffusive flow in the upper parts of the shaft. However, the shaft is not connected to levels above the -540 level, so the water above that level just stands in the shaft and diffusive flow dominates. It thus cannot be seen as an indication of an overall stratification in the abandoned Niederschlema/Alberoda mine.

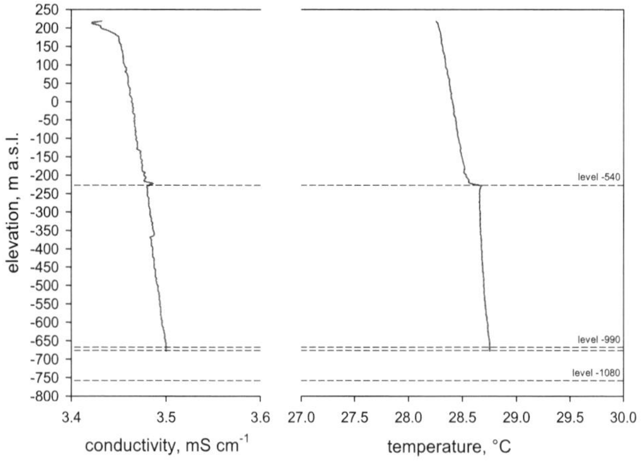

Fig. 109. Electrical conductivity and temperature measurement in the Niederschlema/Alberoda № 371 shaft on 10 March 2004 (modified after Meyer and Schramm [2004] and Jürgen Meyer, pers. comm.).

The course of the stratification could be followed in three shafts in mid-1990 because of repeated temperature measurements (up to 51 weekly to bi-weekly down-shaft measurements). It could be seen that the stratification breaks down when the temperature difference between the upper and lower stratified part falls below a threshold value of about 5–6 K. Once the stratification breaks down, it does not develop again. The reasons for this behaviour are given in section 7.2.1.

As the temperature showed significant fluctuations, turbulent conditions must be assumed to have existed in the observed shafts. Furthermore, the measurements showed that the temperature differences within a single shaft ranged between several K and less than a K and otherwise stayed relatively stable (Fig. 110).

Because of the two tracer tests conducted (1992 and 1995), it became clear that large parts of the mine are extremely well connected hydraulically; effective mean water velocities of up to 8 m min^{-1} were determined.

12.9.4 Results and Interpretation

The measurements of Wolkersdorfer (1996) show that the hydrodynamic characteristics of the seven monitored shafts were very similar during the flooding process. Only one shaft showed differing conditions that could not readily be explained (shaft 372 b above the -1305 level; Fig. 111).

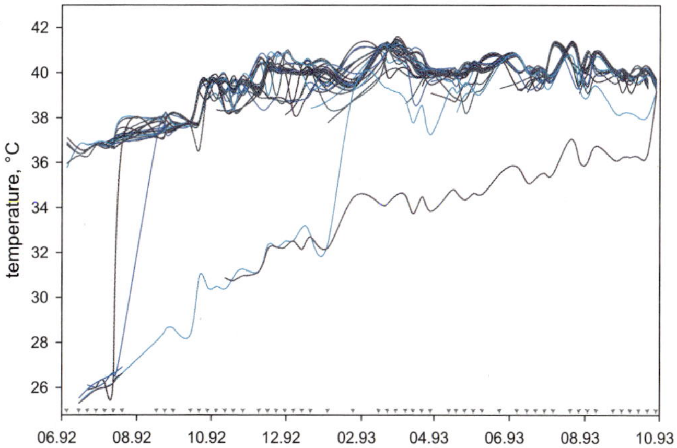

Fig. 110. Temperature–time course in the flooded № 366 II b shaft of the abandoned Niederschlema/Alberoda uranium mine showing a turbulent flow scenario in this shaft. The small triangles above the abscissa with the time scale indicate the individual measuring dates. Each line represents the temperature course in a predefined depth of the shaft (modified and amended after Wolkersdorfer 1996).

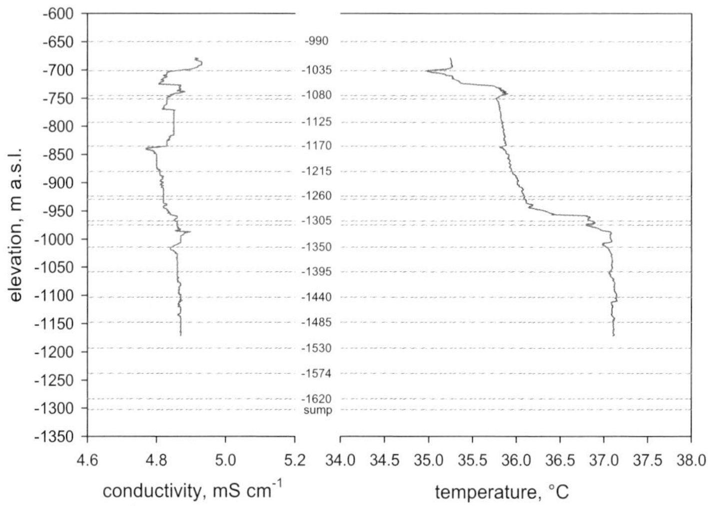

Fig. 111. Electrical conductivity and temperature measurement in the Niederschlema/Alberoda № 372b shaft on 15/16 March 1994.

One of the most prominent similarities is that the changes in the physico-chemical parameters during the flooding process were similar in every shaft. This can be seen especially well in the temperature progression of the shafts №s 366 II b, 371 II b, and 372 b (Fig. 111), but the electrical conductivity and pH measurements show the same trends. All of these shafts were stratified, showing two shaft mine water bodies and a separate water body in the shaft's sump.

The lower shaft water body ranges from the lowermost level in a shaft to the uppermost fully flooded level whilst the upper shaft water body consist of the water body above the last fully flooded level. Both water bodies have distinguishing properties: the lower shaft water body is characterised by similar physico-chemical properties throughout the whole water column, whereas the physico-chemical parameters in the upper shaft water body tend to have an increasing or decreasing trend up to the water surface, depending on the parameter. While the features of the lower shaft water body are typical for convective transport, the upper one indicates diffusive transport. The convective behaviour of the lower shaft water body is also supported by the oscillations in the stationary temperature and electrical conductivity measurements (Uerpmann 1980; Gebhart et al. 1988). Those features can be used to interpret the Niederschlema/Alberoda mine as an open thermosyphon system, as described by Gebhart et al. (1988) and Bau and Torrance (1981a, 1981b).

In both the shafts and adits, the rock temperature, resulting from geothermal processes, affects the mine water. If the rock temperature is higher than the mine water temperature, the mine water is heated, while the water is cooled at the water table and where the rock temperature is below the mine water temperature. The warming of the mine water, usually in the deeper parts of the mine, results in less dense water, so that the water flows upward until it either reaches the water surface, a boundary layer, or a depth where the density of the up flowing water equals that of the water already at that depth. Because of cooling at the water surface or higher levels, the water flows downward in the centre of the shaft, thus inducing a natural convection, also called free convection (see section 7.1). Ideally, laminar convective flow cells would develop in the Niederschlema/Alberoda shafts and adits, but the mine water velocity together with the numerous shaft installations (e.g. guide rails, air tubes, ladders, power supply lines, air ducts) and the surface roughnesses hinder such laminar flow regime. Therefore, it must be concluded that within the Niederschlema/Alberoda mine, the following hydrodynamic situations exist:

- open loops (Fig. 112a),
- open loops with an isothermal reservoir (Fig. 112b),
- turbulent convective flow (Fig. 112d), and
- areas with diffusive transport (Fig. 109)

Neglecting the flow in the matrix, due to its low hydraulic permeabilities, the flooding of the Niederschlema/Alberoda mine (and very likely also in other relatively deep multi-shaft mines), developed in four stages, which are characterised by open loops, open loops with isothermal reservoirs, and finally closed loops (Fig. 112).

Stage I a

Seeping water collects in the sumps and mixes with the water already there. The water flowing into the shaft from the onsetting stations usually has a higher temperature than the water in the sump, as it comes from greater depths. Therefore, it does not mix with the water in the sump and forms a separate layer. If a closed loop has developed, the water flowing through the onsetting station into the shaft can be seen as an entrance flow (Kranawettreiser 1989) and no mixing between the sump and the shaft water body occurs.

Stage I b (Fig. 112a)

Water overflows the sumps and the first level and starts to rise in the shafts. According to the principles of communicating pipes, the water level establishes in accordance with equation 31. The shafts and the first flooded level (excluding the sump) form an open loop with two reservoirs. In that system, the flooded level is the heater (energy source) and the shafts and the water level are the coolers (energy sinks).

As summarised by Bau and Torrance (1981a) and Gebhart et al. (1988), the water flows up- and downward in the vertical workings, driven by convective density driven flow. Even if the interconnected shafts have slightly different temperatures, the overall flow regime is not significantly affected. For the Niederschlema/Alberoda mine, this means that the water in the adits and shafts as well as the workings mix with each other. The water in the sump only mixes with the rest of the water body if the density difference is lower than about 2 kg m^{-3} (Kranawettreiser 1989). The oscillation that can be expected during this type of flow is well demonstrated by the temperature measurements.

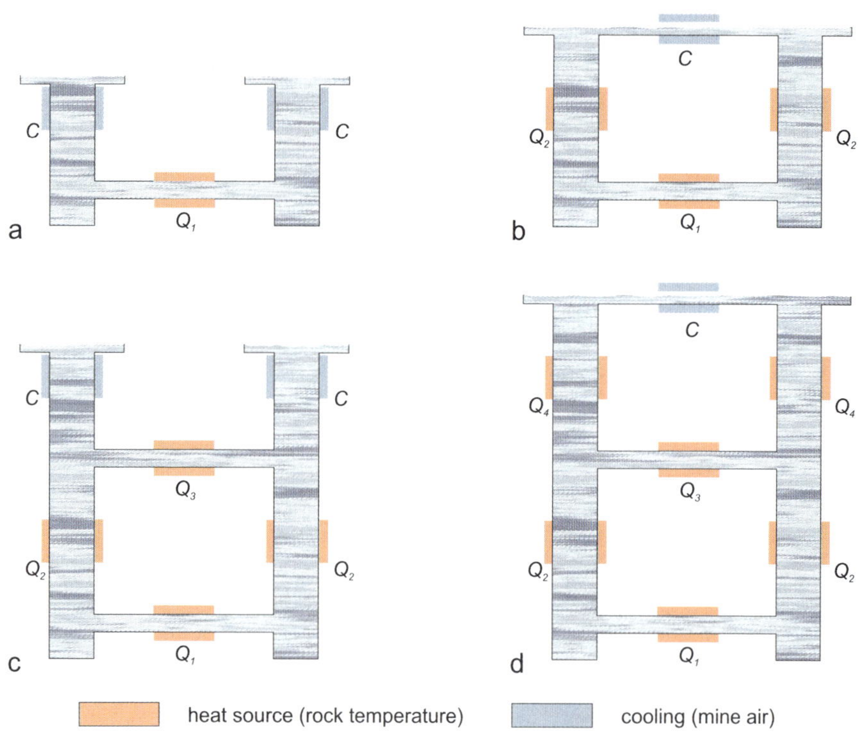

Fig. 112. Schematic representation of the flooding process in an abandoned underground mine as a combination of open and closed thermosyphons with free convection. Explanation see text. Heat flux $Q_1 > Q_2 > Q_3 > Q_4$ (modified after Wolkersdorfer 1996).

Stage II (Fig. 112b)

Starting with the flooding of the next level, the two reservoirs can form a single, near isothermal reservoir. Bau and Torrance (1981a) describe such a situation experimentally and theoretically. After a short initial phase, the water temperature at the flooded level reaches its maximum. Both the maximum temperature and the time until the system reaches a stable mean temperature are a function of the heating rate *Q*. Compared to stage I b, no regular flow builds up; instead, the flow likely changes its direction depending on the initial conditions. If the flooding becomes stationary, the temperature in all parts of the mine will be nearly homogenous, but this is unlikely to occur at a mine like Niederschlema/Alberoda since the rock temperature will not cool enough to match the air temperature.

A behaviour similar to that one described experimentally by Bau and Torrance (1981a) could be observed by a stationary measurement conducted in August 1992 (Fig. 113), when a level overflowed.

Stage III (Fig. 112c)

In stage III, the water rises further in the shafts connected to the second flooded level. Only the water at stations below the on-setting stations is hydraulically connected; the water in the shafts is not. As observed in several measurements, the quasi-isolated water in the shafts cools relatively quickly

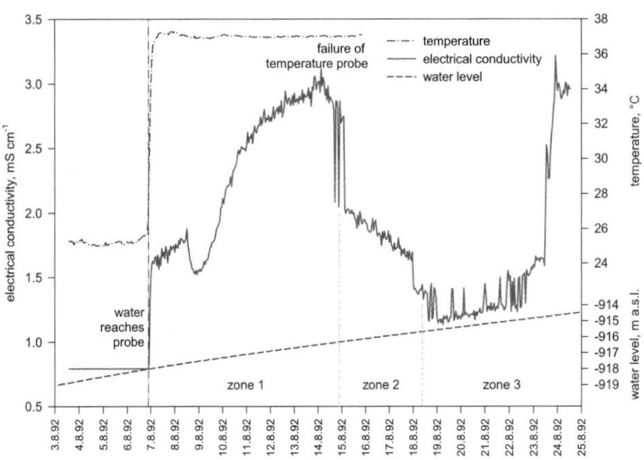

Fig. 113. Temperature and electrical conductivity measurements in the № 366 II b shaft of the abandoned Niederschlema/Alberoda mine during the overflow of the -1260 level (Wolkersdorfer 1996).

and it takes a considerable period of time before the water in the shaft equilibrates with the temperature of the already flooded parts below the on-setting station. This diffusive and convective temperature adaptation between inflowing water (ca. 20 °C) and mine water (ca. 40 °C) is hindered by the different densities of the water as a function of their temperature and mineralisation. In addition, the closed flow field of the lower shaft water body and the geometry of the shaft hinder a complete mixing of the upper and the lower water body. Furthermore, when there is an intermediate layer of thickness $z(\upsilon, g, \rho_1, \rho_2)$ between the upper and lower shaft water bodies, direct interaction between the two layers is nearly impossible (Kranawettreiser 1989).

In a closed loop, temperature as well as velocity varies continuously. In some critical cases, depending on the border conditions, it is even possible that the flow field reverses (Bau and Torrance 1981a). Creveling et al. (1975) and Damerell and Schoenhals (1979) demonstrated that the oscillating flow in closed loops continues to repeat over time.

Stage IV (Fig. 112d)

Stage IV differs from stage II in that the shafts can be connected at more than one level. The loop can incorporate two or three levels, depending on the available energy. Such an effect was observed at the 1992 tracer test, where the tracer was injected into two shafts at different depths. After 5 weeks, a homogenous mixing over 9 levels (about 400 m of vertical distance) was observed. Considering the experiments by Bau and Torrance (1981b) with asymmetrically heated open convection loops, shafts with a colder water temperature than the overall temperature may be included in the fluid loop (in that case, shaft 372 b: Fig. 111).

The final flooding scenario with pumping below the Markus-Semmler-adit level might be concluded at two of the four stages:

- If the flooding stops at stage III, the upper shaft water body is not in convective contact with the rest of the mine water pool. Pumping in such a shaft causes contaminated water from lower parts to flow into the pumping shaft.
- If the flooding stops at stage IV, the last flooded level is completely included in the overall fluid loop, causing the pumped drainage water to be polluted for a long time.

As none of those two scenarios results in a significant water quality improvement, another solution is needed: vertical hermetic sealing of the mine below the last flooded level. When attempting to do this, it is important to either interrupt or to prevent large convective flood loops, which quickly transport the pollutants. Sealing just some of the on-setting stations is not a recommended measure as small dams might quickly fail and therefore would not help to prevent the loop.

To conclude and to point out the impact of such convective loops, Bau and Torrance (1981b), who investigated the advantages of fluid loops in solar water heaters, state:

> Free convection loops provide a means for circulating fluid without the use of pumps.

12.10 Felsendome Rabenstein, Saxony, Germany

12.10.1 Description of Location

Rabenstein/Germany is a suburb of Chemnitz/Saxony and its former limestone mine is located at the western edge of the town. Since 1936, it has served as a popular visitor mine called *Felsendome Rabenstein* (Geißler et al. 1986).

Open-cut mines near the Rabenstein underground mine extracted limestone beginning in early medieval times (Riedel 1993). In 1834, the Rabenstein underground mine opened; 40 years later, the 35 m deep Maschinen Shaft was sunk, connecting the three lowermost levels with each other (15, 22, and 27 m below the surface). At the end of the mine's lifetime, four levels existed: the upper and lower level as well as the 1st and 2nd deep levels (level 3 and 4), extending over an area of 250 × 100 m with caverns ranging from 6–8 m in height (Fig. 114). An underground drift adit in the *Grüne Grotte* (green grotto) and the Maschinen Shaft connected the two deeper levels with the lowest level. To dewater the mine, a 220 m long drainage adit to the Pleiß brook was constructed in 1855, which still dewaters the mine from the lower level, with daily mean discharges between 3 and 40 m^3 and a weighted average of 12 $m^3\ d^{-1}$. Extreme precipitation can induce total discharges up to 120 $m^3\ d^{-1}$. During active mining, the water from the 1st and 2nd deep levels was pumped into the adit. In 1902, six years before the closure of the mine, the two deepest levels were allowed to flood and since then the shaft has been filled with a 15.8 m water column. At two points, the *Grüne Grotte* and the Maschinen Shaft, which are 140 m apart from each other, the flooded part of the mine is accessible. Dietrich (1965) calculated a flooding time of 12 weeks for the two underground levels, which seems to be rather low considering a mean flow of 12 $m^3\ d^{-1}$. Between 2002 and 2005, the electrical conductivity of the mine water was relatively constant at 1,190–1,260 μS cm^{-1}, with a pH of 6.5–8.9, an E_H of 350–510 mV, and a temperature of 7.5–8.7 °C. Slightly different values could be measured at the *Grüne*

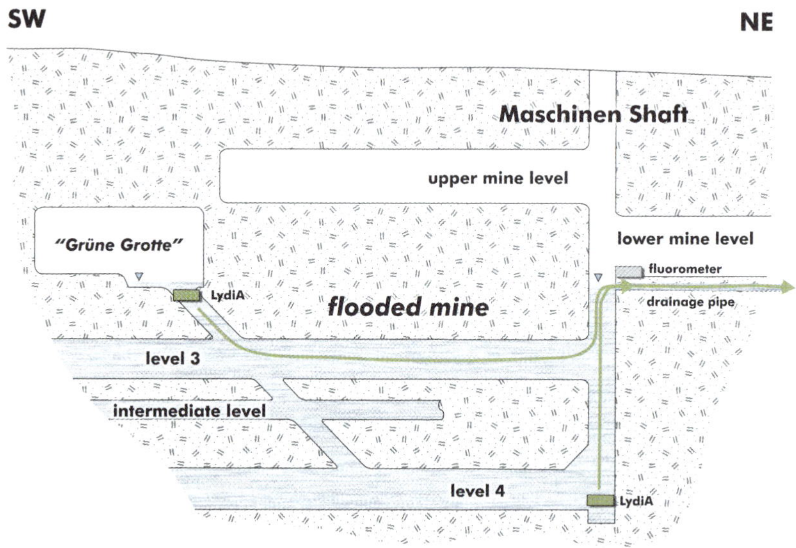

Fig. 114. Sketch of the Felsendome Rabenstein/Germany visitor mine with the two tracer injection sites, the location of the fluorometer, and the likely flow paths of the tracer. LydiA: Tracer injection Apparatus (Fig. 78). Not to scale in the horizontal direction and above the water table.

Grotte, from which the water flows through the flooded part to the Maschinen Shaft: electrical conductivity 800–1,000 µS cm^{-1}, pH 7.1–8.7, E_H 340–510 mV, and a temperature of 4.8–6.8 °C.

Geologically, the mined strata comprise mainly metamorphosed Palaeozoic limestone, a coloured calcite marble, in combination with phyllites, alum shale-like black phyllites, and amphibolites. Pyrite occurs frequently in the limestone and especially in the phyllites, and copper pyrite (chalcopyrite), often oxidised to malachite, can be found in vugs. The whole formation is slightly folded with a northwest striking folding axes, and faults, in a northeastern direction with offsets in the range of 5 m, are common (Dietrich 1965; Freyer et al. 1982; Geißler et al. 1986). Some of the faults are hydraulically active, as water penetrates from the surface to the mine during high precipitation periods (e.g. a fault between the *Blauer Salon* and the *Gnomengang*) and divers have reported that they can regularly see bead-like air bubbles moving in straight lines from the 2nd deep level through the rock into the 1st deep level.

From the available data, the volume of the Maschinen Shaft, the *Grüne Grotte* incline, and the flooded 1st and 2nd deep levels could be roughly cal-

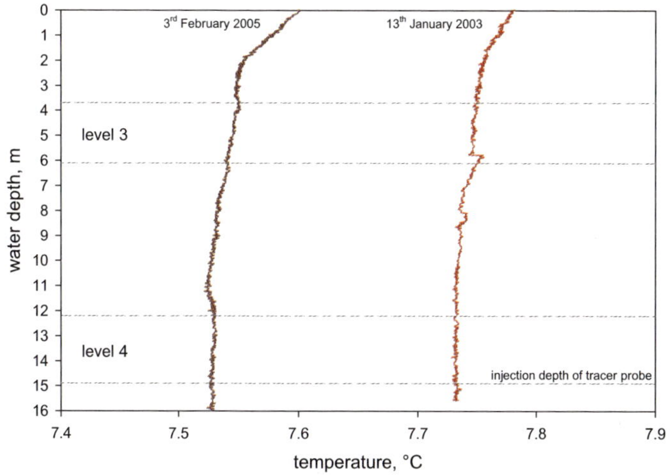

Fig. 115. Temperature–depth profile in the Felsendome Rabenstein machine shaft before and after the tracer test.

culated to be 124, 120, and 2,100 m³. Only the parts that were thought to be directly affected by the tracer test were used in the calculation. The total volume of each level must be another 20–30,000 m³ of volume. From that data, a total flooding time of several years seems more likely than 12 weeks.

12.10.2 Appearance of Stratification

According to the known records and the reports of divers, the flooded and partly backfilled Maschinen Shaft is connected to the 1st and 2nd deep levels at depths of 6 and 12 m below the surface of the water. Several small blind shafts and the incline within the *Grüne Grotte* connect the 1st deep level with the 2nd deep level. No data are available about the shaft's outflow because no precise flow measurements are possible there, but it is estimated at 1–2 L min^{-1}. Only a poorly developed thermal stratification can be found in the Rabenstein Maschinen Shaft. The temperature in the shaft is relatively constant at 7.9 °C and small temperature fluctuations can be observed at the two onsetting stations (Fig. 115).

Such a quasi-stagnant situation is typical for single shaft mines (e.g. Georgi Unterbau, section 14.13) where the geothermal gradient is not high enough to start free convection, nor does a major regional forced convection exist. Water in the shaft flows very slowly and it can be assumed that there is a laminar Poiseuille flow situation in the upper part of the shaft and diffusive flow in the lower part of the shaft.

12.10.3 Investigations conducted

In 2002, the flooded part of the mine was investigated hydrogeologically and the potential for a mine water tracer test was evaluated. The water quality was good and the pH was slightly basic, suggesting that no decomposition of the tracer, sodium fluorescein, would be expected. A discontinuous temperature log in the Maschinen Shaft showed a small temperature variation at the two flooded onsetting stations with no temperature shift. Later, continuous temperature logs revealed a small positive temperature shift of about 0.05 K / 10 m. Therefore, this shaft was selected for a tracer test in a mine with a low vertical flow characteristic and a simple mine geometry, such as in the case of the Georgi Unterbau mine.

Between 2002 and 2005, the mine was visited 16 times and the physico-chemical on-site parameters were measured regularly. One full analysis of the mine water was conducted, showing that the water is a Ca-Mg-SO_4-HCO_3-type water. Five temperature and electrical conductivity logs were conducted in early 2003 and 2005 before and after the tracer test with sodium fluorescein.

On 21 March 2003, the tracer test with sodium fluorescein, scheduled for a duration of about 3 months, was started in the Felsendome Rabenstein mine. With a tracer probe (LydiA; Fig. 63, Fig. 78), 15 g of sodium fluorescein were lowered to a depth of 15 m in the Maschinen Shaft and 250 g at a depth of 2–3 m within the incline of the *Grüne Grotte* (Fig. 114). After 96 weeks, the tracer concentration came down to background values and was therefore brought to an end. Thus, the Felsendome Rabenstein Tracer test has the unique distinction of being the longest lasting tracer test ever conducted in an underground mine (Fig. 116).

To clarify which of the two peaks belongs to which injection place, the vertical distribution of the tracer in the shaft was measured four times in the Maschinen Shaft and once in the *Grüne Grotte* incline.

12.10.4 Results and Interpretation

No relevant stratification could be observed in the Felsendome Rabenstein mine. Neither the temperature nor the electrical conductivity plots show signs of stratification, though small fluctuations in the area of the onsetting stations were observed. While the temperature slightly increases (by about 1.5 K) from bottom to top, the electrical conductivity decreases by about 100 $\mu S\ cm^{-1}$, proving a mainly diffusion-based flow scenario exists in the Maschinen Shaft, with only a small advective component.

Two tracer peaks could be observed, the first one at 65 days, the second one at 350 days after the tracer injection (Fig. 116, Table 27). The first peak came from the tracer injected at a depth of 15 meters in the Maschinen Shaft; the

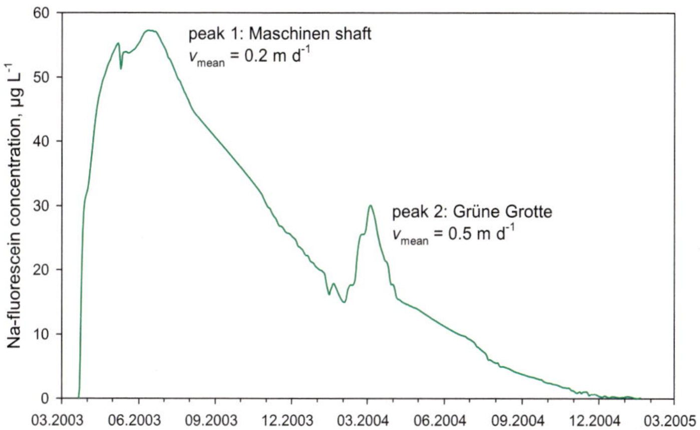

Fig. 116. Breakthrough curve of the 2003–2005 Felsendome/Rabenstein tracer test with Na-fluorescein.

Table 27. Results of the Rabenstein tracer test. Peak 2' is the reconstruction of the travel times for the *Grüne Grotte* tracer from the *Grüne Grotte* to the onsetting station of the 1st deep level at the Maschinen Shaft. It was calculated from the travel times and distances of peak 2 less the distance to the onsetting station and the mean travel times calculated from peak 1 and the mean of the effective and mean velocities of peak 1.

	Effective velocity v_{eff}			Mean velocity v_{mean}		
Peak	Time, d	Distance, m	Velocity, m d^{-1}	Time, d	Distance, m	Velocity, m d^{-1}
1	149	15	0.10	65	15	0.23
2	356	171	0.48	350	171	0.49
2'	320	165	0.53	314	165	0.52

second one was the result of the tracer injected 171 m away in the *Grüne Grotte* incline. From that data, an effective velocity v_{eff} of 0.10 m d^{-1} was calculated for peak 1 and 0.48 m d^{-1} for peak 2. The smaller peak after peak 2 might have been induced by divers training in the flooded underground mine.

In contrast to the first peak, which has an extremely long tail of about 620 days, the second one has a tail of only 120 days, indicating a larger longitudinal dispersion coefficient of $7.1 \cdot 10^{-6}$ m^2 s^{-1} for peak 1 and a smaller one of $1.3 \cdot 10^{-6}$ m^2 s^{-1} for peak 2. From the vertical tracer distribution in the Maschinen Shaft, it became clear that the first peak results from the tracer injected into the Maschinen Shaft and the second one from the tracer in the

Grüne Grotte. The long tail of the first peak shows that the mass transport from the 2nd deep level to the surface is possibly diffusion-based and the flow from the *Grüne Grotte* must be via the connecting adit between the incline and the Maschinen Shaft on the 1st level. If the water had followed the connecting adit at the 2nd level, a tailing similar to peak 1 would have been observed. It can be concluded that the transport from the 2nd level into the region of the 1st deep level is probably diffusion-dominated whereas the shaft above the 1st deep level shows Poiseuille-dominated flow.

12.11 Straßberg, Harz Mountains, Saxony-Anhalt, Germany

12.11.1 Description of Location

The German Straßberg mine is situated in the eastern Harz Mountains approximately 20 km south of Quedlinburg and 6 km west of Harzgerode and was the most important producer of fluorite in the former "German Democratic Republic" (Mohr 1978). At the beginning of mining, silver, copper, and lead were the targets of the miners. From the 18th century until 1990, mining focused on fluorspar, which was mainly found in the deeper parts of the mine (Bartels and Lorenz 1993). Between 1950 and 1970, the *VEB Harzer Spatgrube* joined the three main deposits of the Straßberg mining district by driving two deep adits on the 5th and 9th level (from north to south: Brachmannsberg pit: № 539 Shaft, Straßberg pit: Fluor Shaft and Glasebach pit: Glasebach Shaft). Finally, in 1991, the Straßberg fluorspar mine was closed due to economic and environmental reasons (Kuyumcu and Hartwig 1998) and the flooding of the mine began.

Geologically, the mine is situated in the Mid Harz Fault Zone close to the boundaries of the Lower Devonian Harzgerode zone and the Upper Devonian Tanne zone. Whilst the Harzgerode Zone consists of folded schists, shales, graywacke, limestone, lydites, metabasites, and metabasite tuffs, the Tanne zone comprises graywackes and shales (Schwab 1999). Besides fluorite, the Permian to Cretaceous hydrothermal poly-metallic mineralization of the vein structures comprises approximately eight ore and less than a dozen secondary minerals, among them pyrite, marcasite, chalcopyrite, galena, sphalerite, and arsenopyrite (Kuschka and Franzke 1974; Beck 1981; Junker et al. 1991).

12.11.2 Appearance of Stratification

On 31 May 1991, the drainage water pumps were stopped and the Straßberg and Glasebach underground pits started to flood. Between July 1992 and August 1998, in-situ temperature and electrical conductivity measurements in the № 539 shaft and the Fluor shaft (310 and 147 m deep, respectively) clearly showed that stratification within the water body was taking place (Kahmann and Heinrich 1998; Kindermann 1998; Rüterkamp and Meßer 2000). In the Fluor shaft, three water bodies, separated from each other at the 2nd (328 m a.s.l.) and 5th (243 m a.s.l.) levels, were established. Only two water bodies, separated by the 4th (357 m a.s.l.) level, could be recognised in the № 539 shaft. Evidence for the stratification were differences in temperature, electrical conductivity, and metal concentrations between each of the water bodies, the uppermost always lowest in those parameters, and the lowermost with the highest levels of iron, manganese, and sulphate.

Consequently, in 1993, the DMT – German Mining Technology, proposed to construct three new adits (Brachmannsberg adit, Biwender adit, new Glasebach adit), to drain and treat the lower contaminated mine water within the uppermost water bodies at a water level of 357.7 m a.s.l. (Rüterkamp and Meßer 2000). These three adits were built between 1995 and 1998 and have drained the mine since then. Two provisional active LDS facilities near the Fluor shaft and in the Uhlenbach Valley (close to the entrances of the Brachmannsberg and Biwender adits) are treating the circum-neutral Fe- and Mn-enriched mine water (pH 6.2–8.0, $n = 22.95\%$ conf.) using conventional liming technology. During the year 2007 both LDS treatment facilities will be substituted by a permanent HDS treatment plant.

After completion of the 3-adit system in 1998, the annual water budget of the Straßberg mine increased by almost $2 \cdot 10^6$ m^3. Furthermore, the stratification totally broke down in the № 539 shaft and partly broke down in the Fluor shaft (Rüterkamp and Meßer 2000, appendices 1 and 5), resulting in a generally higher contaminant load than expected. Similar circumstances had already been investigated during the flooding of the Niederschlema/Alberoda mine (Erzgebirge/Germany, Wolkersdorfer et al. 1997a). There, as long as the water level was under a main working level, stratification could be seen in the shafts above the last flooded level. When the main working level was flooded, a new loop was established and the stratification broke down immediately (Wolkersdorfer 1996). Very similar behaviour was observed at the Frazer's Grove/United Kingdom fluorspar mine where the stratification broke down after the water started to drain through the main shaft (Johnson and Younger 2000).

12.11.3 Forced Flow Situation and Tracer Investigations

Historically, the abandoned and flooded Straßberg/Germany fluorspar mine consisted of three mine units, with one central shaft in each: the № 539, Fluor, and Glasebach shafts. In the 2nd half of the 20th century, those units were connected with each other through two deep galleries. Today most of the mine

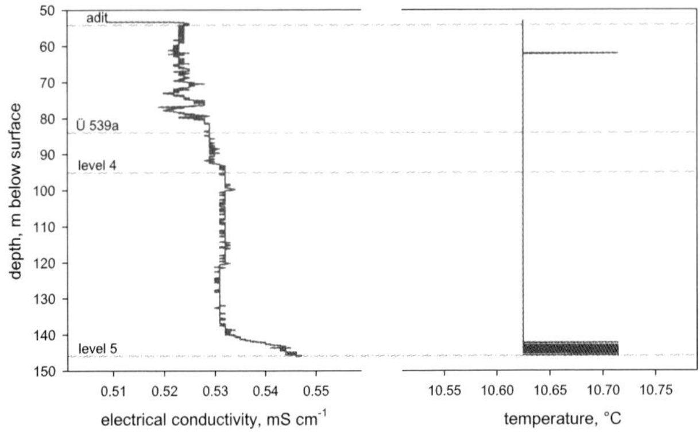

Fig. 117. Temperature and electrical conductivity measurement in the Straßberg/Harz № 539 shaft. Conductivity and temperature both show that there is no stratification in the shaft; there seems to be some minor inflow from the 4th level.

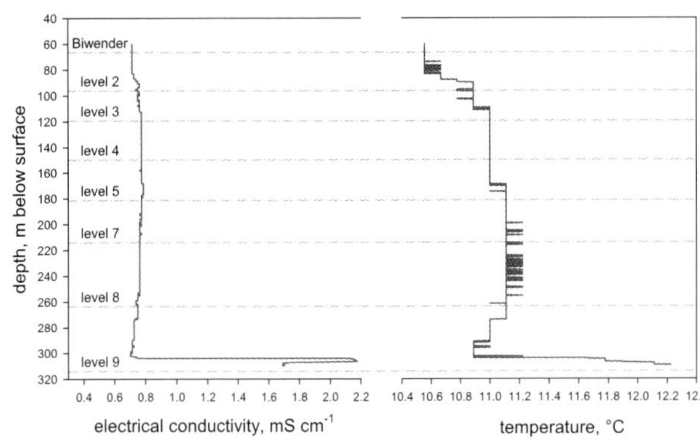

Fig. 118. Temperature and electrical conductivity measurement in the Straßberg/Harz Fluor shaft. Conductivity and temperature show that there is no significant stratification in the shaft except at the 5th level, an indication of water flowing into the shaft from the northern adit, connecting the Brachmannsberg mining field to the Fluor shaft.

water discharges through three drainage adits that were constructed in the 1990s and in which the flow is controlled through Thomson weirs. In June 2002, the water densities in the № 539, Fluor, and Glasebach shafts were 999.468 kg m^{-3}, 999.173 kg m^{-3}, and 999.524 kg m^{-3}, respectively. To prevent forced flow between the three mine units, as the mine operator preferred, the weirs at the outflows should have been set at 357.69 m a.s.l., 357.70 m a.s.l., and 357.59 m a.s.l. In fact, they were set at 357.79 m a.s.l., 357.70 m a.s.l., and 357.76 m a.s.l., forcing the water to flow from the № 539 shaft to the two other shafts.

To investigate the hydraulic conditions, a tracer test was conducted by Wolkersdorfer and Hasche (2001a). At the same time, an extensive water sampling campaign (Wolkersdorfer 2002a) and downhole measurements in the № 539 and Flour shafts were used to support the tracer test (Fig. 117, Fig. 118). A second tracer test at the same location was conducted in 2003 with results similar to the 2000 tracer test.

There were significant differences in the temperature and electrical conductivity profiles in the two shafts. While the № 539 shaft lacked clear stratification, the Fluor shaft was slighlty stratified, as indicated by the temperature profiles and, to a lesser degree, the electrical conductivity. The two tracer tests showed that the water in the Fluor shaft flows upward with a mean effective velocity of ca. $1.4 \cdot 10^{-2}$ m min^{-1}. This explains why the stratification is not as developed as in shafts with smaller water velocities. This can clearly be seen is the shaft's sump at the 9th level and the colder, inflowing water from the Brachmannsberg mining area at the 5th level. Furthermore, there seems to be some inflow from the 3rd level.

The Straßberg/Harz case study clearly shows that stratification in flooded mines is not a static condition but can be disturbed easily. As in the case of the Frazer's grove № 2 shaft, the stratification broke down when the water was allowed to flow freely out of the shaft (in Straßberg, this occurred when the three-adit system became functional). The stratification in the Straßberg mine would have been stable only if the three distinct parts of the mine had been hydraulically separated from each other. Without that hydraulic separation, the water is driven by convection to flow from the northern Brachmannsberg mining field to the central Straßberg and the southern Glasebach mining fields.

12.12 Reiche Zeche, Erzgebirge, Saxony, Germany

12.12.1 Description of Location

The Freiberg/Saxony mining district in Germany once belonged to the richest silver deposits in Europe (Jobst et al. 1994). In 1168 AD, mining started at a small place called Christiansdorf, which later became a part of Freiberg. Due to economic reasons, mining in Freiberg ceased in 1969 and uncontrolled flooding of the mine workings started. Because the deepest drainage adit is the 130 years old *Rothschönberger Stollen*, nearly all the mine water drains through that adit into a northerly direction and discharges 18 km north of Freiberg into the Triebisch and Elbe Rivers. According to different sources, the total flooded mine volume is about 2–5 $\cdot$ 10^6 m^3 (Kolitsch et al. 2001) and the discharge volume is 0.4–1.1 m^3 s^{-1} (Weyer 2003). Since the adit was completed in 1877, the mine water has been discharging into the Triebisch without further treatment, and even after the collapse in 2002, nothing has changed in this regard.

Below the flooding level of *Rothschönberger Stollen* are 494 m of flooded mine workings, which were driven into the polymetallic ore veins of the Erzgebirge (Oelsner 1958). The water penetrates through the flooded shafts, galleries, backfilled veins and open veins and gets enriched in nearly all metals of the periodic table, as can be seen from the chemical analyses (e.g. Baacke 2000). Although the *Rothschönberger Stollen* is the main drainage adit, several other drainage adits exist for isolated parts of the mine, such as the *Königliche Verträgliche Gesellschaft Stollen* and the *Fürstenstollen*, which dewater into the river *Freiberger Mulde* (Becke et al. 1986).

Though numerous studies have been conducted on the Freiberg mine waters, no comprehensive hydrogeological investigation has been done so far. This is mainly because only a small part of the mine is accessible but also because the responsibilities concerning the mine are complex.

Many parts of the Freiberg mining district, especially above the flooding water level, have extremely acid mine waters with pH as low as 2 (Baacke 2000), but it reaches 6–8 at the *Rothschönberger Stollen* portal (Yefimochkina 2004) with an electrical conductivity of 0.7–0.9 mS cm^{-1}. This shows that enough buffer capacity exists to neutralize the water, which mainly comes from the carbonates in the *Braunspatformation* (siderite formation) and the *BiCoNiAg-Formation* (bismuth, cobalt, nickel, silver formation). Geochemically, the discharge water can be classified as a Ca-SO_4-type, and, in some parts of the drainage water system, as a Ca-Mg-SO_4-type, thus showing that calcite and dolomite are the relevant buffering minerals.

Table 28 lists some important parameters of the *Rothschönberger Stollen* drainage waters between 1996 and 2004. Baacke (2000) calculated the average mass loads for some selected parameters for the three relevant discharge adits. If those numbers are used to conservatively estimate the mass loads leaving the mine between its closure in 1969 and 2007, some extreme values can be calculated: 3.3 t of As, 4,500 t of Zn, 1,800 t of Mn, and 117 t of Fe.

Table 28. Selection of mine water parameters from the drainage adit *Rothschönberger Stollen* (from Baacke et al. 1996; Beuge et al. 2001; Yefimochkina 2004). Electrical conductivity (Cond) in µS cm^{-1}, temperature in °C, concentrations in mg L^{-1}. Values are rounded in accordance with the original data.

Parameter	pH	Cond	Temp	Ca	Mg	Cl	SO_4	Zn	As	Cd	Pb	Mn
Min	6.8	620	12	100	24	35	155	2	0.001	0.01	0.002	0.8
Max	7.5	980	14	112	40	54	300	5	0.003	0.03	0.4	2

Some of the existing publications gave a conceptual model of the hydrogeological situation in the flooded and the non-flooded part of the Freiberg mining district (e.g. Baacke and Degner 2000). From a hydrodynamic point of view, the existing conceptual model appears impossible, in that water cannot flow down a shaft against hydrostatic pressure. Water can flow in fluid loops if the hydrodynamic situation allows fluid loops or it can flow through flooded levels in the direction of the lowest hydrodynamic pressure. Moreover, the water flow can split in one of two directions relative to the different hydrostatic pressures at each of the end-points of the possible flows. Tracer tests conducted by the Hydrogeology Department of Freiberg University in other mines (e.g. Wolkersdorfer and Hasche 2001a) clearly showed that the conceptual model of the Freiberg mine needs to be amended (Wolkersdorfer et al. 2007).

12.12.2 Investigations, Stratification, and Interpretation

The only accessible place in the Reiche Zeche mine where the water below the flooding level can be studied is the *Reiche Zeche Schacht*. No other location, so far, is known, where the flooded part of the mine could be investigated. Therefore, the only depth-dependent measurements known are from this location. First, continuous physico-chemical measurements were conducted by Peter (1983) and Zittnan et al. (1990). Their measurements showed that the water in the flooded shaft was stratified at the ½ 10 level (½ 10. Sohle) and the 13 level (13. Sohle); consequently, three mine water bodies could be separated from each other. Though the temperature showed a clear stratification in 1982, this was not true for the chemical composition. Zittnan et al. (1990) also conducted velocity measurements using radioactive tracers and found upward and downward movements of the mine water. Between the water surface and a depth of 412 m, they measured velocities between 0.005 and 0.47 m min^{-1} within 4 convection cells. Interestingly, the cells were not associated with the different levels in the shaft and differed from what was observed by Czolbe et al. (1992) in the Peißen shaft, where water was flowing upwards along the shaft's walls and downward in the middle of the shaft. Flow measurements in 2001 indicated velocities between 0.02 and 0.06 m min^{-1} at 12 different depths and an overall trend of upward movement in the water (Sebastian Kolitsch, pers. comm., the range of that measurement is drawn with two vertical lines in Fig. 119; more than a decade lies between the two velocity measurements). Yet, also "no flow" was measured, thus indicating that turbulent convection cells existed in the shaft. Therefore, it seemed that Zittnan et al. (1990) may have measured part of the upwelling, part of the convection cell, and part of the downwelling one.

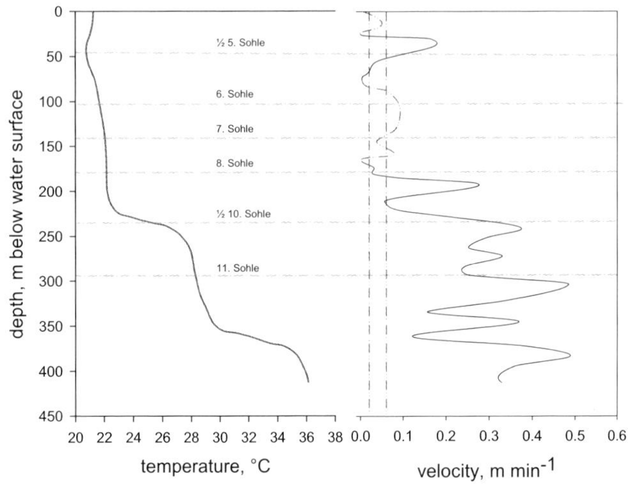

Fig. 119. Temperature and velocity measurement in the Reiche Zeche shaft (modified after Zittnan et al. 1990). Sohle: level. No error-bars were given by the authors.

In April 2002, and in May 2006 the author conducted a tracer test in the *Reiche Zeche Schacht*, based on the results of the previous investigations (Fig. 120). Na-fluorescein was chosen as the tracer, which proved to be stable in the mine water as has been shown by an earlier laboratory investigation with the Reiche Zeche mine water and Na-fluorescein. 50 g of Na-fluorescein were injected into the shaft at a depth of 75 m during the 2002 tracer test, between the ½ 5 and 6 levels (½ 5. and 6. Sohle). From the velocity measurements (ca. 0.06 m min^{-1}), the shaft's cross section (4–5 m^2) and a piston-flow-model, the tracer was expected to reappear after a minimum of 17 hours plus the opening time of the tracer probe (approx. 6–10 hours). Yet, even after 17 days, no tracer was observed at the outflow of the shaft. Three more tracer tests with Na-fluorescein at depths of 40 m, 65 m, and 324 m were conducted in 2006. Only tracer from the injection point at a depth of 40 m, just above the ½ 5 level, reappeared at the shaft's outflow. After seven weeks the tracer test was terminated as no tracers from the 65 m and 324 m deep injection points could be fround (Wolkersdorfer et al. 2007).

Those results comply with the velocity measurements of Zittnan et al. (1990), who observed down welling between the ½ 5 and 6 levels. If that and the tracer tests' results, as discussed in the previous paragraph, represents the real flow situation in the flooded Reiche Zeche shaft – and from all the measurements conducted so far, this hypothesis cannot be rejected – the often published conceptual hydrogeological model of the mine (e.g. Baacke 2000; Baacke and Degner 2000; Bayer 1999) must be reconsidered. The real

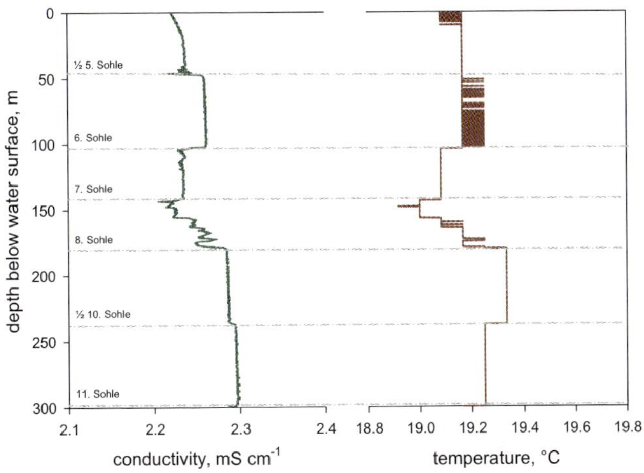

Fig. 120. Temperature and electrical conductivity measurements in the Reiche Zeche shaft (2 February 2001). Sohle: level.

situation in the mine seems to be that the mine water mainly flows from the ½ 5 level (½ 5. Sohle) and that the water below this level forms convective cells that are almost isolated from the upper water body.

12.13 Georgi Unterbau, Tyrol, Austria

12.13.1 Description of Location

The historical silver-barite mines of the famous Kogel mining area can be found 40 km northeast of Innsbruck in Tyrol/Austria, close to Brixlegg. Since medieval times, silver-bearing fahlore, and in the 19th century as well as from 1947 to 1968, barite was mined there. One of these underground mines is the Georgi-Unterbau with the huge *Schwerspat* stope (pit) and its flooded underground works.

In 1887, the miners started to drive the Georgi-Unterbau adit, which opened up rich fahlore and barite resources. Starting in 1900, a 100 m deep two-compartment blind shaft was sunk 320 m southeast of the mine entrance. This shaft opened up the 20, 40, 70, and 100 m main levels and the 10, 75, and 80 m sublevels (Pirkl 1961; Mutschlechner 1984; Krischker 1990; Hanneberg and Schuster 1994) and connected the *14-Nothelfer* stope and the Barbara stope with each other. It was partially to fully flooded in the 1950s and the pre-flooding water make was about 100 L min^{-1} (Hießleitner 1951; Schmidegg 1953); it was pumped out again in 1984 and 1988 for ore prospecting (Krischker 1990). Finally, in 1990, the blind shaft was flooded for the last time and stationary hydraulic conditions exist there today. The Georgi-Unterbau mine is typical of mines that have only one shaft that connects all the flooded levels.

All parts of the mine are within the Devonian Schwaz Dolostone (*Schwazer Dolomit*) of the Northern Tyrol Greywacke area (*Nordtiroler Grauwackenzone*). Typically, the Schwaz Dolostone is a very hard, light white to light grey dolostone, being highly brecciated and fissured in the area investigated. The dolostone hosts silver- and mercury-bearing fahlores (among them schwazite) as well as barite, the mineralisation being bound, but not restricted, to the breccia zones (Pirkl 1961; Wöbking 1982; Arlt and Diamond 1998). Grundmann and Martinek (1994), and Schnorrer (1994, 1996) described 20 ore minerals as characteristic of the Schwaz Dolostone and over 132 minerals for all the rocks in the Schwaz-Brixlegg area.

The first physico-chemical investigation of the flooded blind shaft was conducted in December 2000. It became clear that the mine water was stratified, clearly showing two water bodies separated from each other at the 40 m level. Further investigations, including two tracer tests with 15 μm microspheres and Na-fluorescein, were conducted in August 2001 and February 2002, where 22 L min^{-1} (32 m^3 d^{-1}) and 31 L min^{-1} (45 m^3 d^{-1}) of mine water were flowing out, respectively (Unger 2002; Wackwitz 2002; Wolkersdorfer et al. 2002). About 300 m^3 d^{-1} are coming from other inflows, including 150–200 m^3 from a major fault zone, the Grünwies fault, summing up to 300–400 m^3 d^{-1} of total flow (Table 12).

Based on a piston-flow-model (non-dispersive laminar Poiseuille flow), the mean effective velocity of the mine water in the shaft was calculated to be $4 \cdot 10^{-3}$ m min^{-1}. However, the tracer test resulted in a mean effective velocity of $3 \cdot 10^{-2}$ m min^{-1} above the 40-m level. This difference is typical for all tracer tests conducted by the author so far, whereas the velocity does not fit into the 95% confidence interval of 0.3–1.7 m min^{-1} (Wolkersdorfer 2002b). Similar results for porous media were described by Ren et al. (1996). The low velocity in the Georgi-Unterbau blind shaft is due to the mine being a "single shaft mine", whereas most of the other mines studied are "multiple shaft mines" or include both shafts and galleries. None of the tracers injected below the 40-m level could be retrieved.

12.13.2 Investigations and Appearance of Stratification

Between December 2000 and February 2002, eight temperature and electrical conductivity profiles and two time series were measured in the Georgi-Unterbau blind shaft. Before the tracer test in August 2001, two mine water bodies, separated at the 40-m level, were identified, based on temperatures

Table 29. Exemplified composition of the Georgi-Unterbau mine water (samples BRX-0208-BS0.1 and BRX-0308-BS55 of 2/3 August 2001, filtered through 0.45 µm cellulose acetate filter). Concentrations in mg L^{-1}, depth in m, temp in °C, electrical conductivity in µS cm^{-1}; K, Cl, Ag, and Hg below the detection limits.

depth	temp	pH	cond	Ca	Mg	Sr	Na	Ba
0.15	8.6	7.1	410	60.4	30.5	1.0	0.6	0.08
55	8.3	7.6	411	62.3	30.7	1.0	0.7	0.06

depth	Li	HCO_3^-	SO_4^{2-}	NO_3^-	Sb	Cu	Zn	As
0.15	0.003	214	40.6	4.9	1.0	0.5	0.2	0.2
55	0.002	218	41.0	4.9	1.0	0.6	0.2	0.2

differences. While the upper one had a mean temperature of 8.53±0.01 °C and electrical conductivity of 324±26 µS cm^{-1}, the lower one's mean temperature and electrical conductivity was 8.63±0.03 °C and 332±23 µS cm^{-1}, respectively. In contrast to the temperature, no stratification could be deduced from the electrical conductivity profile, which showed a relatively continuous increase from ca. 335 µS cm^{-1} to ca. 345 µS cm^{-1}. The chemical composition of the mine water behaved similarly (Table 29). After the injection of 200 kg of sodium chloride tracer as a saturated brine in August 2001, the before described stratification significantly changed, albeit the brine was not injected into the blind shaft itself but into a small drop-hole 4 m west from it. Instead of two mine water bodies, four of them appeared to exist, separated at the 20-, 40-, and 70-m levels. Yet, six months later, the previous stratification was restored with a lower temperature (ca. 8.1 °C) and a higher electrical conductivity (ca. 480 µS cm^{-1}).

Interestingly, the stratification pattern in the Georgi-Unterbau blind shaft was stable for at least 19 months, although the temperature difference of 0.1 K between the two mine water bodies is extremely small. On the other hand, the injection of the sodium chloride brine shows that the stratification can easily be destroyed although it had virtually reestablished itself six months thereafter. Unger (2002) numerically modelled a part of the flooded mine and showed that the water seeping through the *Schwerspat* pit into the *14-Nothelfer* pit and from there to the 40-m level is responsible for the temperature drop in the water column above the 40-m level and thus the observed stratification. While the physico-chemical measurements and the chemical analyses of the water indicate good mixing within the upper and lower layers, the results of the tracer test clearly showed restricted mixing be-

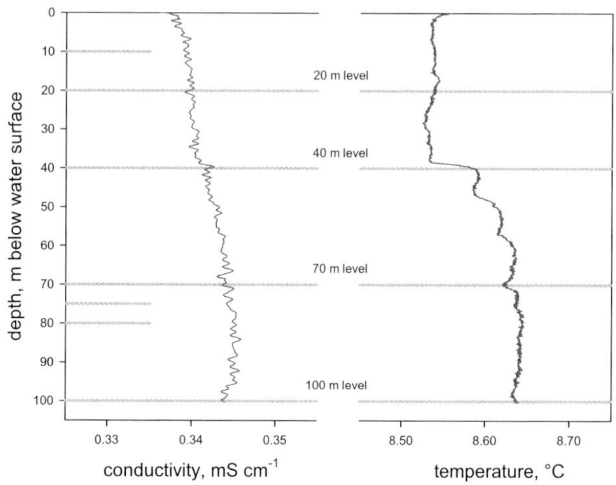

Fig. 121. Temperature and electrical conductivity measurement in August 2001 at the Georgi Unterbau Shaft. The temperature stratification at the 40 m level, even when significantly lower than 1 K, was stable for more than 3 years.

tween the two layers. Furthermore, the temperature and electrical conductivity patterns confirm that there is a turbulent flow regime in the shaft (Fig. 121).

12.13.3 Results and Interpretation

To investigate the situation, two multiple tracer tests were conducted in the mine. Tracers were injected into different depths of the blind shaft and an underground mining lake connected to the shaft. From the tracers injected, only the tracer from the underground mining lake ("Big Lake": green microspheres) and sodium fluorescein from the blind shaft could be detected. A subsequent measurement of the sodium fluorescein concentration in the shaft showed a nearly-normal distribution of the tracer cloud around the tracer's injection depth at 55 m, with tracer concentrations between 200 and 600 $\mu g\ L^{-1}$ between 40 and 80 m depth. This indicates diffusive mass flow was occurring in that part of the shaft. An evaluation of the breakthrough curve of the green microspheres results in a mean flow duration of 52 hours for a distance of 48 m, yielding a mean effective velocity of 0.02 m min^{-1} for the flow between the "Big Lake" and the blind shaft's outflow.

In February 2002, the negative experiences of the first tracer test were considered and the results therefore were better. The sodium fluorescein showed a well-established breakthrough curve within the expected time frame, similar to what was observed in August 2001. During the period of investigation (18 to 24 February 2002), the water temperature of the shaft stayed stable at about 9.2 °C and the electrical conductivity remained around 430 µS cm^{-1}. The outflow of the shaft was also stable at 31.5 L min^{-1} (43–48 m^3 d^{-1}); the pumping capacity of the mini-piston pump (sampling pump) was 0.726 L min^{-1} (1 m^3 d^{-1}). From the 50 g of Na-fluorescein used, only 3.2 g was recovered, which equals a recovery rate of 6.4%. As the breakthrough curve (Fig. 81) shows, the tracer concentration increased after the first peak at 19 February. This might be caused by a density flow of the tracer. The tracer was injected at a relatively high concentration and due to its higher density it first moved downward; later, after dilution, it was transported out of the shaft again. It can therefore be concluded that after a longer time period, more of the tracer would have been discharged and the recovery rate would have been higher.

The peak concentration of the tracer (excluding the background) was about 300 µg L^{-1} at a background concentration of 70 µg L^{-1}, which resulted from the 2001 tracer test. At the end of the tracer test, the tracer concentration was still as high as 13 µg L^{-1}. The calibration of the fluorometer gave a lower detection limit of 7 µg L^{-1} and the maximum concentration design dependent was 7,000 µg L^{-1} (the concentrations in Wolkersdorfer et al. [2002] are wrong as the fluorometer was back calibrated afterwards).

From those results, a mean effective velocity of 0.03 m min^{-1} was calculated, which fits well with the effective velocity calculated from the previous tracer test (0.02 m min^{-1}). This is 7–8 times the velocity calculated ahead of the tracer test (0.004 m min^{-1}) using a piston-flow-model, which is not unusual for mine water tracer tests. So far, the reasons are unclear but the wall roughness of the shafts and adits might play an important role. When comparing the results of the two tracer tests, it is clear that the first tracer test had a larger dispersion than the second one, due to the longer flow path of the first test (48 m) than the second one (10 m).

Unfortunately, the reasons for the stratification in the Georgi Unterbau blind shaft could not be determined by the two tracer tests, because the tracer could not be recovered in the deeper parts of the flooded shaft. It is possible that the deeper parts of the shaft are less hydraulically connected and therefore diffusive flow might dominate. At least from the tracer concentration between the 40- and 70-m levels, diffusive transport can be assumed; otherwise, the tracer distribution around the injection location should have been distributed differently. On the other hand, as the Rabenstein tracer test shows, extremely slow water movements cannot be excluded.

In conclusion, this tracer experiment – like the Rabenstein experiment – showed that no fast convective mixing occurs in the deeper parts of the flooded mine as was observed at the Niederschlema and Straßberg mines.

Therefore, it can be concluded that the missing hydraulic connection between the workings in the deeper parts prevents convective mixing of the mine water. This fact has aroused little attention, though Uerpmann (1980) made a similar observation. It might therefore be concluded that the flooded mine between the 40- and 70-m levels acts as a hydraulic barrier, preventing mass transport between the upper and lower parts of the shaft.

12.14 Salt mines

12.14.1 Description of Locations and Investigations

Germany hosts numerous abandoned salt mines that were flooded in a controlled or uncontrolled manner (Schwandt and Seifert 1999). They have been intensively studied because Germany has a long history of underground salt mining and because Germany's radioactive waste disposal policy, for a long time, focused on these salt mines. The situation in flooded salt mines differs significantly from metal or coal mines, because in addition to thermal stratification, chemical effects have to be considered. The density differences in salt mines are usually greater than in metal mines because the mineralization is greater. This makes salt mines an excellent place to study mine water stratification, but one must remember that such case studies are only transferable to metal and coal mines with restrictions. Of course, there are non-salt mines with chloride contents as high as 130 g L^{-1}, and electricaL conductivities of up to 190 mS cm^{-1} (Coldewey et al. 1999), but such mines are rare. Though this publication focuses on metal and coal mines, some salt mines will be included here because the first monitored controlled flooding of a mine was conducted in the Hope salt mine north of Hannover/Germany. Furthermore, several of the experiences described in papers about that mine flooding were also used in mine water investigations, monitoring programmes, and tracer tests conducted by the author.

Rauche et al. (2002) described how an evaluation of the stratification patterns in a flooded salt mine can be used to estimate the dissolution processes occurring there. In combination with chemical analyses of the mine water and their evaluation in Jänecke-plots (triangular Mg-SO_4-K_2 diagram), it is possible to identify potentially dangerous reactions that may be occurring in the shaft. The anonymous example they give shows a density stratification indicated by a slight jump in the electrical conductivity, though the temperature log clearly showed that convective flow did not seem to dominate the flooded shaft. Because no details are given, their work will not be further discussed here.

All the other examples given here are based on the investigations of Uerpmann (1980), Sander and Herbert (1985), GSF – Gesellschaft für Strahlen-

und Umweltforschung (1986), Herbert (1989), Herbert and Sander (1989), and Czolbe et al. (1992). From the examples in those publications, four salt mines, three in Lower Saxony (Hope and Ronnenberg near Hannover, Beienrode near Wolfsburg) and another one in Saxony-Anhalt (Peißen near Bernburg) are described.

Hope is located 30 km north of Hannover/Germany and was flooded with saturated brine from salt cavern dissolution. Flooding started on 12 March 1984, after the mine was abandoned in January 1984. Before flooding, 16 in-situ measuring installations were installed to monitor the flooding process. Furthermore, an underground dam was constructed to study safety dams for radioactive waste disposal. The whole programme was initiated in a relatively short time period, only six month ahead of the planned flooding. Besides the geochemical, geomechanical, geophysical, and the dam monitoring installations, the aim was to investigate "mass and heat transport in flooded, not backfilled galleries, mine workings, and shafts" (GSF – Gesellschaft für Strahlen- und Umweltforschung 1985; referred sentence translated from German).

Mining in the Hope salt dome began in 1907 with the 521.2 m deep *Adolfsglück* shaft; in 1909, the 628.5 m deep *Hope* shaft was developed. The salt dome has a depth of about 4,000 m, a maximum diameter of 6 km, and consists of different salt types of Permian (Zechstein) age: rock salt, sylvite, kieserite, and anhydrite. A total volume of about $1.6 \cdot 10^6$ m^3 was flooded with saturated NaCl brine (density: 1,200 g m^{-3}) from the Empelde cavern field. During the flooding process and as long as the unflooded parts of the mine were accessible, the brine was regularly analysed by taking several samples underground. Also, the flow path of the brine was controlled and monitored (Herbert 1989). The F&E research project participants conducted the last measurements in October 1988.

The 477 m deep Peißen shaft near Bernburg has a diameter of about 5 m, was developed in 1900, and underwent controlled flooding between 1972 and 1974. It is one of 11 shafts in the salt mines in the Bernburg main anticline and is located within a small syncline, the Peißen syncline. Mining in the Bernburg area started on 31 January 1852 with the Manteuffel shaft, which was where the world's first potassium salt mine was developed (Riemann 1913; Hiltscher 1999). All the mines are within the Zechstein salt formation. Czolbe et al. (1992) investigated the flooded shaft through temperature and density logs as well as chemical analyses and a tracer test with a radioactive γ-tracer. Furthermore, they modelled the hydrodynamic situation of the shaft with the code KASOMO.

The salt mines investigated by Uerpmann (1980) are the Asse I and Asse III (near Wolfenbüttel), the Beienrode shaft near Braunschweig, the Desdemona shaft near Göttingen, and the Deutschland shaft of the flooded Ronnenberg salt mine (Peter Uerpmann, pers. comm.). His work is entitled "Hydrogeological Questions at Radioactive Waste Disposal" and is focused on the safe disposal of such waste. His work does not contain details about the single mines, therefore the missing data were added after personal communication with the author.

12.14.2 Appearance and Interpretation of Stratification

Herbert and Sander (1982), who were responsible for the Hope flooding experiment, had earlier established that the brines in flooded mines were commonly stratified into three layers: an upper layer with slightly saline or fresh water, a middle NaCl-saturated layer with a density of about 1.20 g cm^{-3}, and a lower $MgCl_2$-rich layer with a density of 1.29–1.30 g cm^{-3}. Between those three layers, they found intermediate layers. They concluded that the chemical composition of the rock in contact with the shaft and the geothermal gradient act together to control the formation of the various layers. Furthermore, they observed convective flow in homogenous layers, and no flow between layers with density or temperature gradients (Herbert and Sander 1982; Sander and Herbert 1985). In addition, they found that more than one convective water body and also no-flow and convective flow cells can adjoin each other. Which of those situations occurs is, according to their opinion, a result of the geological setting. The mine geometry was only seen as an additional reason for the stratification. In another paper (Sander and Herbert 1985), they concluded that the $NaCl/MgCl_2$ boundary can be successfully used as a barrier against transport as they noticed that crystalline halite (rock salt) was forming at the boundary. Yet, and this fact is missing in their investigation, such a barrier can only be stable if no forced convection of flow occurs in the shaft.

During the active flooding of the Hope salt mine, numerous physico-chemical measurements were made (GSF – Gesellschaft für Strahlen- und Umweltforschung 1986; Herbert 1989). Fourteen continuous measurements in the Hope, Adolfsglück, and the Abbau 4 well were published. There were many differences and many similarities, of which the stratification in the shafts is the most relevant to this discussion. Unfortunately, the results of the Hope and Adolfsglück shaft cannot be compared directly as the Adolfsglück shaft was backfilled in 1985, so the measurements had to be conducted in a pipe within the backfill. From the differences in the measurements before and after the backfill, it is clear that the post-backfill situation was hydraulically significantly different from the pre-backfill conditions. Stratification was modest in the Adolfsglück shaft prior to backfilling, whereas afterwards, at least two separate mine water bodies could be identified (Fig. 122). Again, as in the case of the wells in the Ruhr area, wells cannot be used to interpret the stratification in a mine as the flow situation in a small diameter well differs significantly from the flow in a large volume shaft. While the Adolfsglück "shaft" stratified after the backfill, this was not the case for the Hope shaft, which was not backfilled (Fig. 123). The interpretation of GSF – Gesellschaft für Strahlen- und Umweltforschung (1986), that the backfill prevents overall turbulent mixing, is definitely true, but this cannot be verified from a measurement in a small diameter well within a backfilled shaft. Some minor stratification was observed during the early stage of the flooding process, but this stratification broke down towards the end of the flooding, leaving just a less-mineralised water lens with a thickness of 30–40 m lying on top of the 180–190 m thick shaft mine water body. The following findings of the "Hope – geochemistry sub project" are relevant to this study of flooded shafts (GSF – Gesellschaft für Strahlen- und Umweltforschung 1986; Herbert 1989):

- a shaft can consist of two or more turbulent mixed water bodies that border on each other, but no-flow layers can also border on turbulent mixed ones;
- water bodies with turbulent flow are characterised by a uniform temperature, electrical conductivity, density, and chemical composition;
- no-flow layers are reflected by concentration and temperature gradients;
- the mass transport is dominated by slow diffusion in no-flow layers, and mainly by fast flow in turbulent mixed layers;

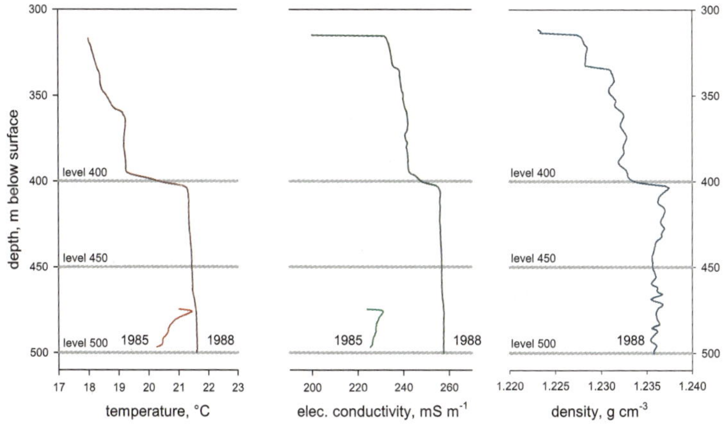

Fig. 122. Comparison of the post- and pre-backfilling situation in the Adolfsglück shaft in 1985 and 1988. The 1988 situation is typical for measurements in small diameter boreholes or wells (modified after GSF – Gesellschaft für Strahlen- und Umweltforschung 1986 and Herbert 1989).

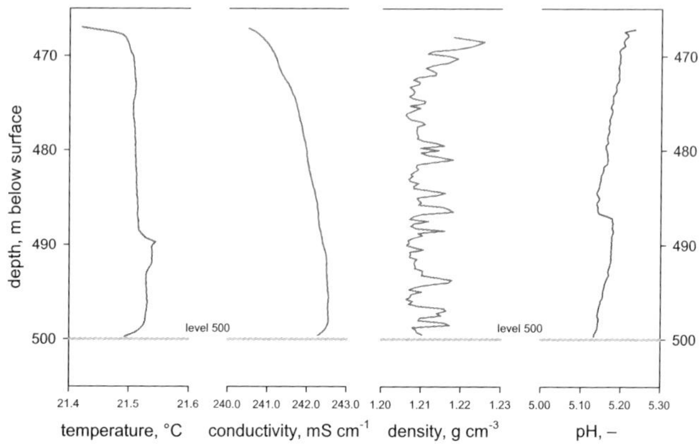

Fig. 123. Results of the multi-parameter measurements in the Hope shaft on 20 September 1985. Only minor stratification can be observed, which – except that the water level rose – did not change significantly until the end of the mine flooding (modified after Herbert 1989).

- boundaries between single layers act like mechanical barriers, resulting in a change in flow direction;
- mass transport across the boundaries is diffusion dominated, even when both zones are mixed by turbulent flow; and
- open and backfilled mine workings show significantly different flow patterns; in open mine workings, the flow is dominated by convective flow while no-flow conditions dominate in backfilled ones.

Density and temperature stratification remained stable for a period of at least 4 months in the Peißen shaft (Czolbe et al. 1992; Fig. 124). Densities of the water and brine ranged between 1.1 g cm^{-3} and 1.3 g cm^{-3} and the temperature ranged between 13.2 and 30.5 °C. A total of four water bodies were observed with no or small intermediate layers. They were characterised by their chemical composition and are similar to the observations of Herbert and Sander (1982): a slightly mineralised "fresh" water layer, a NaCl-layer, a NaCl-KCl-layer and a lowermost $MgCl_2$-layer. Besides the physico-chemical investigations, the investigation included a tracer test. A radioactive γ-tracer was injected at 10 locations in the shaft. The velocities ranged between 0.003 and 1.17 m min^{-1}, which fits well in the overall range of mine water tracer tests (though their velocities are in-situ velocities of the water in the shaft and not the water flowing in the mine). Furthermore, they identified several (at least 5 and a maximum of 8) convective cells within the shaft, which were directly linked to the different mine water bodies. From their measurements,

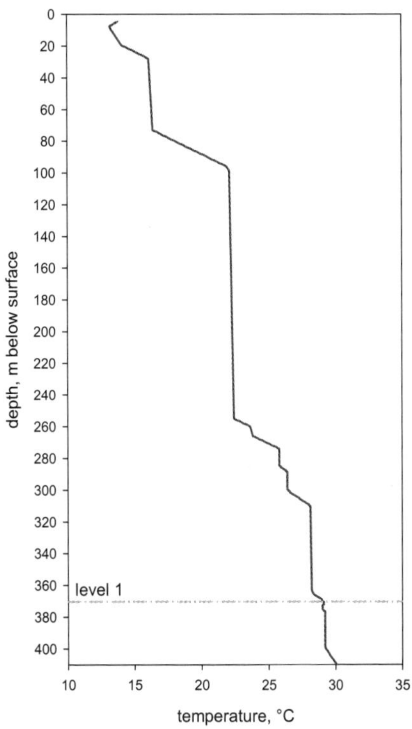

Fig. 124. Temperature profile in the Peißen shaft. The lowermost horizontal line is the mines' 1st level. Each of the "stable" temperature zones represents a convective cell as shown by tracer investigations (modified after Czolbe et al. 1992).

they also concluded that warm water was welling up near the shafts walls and flowing down in the middle of the shaft, as was postulated for the Niederschlema/Alberoda mine (section 12.9). By using the KASOMO hydrodynamic code, originally written to model flow in caverns for radioactive waste disposal, they were able to model and understand their observations. Their measurements are similar to those discussed earlier for single shaft mines, as there is no flow between the Peißen and other shafts. The intermediate layers form relatively stable barriers that cannot be crossed by the different types of water in the shaft. This is supported by the fact that Czolbe et al. (1992) were not able to detect the injected tracers outside of each of the single cells.

Uerpmann (1980) investigated temperature, electrical conductivity, velocity, and chemical compositions of mine waters in five flooded salt mine shafts with depths between ca. 200 and ca. 800 m (the Asse III, Desdemona, Deutschland, Ronnenberg, and Beienrode shafts; Peter Uerpmann, pers. comm.). As in the case studies listed before, he was able to identify different water bodies separated from each other by smaller, intermediate layers of different temperature (Fig. 125). This resulted in staircase profiles, which, according to section 7.1.1, might be due to double-diffusive convection. Due

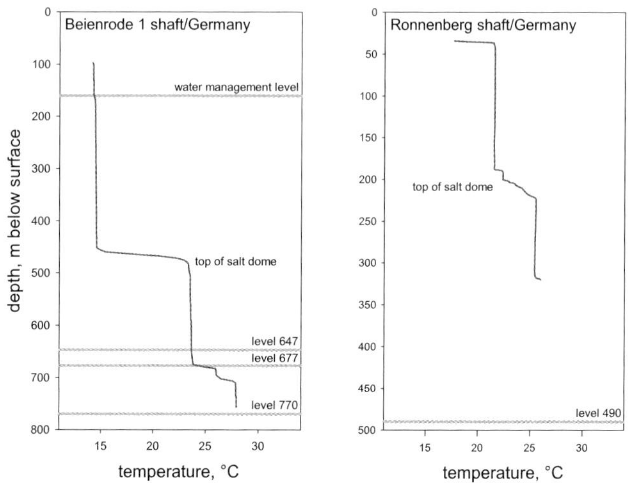

Fig. 125. Temperature–depth profiles in two flooded north-west German salt mines. Left: Beienrode № 1 shaft, right: Ronnenberg shaft (modified after Uerpmann 1980).

to the density differences in flooded salt mines, this effect cannot be excluded as an explanation for the staircase temperature distribution. Uerpmann (1980) also refers to the saline Lake Kivu between Rwanda and the Democratic Republic Congo (Zaïre), which was investigated by Tietze (1978) and where staircase like steps were also detected. An interesting observation made during the flooding of the salt mines was that the water temperatures took only a few weeks to reach nearly the final temperatures observed at the end of the flooding process. This indicates fast convective mixing processes during the first stages of the flooding (see section 7.1) and smaller exchanges later. In a laboratory experiment, Uerpmann (1980) measured mean velocities of 0.06–0.3 m min^{-1} and was able to observe that during his experiment the water stratified and that layers with different physico-chemical properties were separated by one or more distinct intermediate layers. Tracers injected in the two different layers showed relatively good mixing after several days while in the intermediate layers, the tracer distribution and the tracer velocities were significantly less.

13 Literature

Abelin, H. & Birgersson, L. (1985): Migration experiments in the Stripa mine, design and instrumentation. – In: Come, B., Johnston, P. & Müller, A.: Joint CEC/NEA workshop on design and instrumentation of in situ experiments in underground laboratories for radioactive waste disposal CONF 8405165. – p. 180–190, 9 fig.; Rotterdam u.a. (Balkema).

Ackers, P., White, W. R., Perkins, J. A. & Harrison, A. J. M. (1978): Weirs and flumes for flow measurement. – 327 p., 39 plates; Chichester (Wiley & Sons).

Ackman, T. E. & Jones, J. R. (1988): Stream sealing to reduce surface water infiltration into underground mines. – Bureau of Mines Information Circular, IC-9183: 232–239, 6 fig., 3 tab.; Washington.

Ackman, T. E. (2003): An Introduction to the Use of Airborne Technologies for Watershed Characterization in Mined Areas. – Mine Water Env., 22 (2): 62–68, 5 fig.; Berlin.

Adams, N., Carroll, D., Madalinski, K., Rock, S., Wilson, T., Pivetz, B., Anderson, T., Chappell, J., Huling, S., Palmiotti, J. & Sayre, P. (2000): Introduction to Phytoremediation EPA-600/R-99/107. – 72 p., 13 fig., 17 tab.; Cincinnati (U.S. Environmental Protection Agency).

Adams, R. & Younger, P. L. (2001): A Strategy for Modeling Ground Water Rebound in Abandoned Deep Mine Systems. – Ground Water, 39 (2): 249–261, 10 fig.; Worthington.

Adelman, F. L., Bacigalupi, C. M. & Momyer, F. E. (1960): Final report on the Pinot experiment [Colorado]. – Lawrence Radiation Lab. Rept., UCRL-6274: 21; Livermore.

Adlem, C. J. L., Maree, J. P. & McCrindle, R. I. (1994): Water Softening in the Barium Sulphide Process for Sulphate Removal. – Proceedings, 5th International Mine Water Congress, Nottingham, U.K., 2: 601–611, 9 fig., 2 tab.; Nottingham.

Agricola, G. (1556): De re metallica libri XII. – 273 Holzschnitte; Basel.

Ahonen, L. & Tuovinen, O. H. (1995): Bacterial Leaching of Complex Sulfide Ore Samples in Bench-Scale Column Reactors. – Hydrometallurgy, 37 (1): 1–21, 9 fig., 2 tab.; Amsterdam.

Al, T. A. & Blowes, D. W. (1996): Storm-water hydrograph separation of run off from a mine-tailings impoundment formed by thickened tailings discharge at Kidd Creek, Timmins, Ontario. – J. Hydrol., 180 (1–4): 55–78, 15 fig., 2 tab.; Amsterdam.

Aldous, P. J. & Smart, P. L. (1987): Tracing Ground-Water Movement in Abandoned Coal Mined Aquifers Using Fluorescent Dyes. – Ground Water, 26 (2): 172–178, 8 fig., 2 tab.; Worthington.

Aley, T. & Fletcher, M. W. (1976): The water tracer's cookbook. – Missouri Speleology, 16: 1–32, 2 fig., 1 tab.; Columbia.

Al-Fouzan, F., Harbert, W., Dilmore, R., Hammack, R., Sams, J., Veloski, G. & Ackman, T. E. (2004): Methods for Removing Signal Noise from Helicopter Electromagnetic Survey Data. – Mine Water Env., 23 (1): 28–33, 7 fig.; Berlin.

Aljoe, W. W. & Hawkins, J. W. (1993): Neutralization of Acidic Discharges from Abandoned Underground Coal Mines by Alkaline Injection. – Report of Investigations – US Bureau of Mines, 9468: 1–37, 40 fig., 3 tab.; Washington.

Aljoe, W. W. & Hawkins, J. W. (1994): Application of Aquifer Testing in Surface and Underground Coal Mines. – Proceedings, 5th International Mine Water Congress, Nottingham, U.K., 1: 3–21, 8 fig., 1 tab.; Nottingham.

Allen, D. M. & Lepitre, M. E. (2004): Use of Pb, ^{18}O, and ^{2}H Isotopes in Mining-related Environmental Studies. – Mine Water Env., 23 (3): 119–132, 7 fig., 2 tab.; Berlin.

Allen, D. M. & Voormeij, D. A. (2002): Oxygen-18 and Deuterium Fingerprinting of Tailings Seepage at the Sullivan Mine. – Mine Water Env., 21 (4): 168–182, 5 fig., 2 tab.; Berlin.

Alpers, C. N. & Nordstrom, D. K. (1999): Geochemical Modelling of Water-Rock Interactions in Mining Environments. – In: Plumlee, G. S. & Logsdon, M. J.: The Environmental Geochemistry of Mineral Deposits 6A (14). – p. 289–323, 8 fig., 5 tab.; Littleton (Society of Economic Geologists).

Alpers, C. N., Blowes, D. W., Nordstrom, D. K. & Jambor, J. L. (1994): Secondary minerals and acid mine-water chemistry. – In: Jambor, J. L. & Blowes, D. W.: Short Course Handbook on Environmental geochemistry of Sulfide Mine-Wastes 22. – p. 247–270, 4 fig., 12 tab.; Waterloo, Ontario (Mineralogical Association of Canada).

Alpers, C. N., Nordstrom, D. K. & Burchard, J. M. (1992): Compilation and interpretation of water-quality and discharge data for acidic mine waters at Iron Mountain, Shasta County, California, 1940–91. – Water resources investigations report, WRI 91-4160: 173, 9 fig., 5 tab., 1 app.; Reston.

American Public Health Association, American Water Works Association & Water Environment Federation (1998): Standard methods for the examination of water and wastewater. – In: Clesceri, L. S., Eaton, A. D., Franson, M. A. H. & Greenberg, A. E. – 20th edn., Washington (American Public Health Association).

Amezaga, J. M. & Younger, P. L. (2002): Environmental Regulation of Mine Waters in the European Union – D 3 Institutional research – 2 Great Britain. – p. 2.1–2.20, 2 fig.; Newcastle upon Tyne (University of Newcastle).

Amezaga, J. M., Kroll, A., Jarvis, A. P., Rees, B., Puura, E., Miseré, W. M. H., Wings, R. W. M. G., Bajtoš, P., Weber, L., Sárváryné-Szentkatolnay, B., Garzonio, C. A., Wolkersdorfer, Ch., Younger, P. L. & Bowell, R. (2004): Contemporary Reviews of Mine Water Studies in Europe, Part 1. – Mine Water Env., 23 (4): 163–183, 16 fig., 3 tab.; Berlin.

Ammann, G. (1990): Silber, Erz und weißes Gold – Bergbau in Tirol – Tiroler Landesausstellung 1990. – 478 p.; Innsbruck (Tiroler Landesmuseum Ferdinandeum).

Amorim, L. Q., Grandchamp, C. A. P. & Bertachini, A. C. (2001): Dewatering of Iron Ore Mines and Capability to Predict Environmental Impacts – The Experience of MBR. – p. 1–14 [CD-ROM], 9 fig.; Belo Horizonte (Proceedings, IMWA Symposium Belo Horizonte).

Anckorn, J. F. & del Olmo, A. G. (1999): The implications of current European directives and the future of mining in Europe. – In: Fernández Rubio, R.: Mine, Water & Environment II. – p. 791–803, 3 fig., 2 tab.; Sevilla (International Mine Water Association).

Anderson, B. C. (1987): Use of temperature logs to determine intra-well-bore flows, Yucca Mountain, Nevada. – 59 p.; Golden (Master's Thesis Colorado School of Mines).

Angerer, J. & Lechleitner, H. (1893): Bergbau und Hüttenwesen, Gewerbe, Industrie und Handel in Tirol und Vorarlberg. – In: Erzherzog Rudolf: Die österreichisch-ungarische Monarchie in Wort und Bild – Tirol und Vorarlberg. – p. 585–600, 4 fig.; Wien (K.K. Hof- u. Staatsdr.).

Annandale, J. G. (1998): Irrigating with gypsiferous mine water. – p. Afr. J. Sci., 94 (7): 359–360; Johannesburg.

Annandale, J. G., Jovanovic, N. Z., Benade, N. & Tanner, P. D. (1999): Modelling the long-term effect of irrigation with gypsiferous water on soil and water resources. – Agric. Ecosyst. Environ., 76 (2–3): 109–119, 8 fig., 1 tab.; Amsterdam.

Annandale, J. G., Jovanovic, N. Z., Pretorius, J. J. B., Lorentz, S. A., Rethman, N. F. G. & Tanner, P. D. (2001): Gypsiferous mine water use in irrigation on rehabilitated open-cast mine land: Crop Production, soil water and salt balance. – Ecol. Eng., 17 (2–3): 153–164, 3 fig., 4 tab.; Amsterdam.

Annandale, J. G., Jovanovic, N. Z., Tanner, P. D., Benadé, N. & Du Plessis, H. M. (2002): The Sustainability of Irrigation with Gypsiferous Mine Water and Implications for the Mining Industry in South Africa. – Mine Water Env., 21 (2): 81–90, 5 fig., 2 tab.; Berlin.

Anonymous (1901): La fabrique d'absinthe Pernod en feu. – L'illustration – journal universel, 17.8.1901; Paris.

Anonymous (1934): The Henry Process for the Clarification of Polluted Water. – Engineering, 138 (8, 9): 213–215, 293–295, 17 fig.; London.

Anonymous (1935): Wallsend – At their Rising Sun pit the Wallsend & Hebburn Coal Co. Ltd. have erected what is probably the most complete coal preparation plant in the world. – Colliery Engineering; London.

Anonymous (1936): Mine-Sealing Program to Reduce Acid Pollution in Streams. – Engineering news record, 2: 42–43, 3 fig.; New York.

Anonymous (1994): Urteil des Oberverwaltungsgerichts Lüneburg vom 15. Juni 1994 (7 L 5295/92) – Schadensvorsorge bei der Stillegung eines Bergwerksbetriebes. – Z. f. Bergrecht, 135: 277–286; Bonn.

Anonymous (1995): Urteil des Bundesverwaltungsgerichts vom 9. November 1995 (4 C 25.94) – Unternehmerverantwortung bei der teilweisen Betriebseinstellung eines Bergwerks für die Abwehr gemeinschädlicher Einwirkungen – Beseitigung von Sauerwasser im Erzbergwerk Rammelsberg/Harz. – Z. f. Bergrecht, 136: 290–301; Bonn.

Anonymous (1997): Environmental pumping tasks at Wheal Jane. – Mining Magazine, 177 (5): 278–279, 2 fig.; London.

Anonymous (1999a): Scottish Environment Protection Agency to take Action on Mine Water Pollution. – Coal International, 247 (4): 139, 1 fig.; Redhill, Surrey.

Anonymous (1999b): Mining history matters. – Mining Journal, 332 (8523): 198–199; London.

Anonymous (2000): On January 11, Jiangsu Province's Dahuangshan coal mine at Xuzhou was hit by an inrush. – Engineering & Mining Journal, 2001 (3): w10; Prairieville.

Anonymous (2002): NW China Coal Mine Flooded, 15 Trapped. – People's Daily Online, 4.7.2002; Beijing.

Anonymous (2006): On-line Inventory Generates Reports, Enables Tracking Of AML Problems: Office of Surface Mining – 2006 Annual Report. – p. 20; Washington (Office of Surface Mining Reclamation and Enforcement).

Appelo, C. A. J. & Postma, D. (1994): Geochemistry, groundwater and pollution. – 2nd edn., 536 p.; Rotterdam (Balkema).

Aquilina, L., Rose, P., Vaute, L., Brach, M., Gentier, S., Jeannot, R., Jacquot, E., Audigane, P., Tran-Viet, T., Jung, R., Baumgärtner, J., Baria, R. & Gérard, A. (1998): A Tracer Test at the Soultz-Sous-Forêts Hot Dry Rock Geothermal Site. – Proceedings, 23rd Workshop on Geothermal Reservoir Engenieering, SGP-TR-158: 1–8, 3 fig.; Stanford.

Arancibia, E. & Flores, G. (2004): Transición de Chuqui a mina subterránea. – Mineria Chilena, 280 (10): 25, 1 fig.

Aranyakanon, P. & Yongsakul, K. (1981): Offshore tin deposits of Thailand and related mining problems. – In: Halbouty, M. T.: Energy Resources of the Pacific region. – p. 167–178, 9 fig., 1 tab.; Tulsa (American Association of Petroleum Geologists).

Arbeitskreis "Human- und ökotoxikologische Bewertung von Markierungsmitteln in Gewässern" (1997): Human- und ökotoxikologische Bewertung von Markierungsmitteln in Gewässern. – Grundwasser, 2 (2): 61–64, 2 tab.; Heidelberg.

Ardejani, F. D., Singh, R. N., Baafi, E. & Porter, I. (2003): A Finite Element Model to: 1. Predict Groundwater Inflow to Surface Mining Excavations. – Mine Water Env., 22 (1): 31–38, 7 fig., 2 tab.; Berlin.

Ariza, L. M. (1998): River of vitriol. – Sci. Amer., 279 (3): 26–27, 2 fig.; New York.

Arlt, T. & Diamond, L. W. (1998): Composition of tetrahedrite-tennantite and "schwazite" in the Schwaz silver mines, North Tyrol, Austria. – Mineral. Mag., 62 (6): 801–820, 8 fig., 5 tab.; London.

Arnold, I. & Baumgarten, U. (1996): Wege des Wassers – Kohleförderung und Wasserhebung im Lausitzer Braunkohlerevier. – 11 p., 31 fig.; Senftenberg (Lausitzer Braunkohle Aktiengesellschaft).

Arnold, I. & Kuhlmann, K. (1994): Wasserwirtschaft und Bergbau in der Lausitz. – Braunkohle (7): 10–21, 10 fig.; Düsseldorf.

Aurada, K. D. (1970): Berechnung des Wiederanstiegsprozesses im Grubengebäude des Kupferschieferbergbaus in der Mansfelder Mulde mit Hilfe der Monte-Carlo-Methode. – Wasserwirtschaft Wassertechnik, 20 (8): 264–270, 6 fig., 3 tab.; Berlin.

Axler, R., Yokom, S., Tikkanen, C., McDonald, M., Runke, H., Wilcox, D. & Cady, B. (1998): Restoration of a Mine Pit Lake from Aquacultural Nutrient Enrichment. – Restoration Ecology, 6 (1): 1–19, 11 fig., 5 tab.; Cambridge.

Baacke, D. & Degner, T. (2000): Hydrogeochemie, Wärmehaushalt und Strömung des Flutungswassers einer untertägigen Erzgrube. – In: Wippermann, T.: Bergbau und Umwelt – Langfristige geochemische Einflüsse. – p. 163–174, 6 fig., 2 tab.; Berlin (Springer).

Baacke, D. (2000): Geochemisches Verhalten umweltrelevanter Elemente in stillgelegten Polysulfiderzgruben am Beispiel der Grube "Himmelfahrt" in Freiberg/Sachsen. – Freiberger Dissertationen On-line, 35: 139, 72 fig., 62 tab., 7 app.; Freiberg.

Baacke, D., Kluge, A., Martin, M. & Beuge, P. (1996): Geochemische Prozesse beim Stofftransport in Stollnsystemen gefluteter Grubenbaue. – GeoCongress, 2: 1–6, 5 fig., 1 tab.; Köln.

Baas Becking, L. G. M., Kaplan, I. R. & Moore, D. (1960): Limits of the Natural Environment in Terms of pH and Oxidation-Reduction Potentials. – J. of Geol., 68 (3): 243–284, 34 fig., 3 tab.; Chicago.

Bagdy, I. & Sulyokm, P. (1990): Case Example on Acid-Resistent Bulkhead in Hungary. – Int. J. Mine Water, 9 (1–4): 97–109, 7 fig.; Castro Verde.

Baker, A. J. M., Reeves, R. D. & McGrath, S. P. (1991): In Situ Decontamination of Heavy Metal Polluted Soils Using Crops of Metal-Accumulating Plants – A Feasibility Study. – In: Hinchee, R. E. & Olfenbuttel, R. F.: In Situ bioreclamation – Applications and investigations for hydrocarbon and contaminated site remediation. – p. 600–605, 1 fig.; Stoneham (Butterworth-Heinemann).

Balkau, F. (1999): Abandoned mine sites. – In: Natural Resources Management Unit Division „Protection of the Environment and Natural Resources“: Report on the International Round Table on Mining and the Environment ("Berlin II Declaration"). – p. 40–43; Köln (Carl Duisberg Gesellschaft).

Ball, B. R. & Brix, K. V. (2000): Passive Treatment of Metalloids Associated with Acid Rock Drainage. – In: Society for Mining Metallurgy and Exploration: New frontiers in mining: meeting 21st century challenges. – 20 p., 4 fig., 1 tab.; Littleton (Society for Mining, Metallurgy, and Exploration).

Banks, D. (1994): The Abandonment of the Killingdal Sulphide Mine, Norway: a Saga of Acid Mine Drainage and Radioactive Waste Disposal. – Mine Water Env., 13 (3–4): 35–48, 4 fig., 1 tab.; Wollongong.

Banks, D. (2001): A Variable-Volume, Head-Dependent Mine Water Filling Model. – Ground Water, 39 (3): 362–365, 2 fig., 1 tab.; Worthington.

Banks, D. (2004): Geochemical Processes controlling Minewater Pollution: Conference Papers 35. – p. 17–44, 14 fig., 4 tab.; Wien (Umweltbundesamt).

Banks, D., Skarphagen, H., Wiltshire, R. & Jessop, C. (2003): Mine water as a resource – space heating and cooling via use of heat pumps. – Land Contam. Reclam., 11: 191–198, 5 fig.; Richmond.

Banks, D., Younger, P. L., Arnesen, R. T., Iversen, E. R. & Banks, S. B. (1997): Mine-water chemistry: the good, the bad and the ugly. – Environ. Geol., 32 (3): 157–174, 11 fig., 5 tab.; New York, Berlin.

Barnes, I. & Clarke, F. E. (1964): Geochemistry of ground water in mine drainage problems. – US Geological Survey Professional Paper, 473, A: 1–6, 1 fig., 1 tab.; Washington.

Barnes, I., Stuart, W. T. & Fisher, D. W. (1964): Field investigation of mine waters in the northern anthracite field, Pennsylvania. – US Geological Survey Professional Paper, 0473-B: B1–B8, 3 fig., 3 tab.; Reston.

Barry, M. E., Ivey, G. N., Winters, K. B. & Imberger, J. (2001): Measurements of diapycnal diffusivities in stratified fluids. – J. Fluid Mech., 442: 267–291, 11 fig., 2 tab.; Cambridge.

Bartels, C. & Lorenz, E. (1993): Die Grube Glasebach – ein Denkmal des Erz- und Fluoritbergbaus im Ostharz. – Der Anschnitt, 45 (4): 144–158, 10 fig.; Bochum.

Barthel, F. H. (1993): Die Urangewinnung auf dem Gebiet der ehemaligen DDR von 1945 bis 1990. – Geol. Jb., A 142: 335–346, 2 fig., 5 tab.; Hannover.

Barton, C. D. & Karathanasis, A. D. (1999): Renovation of a failed constructed wetland treating acid mine drainage. – Environ. Geol., 39 (1): 39–50, 9 fig., 4 tab.; New York, Berlin.

Batchelor, G. K. (2000): An introduction to fluid dynamics. – 615 p., 172 fig.; Cambridge (Cambridge University Press).

Bates, R. L. & Jackson, J. A. (1987): Glossary of geology. – 3rd edn., 788 p.; Alexandria (American Geological Institute).

Batty, L. C. & Younger, P. L. (2002): Critical role of macrophytes in achieving low iron concentrations in mine water treatment wetlands. – Environ. Sci. Technol., 36 (18): 3997–4002, 6 fig.; Washington.

Batty, L. C. & Younger, P. L. (2003): Effects of External iron Concentration upon Seedling Growth and Uptake of Fe and Phosphate by the Common Reed, *Phragmites Australis* (Cav.) Trin Ex. Steudel. – Ann. Bot., 92 (6): 801–806, 2 fig., 2 tab.; Oxford.

Batty, L. C. (2005): The Potential Importance of Mine Sites for Biodiversity. – Mine Water Env., 24 (2): 101–103; Berlin.

Batty, L. C., Baker, A. J. M. & Wheeler, B. D. (2002): Aluminium and phosphate uptake by *Phragmites australis*: the role of Fe, Mn and Al root plaques. – Ann. Bot., 89 (4): 443–449, 2 fig., 2 tab.; Oxford.

Bau, H. H. & Torrance, K. E. (1981a): Transient and Steady behavior of an open, symmetrically-heated, free convection loop. – Int. J. Heat Mass Transfer, 24: 597–609, 15 fig.; Oxford.

Bau, H. H. & Torrance, K. E. (1981b): On the stability and flow reversal of an asymmetrically heated open convection loop. – J. Fluid Mech., 109: 417–433, 10 fig., 2 tab.; Cambridge.

Bau, H. H. & Torrance, K. E. (1983): On the effects of viscous dissipation and pressure work in free convection loops. – Int. J. Heat Mass Transfer, 26 (5): 727–734, 5 fig., 2 tab.; Oxford.

Baumann, L. (1992): Zur Metallogenie des Erzgebirges sowie Darstellung einiger seiner bedeutenden Lagerstätten. – Schriftenreihe Gesellschaft Deutscher Metallhütten- und Bergleute, 64: 1–30, 17 fig., 2 tab.; Clausthal-Zellerfeld.

Bäumle, R. (2003): Geohydraulic Characterisation of Fractured Rock Flow Regimes Regional Studies in Granite (Lindau, Black Forest, Germany) and Dolomite (Tsumeb Aquifers, Northern Namibia). – 129 p., 61 fig., 17 tab., 4 app.; Karlsruhe (unpubl. Diss. Uni Karlsruhe).

Bäumle, R., Behrens, H., Einsidl, F., Goldschneider, N., Grust, K., Hötzl, H., Käss, W., Kennedy, K., Kinzelbach, W., Kozel, R., Müller, I., Müller, J., Niehren, S., Rossi, P., Schnegg, P.-A., Seiler, K.-P., Witthüser, K., Wohnlich, S. & Zojer, H. (2001): Comparative Tracer Studies in Groundwater. – Beitr. Hydrogeologie, 52: 103–230, 55 fig., 21 tab.; Graz.

Bayer, M. (1999): Die Himmelfahrt Fundgrube – Ein Führer durch das Lehr- und Besucherbergwerk der TU Bergakademie Freiberg. – 144 p., 83 fig., 3 tab.; Freiberg (Eigenverlag TU Bergakademie Freiberg).

Bayless, E. R. & Olyphant, G. A. (1993): Acid-generating salts and their relationship to the chemistry of groundwater and storm runoff at an abandoned mine site in southwestern Indiana, U.S.A. – J. Contam. Hydrol., 12 (4): 313–328, 6 fig., 2 tab.; Amsterdam.

Bear, J. & Bachmat, Y. (1991): Introduction to Modeling of Transport Phenomena in Porous Media. – In: Bear, J.: Theory and Application of Transport in Porous Media 4. – 553 p.; Dordrecht (Kluwer).

Beck (1981): Zwischenbericht "Sulfid-Schwefel". – 18 p., 10 fig.; Rottleberode (VEB Schwerspat).

Becke, A., Douffet, H., Jobst, W., Pforr, H., Sennewald, R., Wächtler, E. & Wagenbreth, O. (1986): Der Freiberger Bergbau – Technische Denkmale und Geschichte. – 1st edn., 382 p., 315 fig., 28 tab.; Leipzig (VEB Deutscher Verlag für Grundstoffindustrie).

Behrens, H., Beims, U., Dieter, H., Dietze, G., Eikmann, T., Grummt, T., Hanisch, H., Henseling, H., Käss, W., Kerndorff, H., Leibundgut, C., Müller-Wegener, U., Rönnefahrt, I., Scharenberg, B., Schleyer, R., Schloz, W. & Tilkes, F. (2001): Toxicological and ecotoxicological assessment of water tracers. – Hydrogeol. J., 9 (3): 321–325, 2 tab.; Berlin.

Bei, X. & China Daily (2004): Chinese Mine Accidents Provide Tragic Lessons. – Mine Water Env., 23 (1): 54; Berlin.

Bekendam, R. F. & Pöttgens, J. J. (1995): Ground movements over the coal mines of southern Limburg, the Netherlands, and their relation to rising mine waters. – IAHS-AISH Publication, 234: 3–12, 6 fig.; The Hague.

Bench, D. W. (2000): PCBs, Mining, and Water Pollution. – Engineering & Mining Journal, 201 (8): 36–38, 1 fig.; Overland.

Benedix, E. H., Casper, S. J., Danert, S., Hübsch, P., Lindner, K. E., Schmiedeknecht, M. & Schubert, R. (1991): Urania Pflanzenreich in vier Bänden – Viren Bakterien Algen Pilze. – 1st edn., 664 p., fig.; Leipzig Jena Berlin (Urania Verlag).

Benkovics, I., Csicsák, J., Csövári, M., Lendvai, Z. & Molnár, J. (1997): Mine Water Treatment – Anion-exchange and Membrane Process. – Proceedings, 6th International Mine Water Association Congress, Bled, Slovenia, 1: 149–157, 6 fig., 2 tab.; Ljubljana.

Benner, S. G., Blowes, D. W. & Ptacek, C. J. (1997): A full-scale porous reactive wall for prevention of acid mine drainage. – Ground Water Monit. Remediat., 17: 99–107, 5 fig., 1 tab.; Dublin.

Bennett, J. C. & Tributsch, H. (1978): Bacterial Leaching Patterns on Pyrite Crystal Surfaces. – J. Bacteriol., 134 (1): 310–317, 3 fig.; Washington.

Benning, H.-J., Mittrach, J. & Heiming, M. (2002): Aufwältigung und Sanierung eines abgeworfenen Schachtes im Bereich einer Deponie. – In: Busch, W., Maas, K., Klapperich, H., Tondera, D., Sroka, A. & Meier, G.: 2. Altbergbau-Kolloquium – Tagungsband. – p. 258–267, 2 fig.; Clausthal-Zellerfeld (Papierflieger).

Benthaus, F. C. & Nitsche, C. (2005): Wiederherstellung des Wasserhaushaltes in den vom Braunkohlenbergbau beeinflussten Gebieten Mitteldeutschlands und der Lausitz. – Wissenschaftliche Mitteilungen, 28: 15–19, 4 fig.; Freiberg.

Berg, M., Tran, H. C., Nguyen, T. C., Pham, H. V., Schertenleib, R. & Giger, W. (2001): Arsenic Contamination of Groundwater and Drinking Water in Vietnam: A Human Health Threat. – Environ. Sci. Technol., 35 (13): 2621–2626, 4 fig., 2 tab.; Washington.

Berk, W. van (1987): Hydrochemische Stoffumsetzung in einem Grundwasserleiter – beeinflußt durch eine Steinkohlebergehalde. – Besondere Mitteilungen zum Deutschen Gewässerkundlichen Jahrbuch, 49: 1–175, 49 fig., 6 tab.; Düsseldorf.

Berk, W. van (2000): Hydrogeochemische Modellierung als Instrument zur Wirkungsanalyse hydrogeochemischer Prozesse. – 270 p., 63 fig., 89 tab.; Aachen (Shaker).

Bernasconi, C. & Betlem, G. (2003): Licences, liability and cross-border pollution. – Environ. Liability, 3: 83–88; London.

Bessler, F. T., Savic, D. A. & Walters, G. A. (2003): Water Reservoir Control with Data Mining. – J. Water Resour. Plan. Manage.-ASCE, 129 (1): 26–34, 8 fig., 4 tab.; Reston.

Betlem, G. & Brans, E. H. P. (2002): The Future Role of Civil Liability for Environmental Damage in the EU. – Yearbook of European Environmental Law, 2: 183–221, 2 tab.; Oxford.

Beuge, P., Kluge, A. & Degner, T. (2001): Prognose geochemischer Auswirkungen der Nachnutzung stillgelegter Bergbau Stollnsysteme – Abschlußbericht über ein Forschungsprojekt gefördert unter dem Az: 15339 von der Deutschen Bundesstiftung Umwelt. – In: Deutsche Bundesstiftung Umwelt. – 88 p., 25 fig., 24 tab., 6 app.; Freiberg (TU Bergakademie Freiberg).

Beyer, W. (1986): Zur mikrobiellen In-situ-Laugung komplexer Sulfiderze des Rammelsbergs bei Goslar. – 118 p., 83 fig., 16 tab.; Berlin (Weidler).

Biaglow, J. A. (1988): Study to determine the effectiveness of the hydraulic mine seal at Big Four Hollow Creek near Lake Hope State Park in southeastern Ohio. – 169 p.; Dayton OH (Master's Thesis).

Biehler, D. & Falck, W. E. (1999): Simulation of the effects of geochemical reactions on groundwater quality during planned flooding of the Königstein uranium mine, Saxony, Germany. – Hydrogeol. J., 7 (3): 284–293, 11 fig., 2 tab.; Berlin.

Bigham, J. M. (1994): Mineralogy of Ochre Deposits Formed by Sulfide Oxidation. – In: Jambor, J. L. & Blowes, D. W.: Short Course Handbook on Environmental geochemistry of Sulfide Mine-Wastes 22. – p. 103–132, 12 fig., 3 tab.; Waterloo, Ontario (Mineralogical Association of Canada).

Birgersson, L., Widen, H., Aagren, T. & Neretnieks, I. (1992): Tracer migration experiments in the Stripa mine 1980-1991 STRIPA-TR-92-25. – 151 p.; Stockholm (Swedish Nuclear Fuel and Waste Management Co.).

Biswas, A. K. (1970): History of hydrology. – 336 p.; Amsterdam u.a. (North-Holland Publishing).

Blanc, G. & Anschutz, P. (1995): New stratification in the hydrothermal brine system of the Atlantis II Deep, Red Sea. – Geology, 23 (6): 543–546, 2 fig., 2 tab.; Boulder.

Bleeker, P. M., Teiga, P. M., Santos, M. H., De Koe, T. & Verkleij, J. A. C. (2003): Ameliorating effects of industrial sugar residue on the Jales gold mine spoil (NE Portugal) using *Holcus lanatus* and *Phaseolus vulgaris* as indicators. – Environ. Pollut., 125 (2): 237–244, 2 fig., 3 tab.; Oxford.

Blockeel, H., Džeroski, S. & Grbović, J. (1999): Simultaneous prediction of multiple chemical parameters of river water quality with TILDE. – Lecture notes in computer science, 1704: 32–40, 1 fig., 3 tab.; Berlin.

Blowes, D. W., Ptacek, C. L., Frind, E. O., Johnson, R. H., Robertson, W. D. & Molson, J. W. (1994): Acid-Neutralization Reactions in inactive Mine Tailings Impoundments and their Effect on the Transport of dissolved Metals. – Proceedings, International Land Reclamation and Mine Drainage Conference, 1: 429–438, 4 fig., 1 tab.; Pittsburgh.

Boehrer, B. (2000): Modal Response of a Deep Stratified Lake: Western Lake Constance. – J. Geophys. Res. C, 105 (12): 28,837–28,845, 9 fig., 1 tab.; Washington.

Boehrer, B., Ilmberger, J. & Münnich, K. O. (2000): Vertical Structure of Currents in Western Lake Constance. – J. Geophys. Res. C, 105 (12): 28,823–28,835, 11 fig., 1 tab.; Washington.

Bogner, J. E. & Doehler, R. W. (1984): Ettringite Formation following Chemical Neutralization Treatment of Acidic Coal Mine Drainage. – Int. J. Mine Water, 3 (1): 13–24, 3 fig.; Granada.

Bohl, W. (1998): Technische Strömungslehre. – 317 p.; Würzburg (Vogel).

Boice, J. D., Mumma, M., Schweitzer, S. & Blot, W. J. (2003): Cancer Mortality in a Texas County with prior Uranium Mining and Milling Activities, 1950–2001. – J. Radiol. Prot., 23 (3): 247–262, 4 fig.; Bristol.

Bolis, J. L., Wildeman, T. R. & Dawson, H. E. (1992): Hydraulic Conductivity of Substrates used for passive Acid Mine Drainage Treatment. – p. 79–89, 2 fig., 1 tab.; Duluth (Proceedings, 9th Annual National Meeting – American Society for Surface Mining and Reclamation).

Bolzicco, J., Ayora, C., Rötting, T. & Carrera, J. (2003): Performance of the Aznalcóllar permeable reactive barrier. – In: Nel, P. J. L.: Mine Water Env. – p. 287–299, 10 fig., 2 tab.; Johannesburg (Proceedings, 8th International Mine Water Association Congress).

Bonfig, K. W. (2002): Technische Durchflussmessung unter Berücksichtigung neuartiger Durchflussmessverfahren. – 3rd edn., 280 p., 221 fig., 17 tab.; Essen (Vulkan).

Bongaerts, J. C. (2002): European Water Law: water policy and water resources management in France: the project de loi sur l'eau. – Europ. Env. Law Rev. (8–9, 10): 239–244; 260–270.

Borden, R. K. (2003): Environmental geochemistry of the Bingham Canyon porphyry copper deposit, Utah. – Environ. Geol., 43 (7): 752–758, 6 fig., 1 tab.; Berlin.

Bosecker, K. (1980): Bakterielles Leaching – Metallgewinnung mit Hilfe von Bakterien. – Metall, 34 (1): 36–40; Berlin.

Bosman, D. J., Clayton, J. A., Maree, J. P. & Adlem, C. J. L. (1990): Removal of Sulphate from Mine Water with Barium Sulphide. – Int. J. Mine Water, 9 (1–4): 149–163, 5 fig., 6 tab.; Castro Verde.

Böttcher, J., Büder, W., Hamann, M., Hamann, S., Linkert, K.-H., Lipp, U., Müller, F., Rochhausen, D., Rosenhahn, L., Schröder, B., Schuppan, W. & Schwarze, K.-H. (1991): Seilfahrt – Auf den Spuren des sächsischen Erzbergbaus. – In: Wismut AG. – 2nd edn., 153 p.; Haltern (Bode).

Boudjemaa, R., Forbes, A. B., Harris, P. M. & Langdell, S. (2003): Multivariate empirical models and their use in metrology. – Reports NPL-CMSC, CMSC 32/03: 55, 30 fig.; Teddington.

Boult, S. (1996): Fluvial Metal Transport near Sources of Acid Mine-Drainage: Relationships of Soluble, Suspended and Deposited Metal. – Mineral. Mag., 60 (399): 325–335, 5 fig., 2 tab.; London.

Bourgeois, W., Gardey, G., Servieres, M. & Stuetz, R. M. (2003): A Chemical Sensor Array based System for Protecting Wastewater Treatment Plants. – Sens. Actuator B-Chem., 91 (1–3): 109–116, 9 fig.; Lausanne.

Bowell, R. J. (2001): Hydrogeochemistry of the Getchell Underground Mine – Part 1: Mine Water Chemistry. – Mine Water Env., 20 (2): 81–97, 10 fig., 7 tab.; Berlin.

Bowell, R. J. (2002): The hydrogeochemical dynamics of mine pit lakes. – Spec. Publ. – Geol. Soc. London, 198: 159–185, 18 fig., 7 tab.; London.

Bowell, R. J. (2003): Pit Lake Systematics: A Special Issue. – Mine Water Env., 22 (4): 167–169; Berlin.

Bowell, R. J. (2004): A review of sulphate removal options for mine waters. – In: Jarvis, A. P., Dudgeon, B. A. & Younger, P. L.: mine water 2004 – Proceedings International Mine Water Association Symposium 2. – p. 75–91, 6 fig., 7 tab.; Newcastle upon Tyne (University of Newcastle).

Bowell, R. J., Barta, J., Gringrich, M., Mansanares, W. & Parshley, J. (1998): Geological Controls on Pit Lake Chemistry: Implications for the assessment of Water Quality in Inactive Open Pits. – In: Nel, P. J. L.: Mine Water and Environmental Impacts 2. – p. 375–386, 5 fig., 1 tab.; Johannesburg (Proceedings International Mine Water Association Symposium).

Bowell, R. J., Connelly, R. J., Ellis, J., Cowan, J., Wood, A., Barta, J. & Edwards, P. (1997): A review of sulfate removal options from mine waters. – Open-File Report – US Geological Survey, OF 97-0496: 14; Reston.

Box, G. E. P. (1976): Science and Statistics. – J. American Statistical Association, 71 (356): 791–799, 3 fig., 1 tab.; Alexandria.

Box, G. E. P. (1979): Robustness in the strategy of scientific model building. – In: Launer, R. L. & Wilkinson, G. N.: Robustness in Statistics. – p. 201–236, 14 fig., 1 tab.; New York (Academic Press).

Box, G. E. P., Hunter, W. G. & Hunter, J. S. (1978): Statistics for Experimenters – An Introduction to Design, Data Analysis, and Model Building. – 653 p., 129 fig., 124 tab.; New York (Wiley).

Brandberg, F., Grundfelt, B., Hoglund, L. O. & Lindgren, M. (1993): Prediction of metal discharges from sulphidic mine tailings covered with soil. – IAHS-AISH Publication, 220: 193–202, 3 fig., 2 tab.; Louvain.

Brandt, A. & Fernando, H. J. S. (1995): Double-Diffusive Convection: Geophysical Monograph 94. – 334 p.; Washington (American Geophysical Union).

Brassington, R. (1999): Field Hydrogeology. – 2nd edn., 248 p., 104 fig., 43 tab., 2 app.; Chichester u.a. (Wiley & Sons).

Brendel, K., Brückner, G., Knitzschke, G., Schwandt, A. & Spilker, M. (1982): Montanhydrologische Aspekte zur Gewährleistung der Bergbausicherheit beim Abbau zechsteinzeitlicher Lagerstätten. – Z. geol. Wiss., 10 (1): 7–31, 13 fig.; Berlin.

Brenner, F. J. (2000): Wildlife and Fishery Considerations in Surface Mine Reclamation. – In: Barnhisel, R. I., Darmody, R. G. & Daniels, W. L.: Reclamation of Drastically Disturbed Lands 15. – p. 399–413, 6 fig., 1 tab.; Madison, Wis. (American Society of Agronomy).

Bretherton, B. (1989): Near-surface acid mine water pools and their implications for mine abandonment, Coeur d'Alene mining district, Idaho. – 152 p., 32 fig., 17 tab.; Moscow (Master's Thesis University of Idaho).

Brewitz, W., Kull, H. & Tebbe, W. (1985): In situ experiments for the determination of macro-permeability and RN-migration in faulted rock formations such as the oolithic iron ore in the Konrad Mine: Design and instrumentation in underground laboratories for radioactive waste disposal. – p. 289–302, 4 fig., 1 tab.; Rotterdam (Balkema).

Briechle, D., Holl-Hagemeir, C., Lögters, C., Mayers-Beecks, E., Milbert, G., Paas, W. & Prüfert, J. (1998): Raw material extraction in the Rhenish lignite district – the effect of the deep open-cast mining on the population and environment. – Chron. Rech. Miniere, 66 (533): 13–29, 19 fig., 1 tab.; Orleans.

Brix, H. (1987a): The applicability of the wastewater treatment plant in Othfresen as scientific documentation of the root-zone method. – Water Sci. Technol., 19 (10): 19–24, 1 fig., 1 tab.; Oxford.

Brix, H. (1987b): Treatment of wastewater in the rhizosphere of wetland plants – the root-zone method. – Water Sci. Technol., 19 (Rio): 107–118, 5 fig., 2 tab.; Oxford.

Brix, H. (2003): Plants used in Constructed Wetlands and their Function. – Proceedings, 1st International Seminar on the Use of Aquatic Macrophytes for Wastewater Treatment in Constructed Wetlands: 81–109, 9 fig., 2 tab.; Lisboa.

Brontë, C. (1993 [1860]): Emma: Unfinished Novels. – p. 98–113; Stroud (Sutton).

Brook, C. & Cresswell, P. J. (1989): Environmental Aspects of Opencast Mining with reference to the Rother Valley Country Park. – Min. Eng.: 191–196, 5 fig.; London.

Brown, A. (1982): A validated theory of underground mine flooding. – Proceedings, 1st International Mine Water Congress, Budapest, Hungary, B: 411–428, 9 fig.; Budapest.

Brown, E. B. & Holley, M. C. (1982): Metal Levels Associated with Tin Dredging and Smelting and Their Effect Upon Intertidal Reef Flats at Ko Phuket, Thailand. – Coral Reefs, 1: 131–137, 9 fig., 2 tab.; Berlin.

Brown, M. C. & Ford, D. C. (1971): Quantitative tracer methods for investigation of karst hydrological systems. – Trans. Cave Research Group of Great Britain, 13 (1): 37–51, 13 fig., 2 tab.; Bridgewater.

Brown, M., Barley, B. & Wood, H. (2002): Minewater Treatment – Technology, Application and Policy. – 500 p.; London (IWA Publishing).

Brunner, H., Strasser, H., Pümpel, T. & Schinner, F. (1993): Mikrobielle Laugung von Kupfer und Zink aus Tetraedrit-haltigem Baryt (Region Kleinkogel/Tirol) durch *Thiobacillus ferrooxidans*. – Arch. f. Lagerst.forsch. Geol. B.-A., 16: 5–12, 6 fig., 3 tab.; Wien.

Bryan, C. G., Hallberg, K. B. & Johnson, D. B. (2004): Microbial populations in surface spoil at the abandoned Mynydd Parys copper mines. – In: Jarvis, A. P., Dudgeon, B. A. & Younger, P. L.: mine water 2004 – Proceedings International

Mine Water Association Symposium 1. – p. 107–112, 3 tab.; Newcastle upon Tyne (University of Newcastle).

Bryan, G. W. & Langston, W. J. (1992): Bioavailability, accumulation and effects of heavy metals in sediments with special reference to United Kingdom estuaries: a review. – Environ. Pollut., 76 (2): 89–131, 9 fig., 5 tab.; Oxford.

Bucksteeg, K. (1986): Sumpfpflanzenkläranlage – Verfahrensvarianten, Betriebserfahrungen, Problem Bodenhydraulik. – GWF, Gas- Wasserfach: Wasser/Abwasser, 127 (9): 429–434, 4 fig., 1 tab.

Büder, W. & Schuppan, W. (1992): Zum Uranerzbergbau im Lagerstättenfeld Schneeberg-Schlema-Alberoda im Westerzgebirge. – Schriftenreihe Gesellschaft Deutscher Metallhütten- und Bergleute, 64: 203–221, 9 fig., 2 tab.; Clausthal-Zellerfeld.

Buhrmann, F., van der Waldt, M., Hanekom, D. & Finlayson, F. (1999): Treatment of industrial wastewater for reuse. – Desalination, 124 (1–3): 263–269, 3 fig., 4 tab.; Amsterdam.

Bundesminister für Wirtschaft (1993): Wismut – Stand der Stillegung und Sanierung. – BMWI Dokumentation, 335: 1–35; Bonn.

Burbey, T. J., Younos, T. & Anderson, E. T. (2000): Hydrologic analysis of discharge sustainability from an abandoned underground coal mine. – J. Am. Water Resour. Assoc., 36 (5): 1161–1172, 10 fig., 4 tab.; Herndon.

Burchi, S. & Mechlem, K. (2005): Groundwater in international law – Compilation of treaties and other legal instuments [sic!]: FAO Legislative Studies 86. – 584 p., 3 tab.; Rome (United Nations Educational Scientific, and Cultural Organization; Food and Agriculture Organization of the United Nations).

Burke, S. P. & Younger, P. L. (2000): Groundwater Rebound in the South Yorkshire Coalfield: A First Approximation using the GRAM Model. – Q. J. Eng. Geol. Hydrogeol., 33 (2): 149–160, 9 fig., 5 tab.; London.

Burnett, J. M., Burnett, M., Ziemkiewicz, P. & Black, D. C. (1995): Pneumatic backfilling of coal combustion residues in underground mines. – In: Hynes, T. P. & Blanchette, M. C.: Sudbury '95 – Mining and the Environment 3 (REHAB 18). – p. 1211–1220, 6 fig.; Ottawa (CANMET).

Busch, W., Maas, K., Klapperich, H., Tondera, D., Sroka, A. & Meier, G. (2002): 2. Altbergbaukolloquium – Tagungsband. – 388 p.; Clausthal-Zellerfeld (Papierflieger).

Cabral, A., Lefebvre, G., Proulx, M., Audet, C., Labbe, M. & Michaud, C. (1997): Use of deinking residues as cover material in the prevention of AMD generation at an abandoned mine site. – Proceedings, International Conference on Tailings and Mine Waste, 4: 257–266, 6 fig., 1 tab.; Rotterdam.

Cacas, M. C., Ledoux, E., de Marsily, G., Tillie, B., Barbreau, A., Durand, E., Feuga, B. & Peaudecerf, P. (1990): Modeling Fracture Flow With a Stochastic Discrete Fracture Network: Calibration and Validation. 1. The Flow Model. – Water Resour. Res., 26 (3): 479–489, 12 fig., 1 tab.; Washington.

Cai, S., Lin, Y., Zhineng, H., Xianzu, Z., Zhalou, Y., Huidong, X., Yuanrong, L., Rongdi, J., Wenhau, Z. & Fangyuan, Z. (1990): Cadmium exposure and health effects among residents in an irrigation area with ore dressing wastewater. – Sci. Total Environ., 90: 67–73, 3 tab.; Amsterdam.

Callaghan, T., Brady, K., Chisholm, W. & Sames, G. (2000): Hydrology of the Appalachian Bituminous Coalbasin. – In: Kleinmann, R. L. P.: Prediction of Water Quality at Surface Coal Mines (3). – p. 36–72, 5 fig., 4 tab.; Morgantown (National Mine Land Reclamation Center).

Calow, R. (1991): Quality Control/Quality Assurance in Geochemical Laboratories. – Explore – Newsletter for the Association of Exploration Geochemists, 72: 23–24; Reno.

Canty, G. A. & Everett, J. (1998): Using tracers to understand the hydrology of an abandoned underground coal mine 15. – p. 62–72, 5 fig., 2 tab.; Princeton (Proceedings, 15th Annual National Meeting – American Society for Surface Mining and Reclamation).

Canty, G. A. & Everett, J. (1999): Remediation of Underground Mine Areas through Treatment with Fly Ash. – 167 p., 30 fig., 43 tab.; Oklahoma City (Oklahoma Conservation Commission).

Carlin, J. F. J. (1999): Tin. – In: Plunkert, P. A. & Jones, T. S.: Metal Prices in the United States through 1998. – p. 159–160, 1 fig., 1 tab.; Washington (US Department of the Interior, US Geolocical Survey).

Carlowitz, H. C. v. (1713): Sylvicultura Oeconomica, oder haußwirthliche Nachricht und Naturmäßige Anweisung zur Wilden Baum-Zucht, nebst Gründlicher Darstellung, wie zuförderst durch Göttliches Benedeyen dem allenthalben und insgemein einreissenden Grossen Holtz-Mangel, vermittelst Säe-Pflantz- und Versetzung vielerhand Bäume zu prospicieren, auch also durch Anflug und Wiederwachs des so wohl guten und schleunig anwachsend, als anderen gewüchsig und nützlichen Holtzes, ganz öde und abgetriebene Holtz-Ländereyen, Plätze und Orte widerum Holzreich, nütz und brauchbar zu machen; Bevorab von Saam-Bäumen und wie der wilde Baum-Saamen zu sammeln, der Grund und Boden zum Säen zuzurichten, solche Saat zu bewerckstelligen, auch der junge Anflug und Wiederwachs zu beachten. Daneben das sogenannte lebendige, oder Schlag-an Ober- und Unter-Holz auffzubringen und zu vermehren, welchen beygefügt die Arten des Tangel- und Laub Holzes, thels deren Eigenschafften und was besagtes Holtz für Saamen trage, auch wie man mit frembden Baum-Gewächsen sich zu verhalten, ferner wie das Holz zu fällen, zu verkohlen, zu äschern und sonst zu nutzen. Alles zu nothdürfftiger Versorgung des Hauß-Bau-Brau-Berg- und Schmeltz-Wesens, und wie eine immerwährende Holtz-Nutzung, Land und Leuten, auch jedem Hauß-Wirthe zu unschätzbaren großen Auffnehmen, pfleglich und füglich zu erziehlen und einzuführen, worbey zugleich eine gründliche Nachricht von den in Churfl. Sächß. Landen Gefundenen Turff Dessen Natürliche Beschaffenheit, grossen Nutzen, Gebrauch und nützlichen

Verkohlung Aus Liebe zu Beförderung des algemeinen Bestens beschrieben. – 432 p.; Leipzig (Braun).

Carlson, A. B., Crilly, P. B. & Rutledge, J. C. (2002): Communication systems – An introduction to signals and noise in electrical communication: McGraw-Hill series in electrical and computer engineering. – 4th edn., 864 p., 479 fig.; Boston (McGraw-Hill).

Carlsson, E. (2004): Sulphide-Rich Tailings Remediated by Soil Cover – Evaluation of cover efficiency and tailings geochemistry, Kristineberg, northern Sweden. – Doctoral Thesis, 44: 56, 20 fig., 7 app.; Lulea.

Carvajal, D. J. & González, A. (2000): Mining Heritage & Closure Mines. – p. 158–165; Rio de Janeiro (CYTED – IMACC – UNIDO).

Cassity, W. D. & Pesic, B. (1999): Interaction of *Thiobacillus ferrooxidans* with arsenite, arsenate and arsenopyrite. – In: Amils, R. & Ballester, A.: Biohydrometallurgy and the environment toward the mining of the 21st century A. – p. 521–532, 10 fig., 1 tab.; Amsterdam (Elsevier).

Castrilli, J. F. (2002): Barriers to Collaboration – Orphaned/Abandoned Mines in Canada (Final Report). – 67 p.; Ontario (National Orphaned/Abandoned Mines Advisory Committee [http://www.abandoned-mines.org/]).

Castro, J. M. & Moore, J. N. (2000): Pit lakes: their characteristics and the potential for their remediation. – Environ. Geol., 39 (11): 1254–1260, 2 fig.; Berlin.

Chałupnik, S. & Wysocka, M. (2000): Removal of radium from mine waters – the experience from the coal mine. – Proceedings, 7th International Mine Water Association Congress: 352–362, 3 fig.; Ustron.

Chałupnik, S., Michalik, B., Wysocka, M., Skubacz, K. & Mielnikow, A. (2001): Contamination of settling ponds and rivers as a result of discharge of radium-bearing waters from Polish coal mines. – J. Environ. Radioact., 54 (1): 85–98, 1 fig., 7 tab.; Oxford.

Chan, G. Y. S., Ye, Z. H. & Wong, M. H. (2003): Comparison of four *Sesbania* Species to remediate Pb/Zn and Cu Mine Tailings. – Environ. Manage., 32 (2): 246–251, 6 tab.; New York.

Chan, T., Reid, J. A. K. & Guvanasen, V. (1987): Numerical modelling of coupled fluid, heat, and solute transport in deformable fractured rock. – In: Tsang, C.-F.: Coupled processes associated with nuclear waste repositories (44). – p. 605–625, 11 fig.; Orlando (Academic Press).

Chaney, R. L., Brown, S. L., Li, Y. M., Angle, J. S., Stuczynski, T. I., Daniels, W. L., Henry, C. L., Siebelec, G., Malik, M., Ryan, J. A. & Compton, H. (2000): Progress in Risk Assessment for Soil Metals, and In-situ Remediation and Phytoextraction of Metals from Hazardous Contaminated Soils. – In: U.S. Environmental Protection Agency: Phytoremediation: State of Science 625/R-01/011b. – p. 1–26, 4 fig., 5 tab.; Boston (U.S. Environmental Protection Agency).

Chansang, H. (1988): Coastal tin mining and marine pollution in Thailand. – Ambio, 17 (3): 223–228, 6 fig., 2 tab.; Stockholm.

Charbonnier, P. (2001): Management of Mining, Quarrying and Ore-Processing Waste in the European Union BRGM/RP-50319-FR. – 79 p., 7 fig., 17 tab., 7 app.; Orleans (BRGM).

Chekan, G. J. (1985): Design of Bulkheads for Controlling Water in Underground Mines. – Bureau of Mines Information Circular, IC-9020: 36, 10 fig., 3 tab., 2 app.; Washington.

Clark, I. D. & Fritz, P. (1999): Environmental Isotopes in Hydrogeology. – 2nd edn., 328 p.; Boca Raton (Lewis).

Clark, L. C. jr., Wolf, R., Granger, D. & Taylor, Z. (1953): Continuous Recording of Blood Oxygen Tensions by Polarography. – J. Appl. Physiol., 6 (3): 189–193, 4 fig., 1 tab.; Bethesda.

Clauser, C., Heitfeld, M., Rosner, P., Sahl, H. & Schetelig, K. (2005): Beispiel Aachener Steinkohlenrevier – Nutzung von Erdwärme in aufgelassenen Bergwerken. – Beratende Ingenieure (6): 14–17, 4 fig.; Wiesbaden.

Clemens, T., Huckinghaus, D., Liedl, R. & Sauter, M. (1999): Simulation of the development of karst aquifers: role of the epikarst. – Int. J. Eart. Sciences (Geol. Rundsch.), 88 (1): 157–163, 6 fig.; Berlin.

Coetzee, H. (1995): Radioactivity and the leakage of radioactive waste associated with Witwatersrand gold and uranium mining. – Proceedings, Uranium-Mining and Hydrogeology, Freiberg, Germany; GeoCongress, 1: 90–99, 6 fig.; Freiberg.

Cohen, R. R. H. & Staub, M. W. (1992): Technical manual for the design and operation of a passive mine drainage system. – 69 p.; Golden (unpublished Report to the US Bureau of Land Reclamation).

Col, M. & Col, C. (2003): Environmental Boron Contamination in Waters of Hisarcik Area in the Kutahya Province of Turkey. – Food Chem. Toxicol., 41 (10): 1417–1420; Oxford.

Coldewey, W. G. & Semrau, L. (1994): Mine Water in the Ruhr Area (Federal Republic of Germany). – Proceedings, 5th International Mine Water Congress, Nottingham, U.K., 2: 613–629, 7 fig.; Nottingham.

Coldewey, W. G., Hewig, R., Richter, R., Rüterkamp, P. & Wedewart, M. (1999): Mittelfristige Entwicklung des Chemismus und der Dichte-Schichtungen von Grubenwässern in Bergwerken und ihre Auswirkungen auf nutzbares Grund- und Oberflächenwasser. – For: Landesoberbergamt Nordrhein-Westfalen. – 83 p., 3 tab., 73 app.; Essen (Deutsche Montan Technologie GmbH).

Collon-Drouaillet, P., Fabriol, R. & Buès, M. (2005): A semi-distributed modelling of the evolution of groundwater quality in flooded mines in the lorraine iron basin. – p. 1–15 [CD-ROM], 10 fig., 3 tab.; Nancy (Proceedings, Post-Mining 2005).

Commonwealth of Pennsylvania (1973): Scarlift Report – Clearfield Creek/Moshannon Watershed SL 1-73:1-101.7. – 333 p., 21 fig.; Harrisburg (Commonwealth of Pennsylvania).

Copa, L. V., Ferderer, D. A., Goklany, I. M. & Kaas, L. M. (1995): Developing Priorities for Verifying and Inventorying Suspected Abandoned Mine Sites on

Public Lands. – In: Hotchkiss, W. R., Downey, J. S., Gutentag, E. D. & Moore, J. E.: Water resources at risk. – p. IMWA.125–140, 10 fig.; Minneapolis (American Institute of Hydrology).

Copper Mines of Tasmania (1995): Mt Lyell Mine Redevelopment – Environmental Management Plan. – 181 p.; Queenstown Tas. (Copper Mines of Tasmania).

Corwin, R. F. (1997): The Self-Potential Method for Environmental and Engineering Applications. – In: Ward, S. H.: Geotechnical and environmental geophysics 1. – 4th edn., S. 127–145, 7 fig.; Tulsa (Society of Exploration Geophysicists).

Coulton, R., Bullen, C., Williams, K., Dey, M. & Jarvis, A. (2004): Active treatment of high salinity mine water. – In: Jarvis, A. P., Dudgeon, B. A. & Younger, P. L.: mine water 2004 – Proceedings International Mine Water Association Symposium 2. – p. 119–124, 3 fig., 6 tab.; Newcastle upon Tyne (University of Newcastle).

Coupland, K., Rowe, O., Hallberg, K. B. & Johnson, D. B. (2004): Biogeochemistry of a subterranean acidic mine water body at an abandoned copper mine. – In: Jarvis, A. P., Dudgeon, B. A. & Younger, P. L.: mine water 2004 – Proceedings International Mine Water Association Symposium 1. – p. 113–119, 4 fig.; Newcastle upon Tyne (University of Newcastle).

Coury, L. (1999): Conductance Measurements – Part 1: Theory. – Current Separations, 18 (3): 91–96, 4 fig., 1 tab.; West Lafayette.

Cravotta, C. A. & Hilgar, G. M. (2000): Considerations for Chemical Monitoring at Coal Mines. – In: Kleinmann, R. L. P.: Prediction of Water Quality at Surface Coal Mines Appendix A. – p. 195–218, 5 fig.; Morgantown (National Mine Land Reclamation Center).

Creveling, H. F., de Paz, J. F., Baladi, J. Y. & Schoenhals, R. J. (1975): Stability characteristics of a single-phase free convection loop. – J. Fluid Mech., 67 (1): 65–84, 10 fig., 1 tab.; Cambridge.

Croxford, S. J., England, A. & Jarvis, A. P. (2004): Application of the PHREEQC geochemical Computer Model during the Design and Operation of UK Mine Water Treatment Schemes. – In: Jarvis, A. P., Dudgeon, B. A. & Younger, P. L.: mine water 2004 – Proceedings International Mine Water Association Symposium 2. – p. 125–134, 8 tab.; Newcastle upon Tyne (University of Newcastle).

Csövári, M., Berta, Z., Csicsák, J., Hideg, J. & Varhegyi, A. (2004): Case Study: Remediation of a former Uranium Mining/Processing Site in Hungary: Conference Papers 35. – p. 81–100, 17 fig., 5 tab.; Wien (Umweltbundesamt).

Cumming, G., Fidler, F. & Vaux, D. L. (2007): Error bars in experimental biology. – J. Cell Biology, 177 (1): 7–11, 7 fig., 1 tab.; New York.

Cunningham, S. D., Shann, J. R., Crowley, D. E. & Anderson, T. A. (1997): Phytoremediation of Contaminated Water and Soil. – ACS Symposium Series, 664: 2–17; Washington.

Cvijić, J. (1893): Das Karstphänomen – Versuch einer morphologischen Monographie. – Geogr. Abh., 5 (3): 217–329, 5 fig., 8 tab.; Wien.

Czolbe, P., Kretzschmar, H.-J., Klafki, M. & Heidenreich, H. (1992): Strömungszellen im gefluteten Salzschacht. – Neue Bergbautechnik, 22 (6): 213–218, 10 fig.; Leipzig.

Dahlkamp, F. J. (1993): Uranium ore deposits. – 460 p.; Berlin (Springer).

Damerell, P. S. & Schoenhals, R. J. (1979): Flow in a Toroidal Thermosyphon with Angular Displacement of Heated and Cooled Sections. – J. Heat Transfer, 101: 672–676, 7 fig.; New York.

Darcy, H. P. G. (1856): Les fontaines publiques de la ville de Dijon – Exposition et application des principes a suivre et des formules a employer dans les questions de distribution d'eau – Ouvrage terminé par un appendice relatif aux fournitures d'eau de plusieurs villes au filtrage des eaux à la fabrication des tuyaux de fonte, de plomb, de tole et de bitume. – 647 p.; Paris (Dalmont).

Davies, G. J., Wireman, M., Stover, B. K. & Anonymous (2000): Mary Murphy's chest pains; resorting to geological angioplasty to study specific metals loading at a high-altitude mine site. – Abstracts with Programs – Geological Society of America, 32: 362; Boulder.

Davis, J. C. (1986): Statistics and Data Analysis in Geology. – 2nd edn., 646 p.; New York u.a. (Wiley & Sons).

Davis, M. W. (1994a): Identification and remediation of a mine flooding problem in Rico, Dolores County, Colorado with a discussion on the use of tracer dyes. – Colorado Geological Survey Open-File Report, 91-1: 15, 1 tab.; Denver.

Davis, M. W. (1994b): The use of tracer dyes for the identification of a mine flooding problem, Rico, Dolores County, Colorado. – Colorado Geological Survey Open-File Report, 91-2: 1–20, 7 fig., 1 tab.; Denver.

Davis, S. N. (1983): Measurement, Prediction, and Hazard Evaluation of Earth Fissuring and Subsidence due to Ground-Water Overdraft. – Office of Water Policy Project, B-092-ARIZ: 44, 12 fig., 1 tab.; Arizona.

de la Vergne, J. (2003): Hard Rock Miner's Handbook. – 3rd edn., 314 p., 13 fig., 98 tab.; Tempe (McIntosh Engineering).

DeHay, K. L. (2003): Assessment and Comparison of 1976–77 and 2002 Water Quality in Mineshafts in the Picher Mining District, Northeastern Oklahoma and Southeastern Kansas. – Water-Resources Investigations Report, 03-4248: 65, 6 fig., 2 tab., 2 app.; Washington.

DeHay, K. L., Andrews, W. J. & Sughru, M. P. (2004): Hydrology and Ground-Water Quality in the Mine Workings within the Picher Mining District, Northeastern Oklahoma, 2002–03. – Scientific Investigations Report, 2004-5043: 62, 24 fig., 4 tab., 4 app.; Washington.

Demchak, J., Skousen, J. G. & McDonald, L. M. (2004): Longevity of Acid Discharges from Underground Mines Located above the Regional Water Table. – J. Environ. Qual., 33 (2): 656–668, 6 fig., 6 tab.; Madison.

Deming, D. (2002): Introduction to Hydrogeology. – 1st edn., 468 p.; Boston u.a. (McGraw-Hill).

Dempsey, B. A. & Jeon, B.-H. (2001): Characteristics of sludge produced from passive treatment of mine drainage. – Geochem.-Explor. Environ. Anal., 1: 89–94, 5 fig., 5 tab.; London.

Diaz, W. D. (1990): Hydrogeochemistry and physical hydrogeology of the Newfoundland zinc mine, Daniel's Harbour, Newfoundland. – 282 p., Refs: 93; illus. incl. 18 tables, 2 plates; St. John's, NF (Master's Thesis).

Dietrich, P. (1965): Zur Geologie des Kalksteinvorkommens von Niederrabenstein. – 17 p., 30 fig., 47 app.; Freiberg (unpubl. Thesis Bergakademie Freiberg).

Dinelli, E. & Tateo, F. (2002): Different types of fine-grained sediments associated with acid mine drainage in the Libiola Fe-Cu mine area (Ligurian Apennines, Italy). – Appl. Geochem., 17 (8): 1081–1092, 4 fig., 4 tab.; Oxford.

Diz, H. R. (1998): The Selective Oxide System™: Acid Drainage Treatment that avoids the Formation of Sludge. – Mine Water Env., 17 (1): 1–7, 3 fig.; Lakewood, CO.

Dizdarevič, T. & Režun, B. (1998): Environmental Problems related to the closing down of the Idrija Mercury Mine 2. – p. 367–374, 4 fig.; Johannesburg (Proceedings International Mine Water Association Symposium).

Dobler, L. (1999): Der Einfluß der Bergbaugschichte im Ostharz auf die Schwermetalltiefengradienten in historischen Sedimenten und die fluviale Schwermetalldispersion in den Einzugsgebieten von Bode und Selke im Harz. – 120 p., 22 fig., 41 tab., 9 maps, 24 app.; Halle (unpubl. Diss. Univ. Halle-Wittenberg).

Doherty, J. (1990): The interpretation of pump-test data from a disused underground mine. – J. Hydrol., 114 (1–2): 109–123, 7 fig., 2 tab.; Amsterdam.

Domenico, P. A. & Schwartz, F. W. (1998): Physical and Chemical Hydrogeology. – 2nd edn., 506 p.; New York u.a. (Wiley & Sons).

Doornbos, M. H. (1989): Hydrogeologic controls upon the water quality of the acid mine drainage in the Sand Coulee-Stochett area, Montana. – 103 p., Refs: 57; Butte (Master's Thesis Montana College of Mineral Science & Technology).

Döring, M. (1993): Unterirdische Wasserkraftwerke im Bergbau. – Wasserwirtschaft, 83 (5): 272–278, 6 fig.; Stuttgart.

Döring, M. (1999): Eisen und Silber – Wasser und Wald. – 226 p.; Kreuztal (wielandschmiede).

Douglas, G. E., John, D. M., Williamson, D. B. & Reid, G. (1998): The aquatic algae associated with mining areas in Peninsula Malaysia and Sarawak: their composition, diversity and distribution. – Nova Hedwigia, 67 (1–2): 189–211, 9 fig., 4 tab.; Berlin.

Dowdall, M., Vicat, K., Frearson, I., Gerland, S., Lind, B. & Shaw, G. (2004): Assessment of the radiological impacts of historical coal mining operations on the environment of Ny-Alesund, Svalbard. – J. Environ. Radioact., 71 (2): 101–114, 5 fig., 5 tab.; Oxford.

Downing, B. W. & Mills, C. (1998): Quality Assurance/Quality Control for Acid Mine Drainage. – 1 p., 7 fig. (http://www.enviromine.com/ard/Acid-Base%20 Accounting/Quality.htm 5.1.2005).

Doyle, G. A. & Runnells, D. D. (1997): Physical limnology of existing mine pit lakes. – Mining Engineering, 49 (12): 76–80, 5 fig., 1 tab.; Littleton.

Drazin, P. G. (2002): Introduction to Hydrodynamic Stability: Cambridge texts in applied mathematics 32. – 1st edn., 258 p., 78 fig.; Cambridge (Cambridge University Press).

Driscoll, F. G. (1987): Groundwater and Wells. – 2nd edn., 1089 p.; Minnesota (Johnson Division).

Droste, R. L. (1997): Theory and Practice of Water and Wastewater Treatment. – 800 p.; New York (Wiley & Sons).

Du Plessis, H. M. (1983): Using Lime Treated Acid-Mine Water for Irrigation. – Water Sci. Technol., 15 (2): 145–154, 5 fig., 3 tab.; Oxford.

Duane, M. J., Pigozzi, G. & Harris, C. (1997): Geochemistry of some deep gold mine waters from the western portion of the Witswatersrand Basin, South Africa. – J. Afr. Earth Sci., 24 (1–2): 105–123, 6 fig., 2 tab.; Oxford.

Duarte, J. C., Estrada, P., Beaumont, H., Sitima, M. & Pereira, P. (1990): Biotreatment of Tailings for Metal recovery. – Int. J. Mine Water, 9 (1–4): 193–206, 4 fig., 3 tab.; Castro Verde.

Duffus, J. H. (2002): "Heavy Metals" – A meaningless term? – Pure Appl. Chem., 74 (5): 793–807, 3 fig., 4 tab.; Oxford.

Dumpleton, S., Robins, N. S., Walker, J. A. & Merrin, P. D. (2001): Mine water rebound in South Nottinghamshire: risk evaluation using 3-D visualization and predictive modelling. – Q. J. Eng. Geol. Hydrogeol., 34 (3): 307–319, 6 fig., 3 tab.; London.

Dunham, K. C. (1990 [1948]): Geology of the Northern Pennine Orefield – Tyne to Stainmore. – In: British Geological Survey: Economic Memoir of the British Geological Survey – England and Wales 1. – 2nd edn., 299 p., 46 fig., 75 tab., 10 plates; London (Her Majesty's Stationery Office).

Durgin, W. W. (1981): Flow: Its Measurement and Control in Science and Industry 2. – 867 p.; Triangle Park NC (ISA, Research).

Dushenkov, S., Vasudev, D., Kapulnik, Y., Gleba, D., Fleisher, D., Ting, K. C. & Ensley, B. (1997): Removal of Uranium from Water Using Terrestrial Plants. – Environ. Sci. Technol., 31 (12): 3468–3474, 9 fig., 2 tab.; Washington.

Dzombak, D. A. & Morel, F. M. M. (1990): Surface Complexation Modeling – Hydrous Ferric Oxides. – 393 p.; New York (Wiley).

Ebeling, W. (1986): Betriebserfahrungen mit dem Wurzelraumverfahren. – Schriftenr. WAR, 26: 31–49, 6 fig., 1 tab.; Darmstadt.

Ebena, G. (2003): Sulfidic mine waste microorganisms in an ecological context. – Linköping studies in science and technology, 841: 51, 8 fig., 1 tab., 6 app.; Linköping.

Ebner, H. G. & Schwarz, W. (1973): Untersuchungen über die Laugung von Uranerzen mit Hilfe von Bakterien. – Erzmetall, 26 (10): 484–490, 3 fig., 7 tab.; Clausthal-Zellerfeld.

Eckhart, E. & Riepler, F. (2004): Sicherungsmaßnahmen in den Altbergbauen Nassereith (Tirol) und Schlaining (Burgenland) unter besonderer Berücksichtigung der hydrogeologischen Verhältnisse. – In: Meier, G., Sroka, A., Löbel, K.-H., Klapperich, H., Tondera, D., Busch, W., Wagner, H. & Randjbar, B.: 4. Altbergbau-Kolloquium – Tagungsband. – p. 154–170, 15 fig.; Essen (Glückauf).

Edwards, K. J., Bond, P. L., Druschel, G. K., McGuire, M. M., Hamers, R. J. & Banfield, J. F. (2000): Geochemical and biological aspects of sulfide mineral dissolution: lessons from Iron Mountain, California. – Chem. Geol., 169: 383–397, 11 fig.; Amsterdam.

Edwards, P. J., Bolton, C. P., Ranson, C. M. & Smith, A. C. (1997): The River Pelenna minewater treatment project. – In: Younger, P. L.: Minewater Treatment Using Wetlands. – p. 17–32; London (Chartered Institution of Water and Environmental Management).

Elander, P., Lindvall, M. & Hakansson, K. (1998): MiMi – Prevention and control of pollution from mining waste products: State-of-the-art-report MiMi 1998:2. – 29 p.; Linköping (MiMi Print).

Ellis, D. V., Poling, G. W. & Baer, R. L. (1995): Submarine Tailings Disposal (STD) for Mines – An Introduction. – Mar. Geores. Geotechnol., 13 (1–2): 3–18, 4 fig.; London.

Eloff, E., Greben, H. A., Maree, J. P., Radebe, B. V. & Gomes, R. E. (2003): Biological sulphate removal using hydrogen as the energy source. – In: Nel, P. J. L.: Mine Water and the Environment. – p. 99–108, 3 fig., 5 tab.; Johannesburg (Proceedings, 8th International Mine Water Association Congress).

Enderle, G. J. & Friedrich, K. (1995): East-German Uranium Miners (Wismut) – Exposure Conditions and Health Consequences. – Stem Cells, 13 (Suppl. 1): 78–89, 9 fig., 2 tab.; Dayton.

Environment Agency (1997): Use and Design of Oil Separators in Surface Water Drainage Systems. – Pollution prevention guidelines, PPG 3: 1–4; Almondsbury.

Environmental Protection Agency (2001): Environmental Protection Agency Annual Report and Account 2000. – 84 p.; Wexford (Environmental Protection Agency).

Erickson, P. M., Kleinmann, R. L. P., Posluszny, E. T. & Leonard-Mayer, P. J. (1982): Hydrogeochemistry of a large mine pool. – Proceedings, 1st International Mine Water Congress, Budapest, Hungary, A: 27–42, 7 fig., 2 tab.; Budapest.

ERMITE Consortium, [eds. Younger, P. & Wolkersdorfer, Ch.]. (2004): Mining Impacts on the Fresh Water Environment: Technical and Managerial Guidelines for Catchment Scale Management. – Mine Water Env., 23 (Supplement 1): S2–S80, 28 fig., 6 tab.; Berlin.

Ernst, W. H. O. (1996): Bioavailability of heavy metals and decontamination of soils by plants. – Appl. Geochem., 11 (1 & 2): 163–167, 4 fig.; Oxford.

Ernst, W. H. O. (2004): Phytoextraction – options and impossibilities. – Abstracts, Geomicrobiology in remediation of mine waste (3. Jenaer Sanierungskolloquium): 11:30–12:10; Jena.

European Commission (2000): Directive 2000/60/EC of the European Parliament and of the Council of 23 October 2000 establishing a framework for Community action in the field of water policy. – Official Journal of the European Communities, L 327: 1–72, 2 fig.; Luxembourg.

Evangelou, V. P. (1995): Pyrite oxidation and its control – solution chemistry, surface chemistry, acid mine drainage (AMD), molecular oxidation mechanisms, microbial role, kinetics, control, ameliorites and limitations, microencapsulation. – 293 p.; Boca Raton (CRC Press).

Exner, C. (1952): Zur Geologie und Petrographie des Radhausberg-Unterbaustollens bei Badgastein. – Tschermaks Min. u. Petr. Mitt., 3 (1): 79–81; Wien.

Fahle, W. & Arnold, I. (1996): Der Einsatz von Dichtwänden im Lausitzer Braunkohlerevier. – Braunkohle, 48 (4): 407–412, 9 fig.; Düsseldorf.

Faust, K. M., Denton, R. A. & Plate, E. J. (1983): Schichtungseffekte in Flußstauhaltungen: Laborversuche. – Wasserwirtschaft, 73 (11): 386–392, 17 fig.; Stuttgart.

Fauville, A., Mayer, B., Frommichen, R., Friese, K. & Veizer, J. (2004): Chemical and isotopic evidence for accelerated bacterial sulphate reduction in acid mining lakes after addition of organic carbon: laboratory batch experiments. – Chem. Geol., 204 (3–4): 325–344; Amsterdam.

Fawcett, R. J., Hibberd, S. & Singh, R. N. (1984): An Appraisal of Mathematical Models to predict Water Inflows into underground Coal Workings. – Int. J. Mine Water, 3 (2): 33–54, 5 fig.; Granada.

Featherstone, A. M. & O'Grady, B. V. (1997): Removal of Dissolved Copper and Iron at the Freshwater-Saltwater Interface of an Acid Mine Stream. – Mar. Pollut. Bull., 34 (5): 332–337, 4 fig., 1 tab.; Oxford-New York.

Fedotov, S., Bashkirtseva, I. & Ryashko, L. (2002): Simple stochastic model for laminar-to-turbulent subcritical transition. – Annual Research Briefs, 2002: 397–404, 4 fig.; Stanford.

Felegy, E. W., Johnson, L. H. & Westfield, J. (1948): Acid mine-water in the anthracite region of Pennsylvania. – Tech. Paper, 710: 49, 13 fig., 11 tab.; Washington, DC.

Fellows, P. A. (1937): Sealing Projects – Sharply Reduce Stream Pollution From Abandoned Mines. – Coal Age, 42: 158–161, 13 fig., 2 tab.; New York.

Fernández Rubio, R. (2004): Sustainable mining – Environmental assets. – In: Jarvis, A. P., Dudgeon, B. A. & Younger, P. L.: mine water 2004 – Proceedings International Mine Water Association Symposium 2. – p. 251–271, 37 fig.; Newcastle upon Tyne (University of Newcastle).

Fernández-Rubio, R. & Fernandez Lorca, D. (1993): Mine Water drainage. – Mine Water Env., 12 (1–4): 107–130, 22 fig., 1 tab.; Wollongong.

Fernández-Rubio, R., Fernández-Lorca, S. & Esteban Arlegui, J. (1987): Preventive Techniques for Controlling Acid Water in Underground Mines by Flooding. – Int. J. Mine Water, 6 (3): 39–52, 11 fig.; Budapest.

Fetter, C. W. (1988): Applied hydrogeology. – 2nd edn., 592 p.; Columbus u.a. (Merrill).

Ficklin, W. H., Mosier & E. L. (1999): Field Methods for Sampling and Analysis of Environmental Samples for Unstable and Selected Stable Constituents. – In: Plumlee, G. S. & Logsdon, M. J.: The Environmental Geochemistry of Mineral Deposits 6A (12). – p. 249–264, 2 fig., 3 tab.; Littleton (Society of Economic Geologists).

Ficklin, W. H., Plumlee, G. S., Smith, K. S. & McHugh, J. B. (1992): Geochemical classification of mine drainages and natural drainages in mineralized areas. – Proceedings, International Symposium on Water-Rock Interaction, 7: 381–384, 2 fig., 2 tab.; Park City.

Field, M. S. (2002): The QTRACER2 Program for Tracer-Breakthrough Curve Analysis for Tracer Tests in Karstic Aquifers and Other Hydrologic Systems EPA/600/R-02/001. – 179 p., 51 fig., 13 tab.; Washington (U.S. Environmental Protection Agency – Office of Research and Development).

Field, M. S. (2003): Tracer-Test Planning Using the Efficient Hydrologic Tracer-Test Design (EHTD) Program EPA/600/R-03/034. – 175 p., 32 fig., 34 tab.; Washington (U.S. Environmental Protection Agency – Office of Research and Development; National Center for Environmental Assessment).

Fisher, J. & Greenbank, L. (2000): The Rogerley Mine, Weardale County Durham, England. – Rocks and Minerals, 75 (1): 54–62, 16 fig.; Peekskill.

Fishman, M. J. & Friedman, L. C. (1989): Methods for determination of inorganic substances in water and fluvial sediments. – Techniques of Water-Resources Investigations, TWRI 05-A1 (3): 545, 16 tab.; Reston.

Fitzpatrick, R. W. & Self, P. G. (1997): Iron Oxyhydroxides, Sulfides and Oxyhydroxysulfates as Indicators of Acid Sulfate Weathering Environments. – Advances in Geoecology, 30: 227–240, 2 fig., 2 tab.; Reiskirchen.

Flach, D. & Heick, C. (1986): Untersuchungen zum mikroseismischen Verhalten eines mit Lauge gefluteten Salzbergwerkes am Beispiel Hope. – gsf-Bericht (20): 130–195, 39 fig.; München.

Flynn, R. M., Schnegg, P.-A., Costa, R., Mallen, G. & Zwahlen, F. (2005): Identification of zones of preferential groundwater tracer transport using a mobile downhole fluorometer. – Hydrogeol. J., 13 (2): 366–377, 9 fig., 1 tab.; Berlin.

Fofonoff, N. P. & Millard Jr., R. C. (1983): Algorithms for computation of fundamental properties of seawater. – UNESCO Tech. Pap. Mar. Sci., 44: 53, 13 fig.; Paris.

Fontaine, R. C., Davis, A. & Fennemore, G. G. (2003): The Comprehensive Realistic Yearly Pit Transient Infilling Code (CRYPTIC): A Novel Pit Lake Analytical Solution. – Mine Water Env., 22 (4): 187–193, 8 fig., 2 tab.; Berlin.

Foreman, J. W. (1971): Deep Mine Sealing. – Proceedings, 1st Acid Mine Drainage Workshop, 1: 19–45, 21 fig.; Athens, Ohio.

Foreman, J. W. (1972): Evaluation of mine sealing in Butler County, Pennsylvania. – Papers Presented Before the Symposium on Coal Mine Drainage Research, 4: 83–95, 4 fig., 3 tab.; Washington.

Freeze, R. A. & Cherry, J. A. (1979): Groundwater. – 604 p.; Englewood Cliffs (Prentice Hall).

Freyer, G., Geissler, E., Hoth, K. & Tran, T. N. (1982): Die Altersstellung des Karbonatgesteinshorizonts Rabenstein-Auerswalde und ihre Bedeutung für die Geologie des südöstlichen Schiefermantels des sächsischen Granulitgebirges. – Z. geol. Wiss., 10: 1403–1424, 6 fig., 1 tab., 4 plates; Berlin.

Fried, J. J. (1972): Miscible pollutions of groundwater – a study in methodology. – In: Biswas, A. K.: Modelling of water resources systems 2. – p. 520–529, 3 fig.; Montreal (Harvest House).

Friese, K., Wendt-Potthoff, K., Zachmann, D. W., Fauville, A., Mayer, B. & Veizer, J. (1998): Biochemistry of iron and sulfur in sediments of an acidic mining lake in Lusatia, Germany. – Water Air Soil Pollut., 108: 231–247, 8 fig., 2 tab.; Dortrecht.

Frigaard, I. A. & Nouar, C. (2003): Predicting Transition to Turbulence in Well Construction Flows. – 8th SPE Latin American and Caribbean Petroleum Engineering Conference, SPE 81150: 1–9, 6 fig.; Port-of-Spain.

Frostad, S., Klein, B. & Lawrence, R. W. (2002): Evaluation of Laboratory Kinetic Test Methods for Measuring Rates of Weathering. – Mine Water Env., 21 (4): 183–192, 4 fig., 6 tab.; Berlin.

Frydrych, D., Maryška, J., Mužák, J. & Severýn, O. (2000): Mathematical Modelling of a Uranium Mine Flooding. – In: Handlovičová, A.: Algoritmy 2000 – Proceedings of Contributed Papers and Posters 15. – p. 110–119, 6 fig.; Bratislava (Slovenská Technická Univerzita).

Fuge, R., Pearce, F. M., Pearce, N. J. G. & Perkins, W. T. (1994): Acid mine drainage in Wales and influence of ochre precipitation on water chemistry. – ACS Symposium Series, 550: 261–274, 3 fig., 4 tab.; Washington.

Furrer, G., Phillips, B. L., Ulrich, K.-U., Pöthig, R. & Casey, W. H. (2002): The Origin of Aluminum Flocs in Polluted Streams. – Science, 297 (5590): 2245–2247, 3 fig., 1 tab. + 3 fig., 1 tab.; Washington.

Fyson, A. & Nixdorf, B. (2005): The Lignite Pit Lakes of Lusatia, Germany – Chemistry, Ecology and Acidity Removal Strategies. – p. 1–20, 6 fig., 5 tab.; Princeton (Proceedings, Annual National Meeting – American Society for Surface Mining and Reclamation).

Fyson, A., Nixdorf, B., Kalin, M. & Steinberg, C. E. W. (1997): Mesocosm Studies to Assess Acidity Removal from Acidic Mining Lakes through Controlled Eutrophication. – Ecol. Eng., 10: 229–245, 5 fig., 1 tab.; Amsterdam.

Fytas, K. & Bousquet, P. (2002): Silicate Micro-encapsulation of Pyrite to Prevent Acid Mine Drainage. – Bull. Can. Inst. Min. & Met., 95 (1063): 96–99, 6 fig., 3 tab.; Calgary.

Galinsky, G., Leistner, J., Scheuermann, G., Ebert, S. & Förderverein Drei-Brüder-Schacht e.V. (2001): Kavernenkraftwerk Drei-Brüder-Schacht – Geschichte und Überlegungen zur Rekonstruktion. – 2nd edn., 72 p., 70 fig., 1 tab.; Freiberg (Saxonia Standortentwicklungs- und -verwaltungsgesellschaft).

Galloway, D. L. & Erickson, J. R. (1985): Tracer test for evaluating nonpumping intraborehole flow in fractured media. – Trans. Amer. Nucl. Soc., 50: 192–193; La Grange Park.

Gammons, C. H., Poulson, S. R., Metesh, J. J., Duaime, T. E. & Henne, A. R. (2003): Geochemistry and Isotopic Composition of H_2S-rich Water in Flooded Underground Mine Workings, Butte, Montana, USA. – Mine Water Env., 22 (3): 141–148, 3 fig., 3 tab.; Berlin.

Gatzweiler, R. & Mager, D. (1993): Altlasten des Uranbergbaus – Der Sanierungsfall Wismut. – Die Geowissenschaften, 11 (5–6): 164–172, 10 fig., 1 tab.; Berlin.

Gatzweiler, R., Hähne, R. E. M., Meyer, J. & Snagovsky, S. (1997): Prognosis of the Flooding of Uranium mining Sites in East Germany with the Help of numerical Box-Modeling. – Proceedings, 6th International Mine Water Association Congress, Bled, Slovenia, 1: 57–63, 2 tab.; Ljubljana.

Gatzweiler, R., Jakubick, A. T., Meyer, J., Paul, M. & Schreyer, J. (2002): Flooding of the WISMUT mines – learning by doing or applying a comprehensive systematic approach? – In: Merkel, B. J., Planer-Friedrich, B. & Wolkersdorfer, Ch.: Uranium in the Aquatic Environment. – p. 745–754; Heidelberg (Springer).

Gebhart, B., Jaluria, Y., Mahajan, R. L. & Sammakia, B. (1988): Buoyancy-Induced Flows and Transport. – 1001 p.; Berlin (Springer).

Geisenheimer, P. (1913): Die Wasserversorgung des Oberschlesischen Industriebezirks. – 97 p., 29 fig., 30 tab., 1 Kart.; Kattowitz (Selbstverl. des Oberschlesischen Berg- und Hüttenmännischen Vereins).

Geissler, E., Nhuan, T. T. & Strauss, R. (1986): Die Rabensteiner unterirdischen Felsendome – zur Geologie und Geschichte des Schaubergwerkes. – 4th edn., 38 p., 20 fig.; Karl-Marx-Stadt [Chemnitz] (Städtische Museen).

Geletneky, J. W. (2002): Hydrogeologische/Hydrologische Untersuchung einer Prä-Flutungssituation am Beispiel des Gessentals im ehemaligen ostthüringischen Uranbergbaugebiet. – 262 p., 75 fig., 28 tab., 9 app.; Jena (unpubl. Diss. Universität Jena).

Geller, W., Friese, K., Herzsprung, P., Kringel, R. & Schultze, M. (2000b): Limnology of sulphur-acidic mining lakes. II Chemical properties: The main constituents and buffering systems. – Verh. Internat. Verein. Limnol., 27: 2475–2479, 5 fig., 3 tab.; Stuttgart.

Geller, W., Friese, K., Herzsprung, P., Kringel, R., Schimmele, M., Schultze, M., Wendt-Potthoff, K. & Wölfl, S. (2000a): Geogene Versauerung von Tagebaurestgewässern. – In: Guderian, R. & Gunkel, G.: Handbuch der Umweltveränderlichkeiten und Ökotoxikologie 3A (3.5). – p. 352–380, 8 fig., 2 tab.; Heidelberg (Springer).

Geller, W., Klapper, H. & Salomons, W. (1998): Acidic Mining Lakes – Acid Mine Drainage, Limnology and Reclamation. – 435 p., 155 fig., 55 tab.; Berlin (Springer).

Genser, C. (1933): Radioaktive Heilquellen in Deutschland. – Z. dt. geol. Ges., 85: 482–495, 3 fig., 1 plates; Berlin.

Georgaki, I., Dudeney, A. W. L. & Monhemius, A. J. (2004): Characterisation of Iron-rich Sludge: Correlations Between Reactivity, Density and Structure. – Miner. Eng., 17 (2): 305–316, 13 fig., 1 tab.; Oxford.

Georgescu, D., Stoica, L., Radulescu, R. & Filip, D. (1995): Radium Content Reduction from Radioactive Effluents resulted in Uranium ores Mining and Processing. – Proceedings, Uranium-Mining and Hydrogeology, Freiberg, Germany; GeoCongress, 1: 197–206, 8 fig., 3 tab.; Freiberg.

Gerth, A., Hebner, A., Kiessig, G. & Zellmer, A. (2005): Passive Treatment of Minewater at the Schlema-Alberoda Site. – In: Merkel, B. J. & Hasche-Berger, A.: Uranium in the Environment. – p. 409–414, 2 fig.; Heidelberg (Springer).

Gertsch, R. E. & Bullock, R. L. (1998): Techniques in Underground Mining – Selections From Underground Mining Handbook. – 823 p.; Littleton (Society for Mining, Metallurgy, and Exploration).

Geyh, M. A. & Schleicher, H. (1990): Absolute Age Determination – Physical and Chemical Dating Methods and Their Application. – 503 p., 146 fig.; Berlin (Springer).

Gibert, O., Rötting, T., De Pablo, J., Cortina, J. L., Bolzicco, J., Carrera, J. & Ayora, C. (2004): Metal Retention Mechanisms for the Aznalcóllar permeable reactive Barrier (SW Spain). – In: Jarvis, A. P., Dudgeon, B. A. & Younger, P. L.: mine water 2004 – Proceedings International Mine Water Association Symposium 2. – p. 61–68, 7 fig.; Newcastle upon Tyne (University of Newcastle).

Giese, S., Grubert, A. & Hübner, C. (1994): Ergebnisse geophysikalischer Untersuchungen des Carolinen-Gangzuges (Bergbaurevier Freiamt-Sexau, Mittlerer Schwarzwald) und des darauf umgegangenen historischen Bergbaus. – Abh. geol. Landesamt Baden-Württemberg, 14: 343–360, 6 fig., 1 tab.; Freiburg.

Gillmore, G. K., Grattan, J., Pyatt, F. B., Phillips, P. S. & Pearce, G. (2002): Radon, water and abandoned metalliferous mines in the UK: Environmental and Human Health implications. – In: Merkel, B. J., Planer-Friedrich, B. & Wolkersdorfer, Ch.: Uranium in the Aquatic Environment. – p. 65–76, 2 tab.; Heidelberg (Springer).

Girts, M.A. & Kleinmann, R. L. P. (1986): Constructed wetlands for the treatment of acid mine drainage: a preliminary review. – p. 165–171– Proceedings, 1986 Symposium on Mining, Hydrology, Sedimentology, and Reclamation; University of Kentucky, Lexington

Gisman, S. & Szczypa, H. (1982): Trends in Utilization of Mine Waters in the Upper-Silesian Coal Basin. – Proceedings, 1st International Mine Water Congress, Budapest, Hungary, D: 17–23, 1 tab.; Budapest.

Goette, H. (1934): Das Henry-Verfahren zum Klären von Ab- und Schlammwässern. – Technische Blätter – Wochenschr. z. Dtsch. Bergwerksztg., 42: 683–684; Essen.

Goetz, D., Roussel, V. & Tincelin, E. (1994): Cave Ins Linked to Mine Flooding. – Proceedings, 5th International Mine Water Congress, Nottingham, U.K., 1: 29–42, 5 fig., 4 tab.; Nottingham.

Goldbrunner, J. E., Ramspacher, P., Zojer, H., Zötl, J. G., Moser, H., Rauert, W. & Stichler, W. (1982): Die Anwendung natürlicher und künstlicher Tracer in einem hochalpinen Magnesitbergbau. – Beitr. Geol. Schweiz – Hydrol., 28 (II): 407–422, 4 fig.; Bern.

Gombkötö, I., Kováčs, F., Böhm, J. & Debreczeni, Á. (2002): Removal of Uranium from contaminated Groundwater at a former Uranium Mining Site. – p. 97–102, 7 fig.; Karlsruhe (Proceedings, First Image-Train Cluster-Meeting).

Goodman, A. E., Khalid, A. M. & Ralph, B. J. (1981): Microbial ecology of Rum Jungle II: environmental study of two flooded opencuts and smaller, associated water bodies. – 61 p., 10 fig., 36 tab.; Lucas Heights (Australian Atomic Energy Commission Research Establishment).

Gospodarič, R. & Habič, P. (1976): Underground Water tracing – Investigations in Slovenia 1972–1975. – 309 p., 85 fig., 46 tab., 11 plates; Ljubljana (Institute Karst Research).

Gottschalk, V. H. & Buehler, H. A. (1912): Oxidation of Sulphides. – Econ. Geol., 7: 15–34, 1 fig., 3 tab.; Lancaster.

Gourdon, R. & Funtowicz, N. (1995): Bioleaching of metals from industrial contaminated soil using sulphuric acid produced by bacterial activity – a feasibility study. – Soil & Environment, 5: 1049–1056, 5 fig., 4 tab.; Dordrecht.

Government of Ireland (1997): Sustainable Development – A Strategy for Ireland. – 258 p., 48 fig., 48 tab.; Dublin (Government Publications).

Grandchamp, C. A. P., Sperling, E. v. & Almeida, D. C. d. (2001): Águas Claras Mine – A successful dewatering story. – p. 1–8 [CD-ROM], 6 fig., 2 tab.; Belo Horizonte (Proceedings, IMWA Symposium Belo Horizonte).

Grapes, T. R. & Connelly, R. J. (1998): Evaluation of Mine Water Inflows and associated Impacts at the Olympias mine, Halkidiki, Greece 1. – p. 13–23, 2 fig.; Johannesburg (Proceedings International Mine Water Association Symposium).

Graps, A. (1995): An Introduction to Wavelets. – IEEE Computational Science & Engineering, 2 (2): 1–18, 8 fig.; Los Alamitos.

Gray, D. M. (1970): Handbook on the Principles of Hydrology: A water information center publication. – 691 p.; Ottawa (Canadian Committee for the International Hydrological Decade).

Gray, D. M., McKay, G. A. & Wigham, J. M. (1970): Energy, Evaporation and Evapotranspiration. – In: Gray, D. M.: Handbook on the Principles of Hydrology (III). – p. 3.1–3.66, 26 fig., 13 tab.; Ottawa (Canadian Committee for the International Hydrological Decade).

Griffin, B. C., Ahlgren, K., Coleman, M., Jurgensmeyer, V., Littlefield, R., Martin, T., McCaleb, N., Roberts, L., Seybolt, D. & Smith, M. (2000): Final Report – Governor Frank Keating's Tar Creek Superfund Task Force. – 30 p., 16 fig., 11 tab.; Oklahoma City (Office of the Secretary of Environment).

Grimes, D. J., Ficklin, W. H., Meier, A. L. & McHugh, J. B. (1995): Anomalous gold, antimony, arsenic, and tungsten in ground water and alluvium around disseminated gold deposits along the Getchell Trend, Humboldt County, Nevada. – J. Geochem. Explor., 52 (3): 351–371, 13 fig., 1 tab.; Amsterdam-New York.

Grisak, G. E. & Pickens, J. F. (1981): An analytical Solution for solute Transport through fractured Media with Matrix Diffusion. – J. Hydrol., 52 (1–2): 47–57, 7 fig., 1 tab.; Amsterdam.

Grundmann, G. & Martinek, K.-P. (1994): Erzminerale und Gangarten des Bergbaugebietes Schwaz-Brixlegg. – Mineralien Magazin, Lapis, 19 (7–8): 28–40, 44 fig.; München.

Grünewald, U. (2001): Water resources management in river catchments influenced by lignite mining. – Ecol. Eng., 17 (2–3): 143–152, 7 fig., 2 tab.; Amsterdam.

GSF – Gesellschaft für Strahlen- und Umweltforschung (1985): Untersuchungen endlagerrelevanter Vorgänge vor, während und nach der Flutung des Kalisalzbergwerkes Hope. 1. Bericht. – gsf-Bericht, 12/85: 286; Neuherberg.

GSF – Gesellschaft für Strahlen- und Umweltforschung (1986): Untersuchungen endlagerrelevanter Vorgänge vor, während und nach der Flutung des Kalisalzbergwerkes Hope. Statusseminar zum FE-Projekt Hope am 31. Oktober 1985 in Braunschweig. – gsf-Bericht, 20/86: 286; Neuherberg.

Gulati, M. S., Lipman, S. C. & Strobel, C. J. (1978): Tritium Tracer Survey at the Geysers. – In: Geothermal Resource Council: Geothermal Resource Council Annual Meeting: Geothermal Energy – A Novelty Becomes Resource, Hilo, Hawaii, Jul 25–27 1978 2. – p. 237–239, 1 tab.; Davis, Calif. (Geothermal Resource Council).

Gundelwein, A. (2001): Die Roeder'sche Zeit – Das Ende des "Mittelalters" am Rammelsberg. – In: Roseneck, R.: Der Rammelsberg – Tausend Jahre Mensch-Natur-Technik 2. – p. 172–181, 8 fig.; Goslar (Goslarsche Zeitung).

Günther, W., Eibner, C., Lippert, A. & Paar, W. ([ca. 1993]): 5000 Jahre Kupferbergbau Mühlbach am Hochkönig-Bischofshofen. – 396 p.; Bischofshofen (Eigenverlag Marktgemeinde Bischofshofen).

Gurrieri, J. T. (1998): Distribution of metals in water and sediment and effects on aquatic biota in the upper Stillwater River Basin, Montana. – J. Geochem. Explor., 64: 83–100, 6 fig., 3 tab.; Amsterdam.

Gusek, J. J. & Wildeman, T. R. (2002): Passive Treatment of Aluminum-Bearing Acid Rock Drainage. – Proceedings, West Virginia Surface Mine Drainage Task Force Symposium, 23: 1–9 [CD-ROM], 3 fig., 1 tab.; Morgantown.

Gusek, J. J. (2002): Sulfate-Reducing Bioreactor Design and Operating Issues – Is this the Passive Treatment Technology for your Mine Drainage? – p. 1–14 [CD-

ROM], 4 fig.; Park City (Proceedings, Annual Conference – National Association of Abandoned Mine Land Programs).

Häge, K. (1995): Methoden und Erfahrungen zur Vermeidung von Umweltauswirkungen im Lausitzer Braunkohlenrevier. – Braunkohle (7): 4–8, 6 fig.; Düsseldorf.

Hagen, M. & Lange, G. (1995): Der Flutungsprozess ehemaliger Uranerzgruben in Ostdeutschland als Sanierungsschwerpunkt der Wismut GmbH. – Erzmetall, 48 (11): 790–804, 8 fig.; Weinheim.

Hallberg, K. B. & Johnson, D. B. (2005): Mine Water Microbiology. – Mine Water Env., 24 (1): 28–32, 2 fig., 1 tab.; Berlin.

Hámar, T., D'Allessandro, M. & Bidaglio, G. (2002): Legislation on Mining Waste Management in Central and Eastern European Candidate Countries EUR 20545 EN. – 196 p.; Ispra (Joint Research Centre of the European Commission).

Hamilton, Q. U. I., Lamb, H. M., Hallett, C. & Proctor, J. A. (1997): Passive Treatment Systems for the Remediation of Acid Mine Drainage Wheal Jane, Cornwall, UK. – In: Younger, P. L.: Minewater Treatment Using Wetlands. – p. 33–56, 9 fig., 2 tab.; London (Chartered Institution of Water and Environmental Management).

Hamilton, R. M., Bowen, G. G., Postlethwaite, N. A. & Dussek, C. J. (1994b): The abandonment of Wheal Jane, a tin mine in South West England. – Proceedings, 5th International Mine Water Congress, Nottingham, U.K., 2: 543–551, 4 fig., 2 tab.; Nottingham.

Hamilton, R. M., Taberham, J., Waite, R. R. J., Cambridge, M., Coulton, R. H. & Hallewell, M. P. (1994a): The development of a temporary treatment solution for the acid mine water discharge at Wheal Jane. – Proceedings, 5th International Mine Water Congress, Nottingham, U.K., 2: 643–656, 7 fig., 5 tab.; Nottingham.

Hamm, V., Fabriol, R., Vaute, L., Fourniguet, G. & Baubron, J.-C. (2005): Modelling the impact of back-filling material on the water quality of flooded iron mines (Lorraine, France). – p. 1–12 [CD-ROM], 7 fig., 2 tab.; Nancy (Proceedings, Post-Mining 2005).

Hammack, R. W., Love, E. I., Veloski, G. A., Ackman, T. E. & Harbert, W. (2003b): Using Helicopter Electromagnetic Surveys to Identify Environmental Problems at Coal Mines. – Mine Water Env., 22 (2): 80–84, 7 fig.; Berlin.

Hammack, R. W., Sams, J. I., Veloski, G. A. & Mabie, J. S. (2003a): Geophysical Investigation of the Sulphur Bank Mercury Mine Superfund Site, Lake County, California. – Mine Water Env., 22 (2): 69–79, 15 fig.; Berlin.

Hanneberg, A. & Schuster, H. (1994): Geschichte des Bergbaus in Schwaz und Brixlegg. – Mineralien Magazin, Lapis, 19 (7–8): 13–21, 16 fig., 3 tab.; München.

Hansen, C. (1998): Modellierung der hydrogeochemischen Entwicklung saurer Grubenwässer – Flutung der Schwefelkieslagerstätte Einheit bei Elbingerode (Harz). – 137 p., 76 fig., 6 tab.; Clausthal-Zellerfeld (unpubl. Diplomarbeit TU Clausthal).

Hanzlík, J. & Vydra, J. (1985): Liquidation of abandoned Mine Excavations by Flooding. – Proceedings, 2nd International Mine Water Association Congress, 2: 953–965, 6 fig., 2 tab.; Granada.

Harries, J. R. (1990): Acid Drainage from Waste rock Dumps at Mine Sites (Australia and Scandinavia) ANSTO/E692. – 30 p., 21 fig., 6 tab.; Lucas Heights (Australian Nuclear Science and Technology Organisation).

Harris, D. (2004): Post-mining regeneration at the Garden of Eden. – Mining Environmental Management, 12 (1): 8–11, 3 fig.; London.

Harvey, R. W., George, L. H., Smith, R. L. & LeBlanc, D. R. (1989): Transport of microspheres and indigenous bacteria through a sandy aquifer; results of natural- and forced-gradient tracer experiments. – Environ. Sci. Technol., 23 (1): 51–56, 5 fig., 3 tab.; Washington.

Harvey, R. W., Kinner, N. E., McDonald, D., Metge, D. W. & Bunn, A. L. (1993): Role of physical heterogeneity in the interpretation of small-scale laboratory and field observations of bacteria, microbial-sized microsphere, and bromide transport through aquifer sediments. – Water Resour. Res., 29 (8): 2713–2721, 2 fig., 2 tab.; Washington.

Hasche, A. & Wolkersdorfer, Ch. (2004): Mine Water Treatment with a Pilot Scale RAPS-System. – Wissenschaftliche Mitteilungen, 25: 93–99, 8 fig., 1 tab.; Freiberg.

Hasche, A. (2001): Multitracerversuch in einem gefluteten Bergwerk – Vorbereitung, Durchführung, Auswertung. – 99 p., 16 fig., 29 tab., 36 app., 2 app.; Freiberg (unpubl. Diploma Thesis TU Bergakademie Freiberg).

Hasse, J. (2000): Flutung des Metallerzbergwerkes Meggen. – In: Wismut GmbH: Proceedings Wismut 2000 – Bergbausanierung Internationale Konferenz. – 9 p., 7 fig.; Chemnitz (Eigenverlag Wismut GmbH).

Hause, D. R. & Willison, L. R. (1986): Deep Mine Abandonment Sealing and Underground Treatment to Prelude Acid Mine Drainage. – Proceedings, West Virginia Surface Mine Drainage Task Force Symposium, 7 (Paper 14): 14, 13 fig.; Morgantown.

Hawkins, J. W. & Aljoe, W. W. (1992): Pseudokarst Groundwater Hydrologic Characteristics of a Mine Spoil Aquifer. – Mine Water Env., 11 (2): 37–52, 11 fig., 1 tab.; Wollongong.

Hawkins, J. W. (1994): Statistical characteristics of coal-mine discharges on western Pennsylvania remining sites. – Water Resour. Bull., 30 (5): 861–869, 5 fig., 3 tab.; Bethesda.

Hawkins, J. W. (1995): Characterization and Effectiveness of Remining Abandoned Coal Mines in Pennsylvania. – Report of Investigations – US Bureau of Mines, RI 9562: 1–38, 13 fig., 14 tab.; Washington.

Hawkins, J. W. (1998): Remining. – In: The Pennsylvanian Department of Environmental Protection: Coal mine drainage prediction and pollution prevention in Pennsylvania (17). – p. 17.1–17.6, 3 fig.; Harrisburg (Pennsylvanian Bureau of Mining and Reclamation).

Hayek, E. & Wierer, H. (1951): Über den Arsengehalt von Stollenwässern in der Umgebung Badgasteins. – Tschermaks Min. u. Petr. Mitt., 2 (3): 290–291; Wien.

Hedin, R. S. (2003): Recovery of marketable iron oxide from mine drainage in the USA. – Land Contam. Reclam., 11 (2): 93–97, 6 fig., 2 tab.; Richmond.

Hedin, R. S., Nairn, R. W. & Kleinmann, R. L. P. (1994a): Passive Treatment of Coal Mine Drainage. – Bureau of Mines Information Circular, IC-9389: 1–35, 13 fig., 19 tab.; Washington.

Hedin, R. S., Stafford, S. L. & Weaver, T. J. (2005): Acid Mine Drainage Flowing from Abandoned Gas Wells. – Mine Water Env., 24 (2): 104–106, 1 fig., 2 tab.; Berlin.

Hedin, R. S., Watzlaf, G. R. & Nairn, R. W. (1994b): Passive treatment of acid-mine drainage with limestone. – J. Environ. Qual., 23 (6): 1338–1345, 3 fig., 6 tab.; Madison.

Heide, B. & Hasse, J. (1997): Stillegung des Metallerzbergwerks Meggen nach 140 Jahren aktivem Bergbau. – Glückauf, 133: 233–239, 6 fig., 4 tab.; Essen.

Heide, B. & Hasse, J. (1998): Der Anfang und das Ende des Metallerzbergwerkes Meggen. – Bergbau, 49 (3): 115–122, 6 fig., 4 tab.; Gelsenkirchen.

Heidenreich, H., Klafki, M., Kretzschmar, H. J. & Ziegert, H. R. (1991): Strömungsmechanisch-mathematische Simulation eines gefluteten Endlagers. – Kernenergie, 34 (6): 270–275, 7 fig.; Berlin.

Heitfeld, K.-H., Heitfeld, M., Rosner, P., Sahl, H. & Schetelig, K. (2002): Mine Water Recovery in the Coal Mining District of Aachen – Impacts and Measures to Control Potential Risks. – In: Merkel, B. J., Planer-Friedrich, B. & Wolkersdorfer, Ch.: Uranium in the Aquatic Environment. – p. 1027–1036, 6 fig.; Heidelberg (Springer).

Helling, C., Nitzsche, O., Ullrich, B. & Merkel, B. J. (1998): The Influence of municipal Waste on Uranium Mill Tailings: A hydrogeochemical Study on a mixed Tailings Site. – Mine Water Env., 17 (1): 41–51, 3 fig., 7 tab.; Lakewood, CO.

Henton, M. P. (1976): Predicting the quality of mine water discharges. – Effluent & Water Treatment Journal, 16 (11): 568, 572, 1 tab.; London.

Hentschel, T., Hruschka, F. & Priester, M. (2002): Global Report on Artisanal & Small-Scale Mining. – Mining, Minerals and Sustainable Development, 70: 1–67, 16 fig., 7 tab.; London.

Herbert Jr., R. B. (1999): MiMi – Sulfide oxidation in mine waste deposits – A review with emphasis on dysoxic weathering. – 45 p., 10 fig., 3 tab.; Luleå (MiMi Print).

Herbert, H.-J. & Sander, W. (1982): Mineralogisch-chemische Prozesse in laugenerfüllten Schächten des Zechsteinsalinars. – Fortsch. d. Min., 60: 96–98; Stuttgart.

Herbert, H.-J. & Sander, W. (1989): Verfahren zur verfälschungsfreien Messung und Probennahme von hochkonzentrierten Salzlösungen im Untertagebereich. – Kali und Steinsalz, 10 (4/5): 137–141, 3 fig.; Kassel.

Herbert, H.-J. (1989): Geochemische Vorgänge bei der Flutung des Kalisalzbergwerkes Hope – Abschlußbericht des Teilvorhabens Geochemie des FE-Vorhabens Hope. – gsf-Bericht, 5/89: 62, 4 tab., 72 fig. im Anhang; Neuherberg.

Herschy, R. W. (1995): Streamflow measurement. – 2nd edn., 524 p.; London (Spon).

Herzsprung, P., Friese, K., Packroff, G., Schimmele, M., Wendt-Potthoff, K. & Winkler, M. (1998): Vertical and Annual Distribution of Ferric and Ferrous Iron in Acidic Mining Lakes. – Acta Hydrochim. Hydrobiol., 26 (5): 253–262, 9 fig., 7 tab.; Weinheim.

Hettler, J. (1995): Bergbau und Umwelt in Papua-Neuguinea – Die Ok Tedi-Mine und das Fly River-Flußökosystem. – Berliner geowissenschaftliche Abhandlungen: Reihe A, 176: 110, 29 fig., 17 tab., 8 app.; Berlin.

Hewett, R. (2003): Data Mining for generating Predictive Models of local Hydrology. – Appl. Intell., 19 (3): 157–170, 11 fig.; Dordrecht.

Hiessleitner, G. (1951): Geologischer Gutachtensbericht über den Schwerspatbergbau Großkogl bei Brixlegg Archiv Montanwerke Brixlegg. – 26 p., 5 fig., 5 tab., 8 Beil.; Brixlegg (Montanwerke G.m.b.H.).

Hiltscher, A. (1999): 140 Jahre Salzbergbau in der Region Anhalt – Ein Überblick. – Der Aufschluss, Sonderband: 103–116, 13 fig., 3 tab.; Heidelberg.

Himmelsbach, T. & Wendland, E. (1999): Schwermetalltransport in Sandsteinen unter Bedingungen einer hochsalinaren Porenwasserlösung – Laborversuch und Modellierung. – Grundwasser, 4 (3): 103–112; Berlin.

Himmelsbach, T., Hötzl, H., Käss, W., Leibundgut, C., Małoszewski, P., Meyer, T., Moser, H., Rajner, V., Rank, D., Stichler, W., Trimborn, P. & Veulliet, E. (1992): Fractured Rock – Test site Lindau/Southern Black Forest (Germany). – Steir. Beitr. Hydrogeol., 43: 159–228, 28 fig., 9 tab.; Graz.

Hiscock, K. M. (2005): Hydrogeology – Principles and Practice. – 389 p., 266 fig., 60 tab.; Oxford (Blackwell).

Hjulström, F. H. (1935): Studies of the morphological activity of rivers as illustrated by the river Fyris. – Bull. Geol. Ass. Uppsala, 25: 221–525, 51 fig., 38 tab., 8 plates; Uppsala.

Hobba, W. (1987) Underground Coal Mines as Sources of Water for Public Supply in Northern Upshur County, West Virginia. – US Geological Survey Water Resource Investigation, WRI 84–4115.

Hodgkinson, R. A. & Armstrong, A. C. (1996): Field studies of runoff processes on restored land in South Wales and the design of channels for erosion control. – In: Anderson, M. G. & Brooks, S. M.: British Geomorphological Research Group Symposia Series – Advances in hillslope processes 12 (27). – p. 613–633, 11 fig., 6 tab.; Chichester – New York (Wiley).

Hoehn, E., Eikenberg, J., Fierz, T., Drost, W. & Reichlmayr, E. (1998): The Grimsel Migration Experiment – field injection–withdrawal experiments in fractured rock with sorbing tracers. – J. Contam. Hydrol., 34 (1–2): 85–106, 9 fig., 4 tab.; Amsterdam.

Höglund, L. O., Herbert, R., Lövgren, L., Öhlander, B., Neretnieks, I., Moreno, L., Malmström, M. E., Elander, P., Lindvall, M. & Lindström, B. (2004): MiMi – Performance Assessment Main Report. – 345 p.; Luleå (MiMi Print).

Holgate, R. (1991): Prehistoric Flint Mines: Shire archaeology 67. – 56 p.; Princes Risborough (Shire).

Hölting, B. & Coldewey, W. G. (2005): Hydrogeologie – Einführung in die Allgemeine und Angewandte Hydrogeologie. – 6th edn., 326 p., 118 fig., 69 tab.; München (Elsevier).

Hopkins, W. B. & Simpson, T. A. (1960): Montana Earthquakes Noted in Pennsylvania Mine-Water Pools. – Trans. Amer. Geophys. Union, 41 (2): 435–436; Washington.

Horne, R. N., Johns, R. A., Adams, M. C., Moore, J. N. & Stiger, S. G. (1987): The use of tracers to analyze the effects of reinjection into fractured geothermal reservoirs. – p. 37–52; Washington (Proceedings 5th geothermal program review).

Hoskin, W. (2003): Abandoned Mine Sites: Problems, Issues and Options. – In: Puura, E., Marmo, L. & D'Alessandro, M.: Workshop on Mine and Quarry Waste – The Burden from the Past. – p. 53–55; Luxembourg (Office for Official Publications of the European Communities).

Hoskin, W. M. A. (2000): Mine Closure – The 21st Century Approach. – In: Villas Bôas, R. C. & Barreto, M. L.: Mine Closure in Iberoamerica. – p. 109–116; Rio de Janeiro (CYTED – IMACC – UNIDO).

Hötzl, H. & Werner, A. (1992): Tracer Hydrology – 6th International Symposium on Water Tracing. – 464 p.; Rotterdam (Balkema).

Hounslow, A. W. (1995): Water Quality Data – Analysis and Interpretation. – 397 p., 88 fig., 73 tab.; London (Lewis Publishers).

Houston, K. S., Milionis, P. N., Eppley, R. L., Harrington, J. M. & Harrington, J. G. (2005): Field Demonstration of In-Situ Treatment and Prevention of Acid Mine Drainage in the Abandoned Tide Mine, Indiana County, Pennsylvania. – Proceedings, West Virginia Surface Mine Drainage Task Force Symposium, 26: 1–9, 6 fig.; Morgantown.

Hudson, N. W. (1993): Field measurement of soil erosion and runoff: Food and Agriculture Organization of the United Nations soils bulletin 68. – 139 p.; Rome (Food and Agriculture Organization of the United Nations).

Hunt, D. J. & Reddish, D. J. (1997): The stability implications of groundwater recharge upon shallow, abandoned mineworkings in the UK. – In: Marinos, P. G., Koukis, G. C., Tsiambaos, G. C. & Stournaras, G. C.: Engineering geology and the environment 3. – p. 2425–2430, 3 fig.; Rotterdam (Balkema).

Hunt, L. E. & Howard, A. G. (1994): Arsenic Speciation and Distribution in the Carnon Estuary Following the Acute Discharge of Contaminated Water from a Disused Mine. – Mar. Pollut. Bull., 28 (1): 33–38, 5 fig., 1 tab.; Oxford.

Integrated Geotechnical Engineering Services Specialists in Unsaturated Zone Hydrology (2003): Evaluation of the Long-Term Performance of Dry Cover

Systems Final Report. – For: International Network for Acid Prevention. – 45 p., 15 fig., 4 tab., 9 app.; Saskatoon (Integrated Geotechnical Engineering Services Specialists in Unsaturated Zone Hydrology).

International Institute for Environment and Development (2002): Mining for the Future – Appendix C: Abandoned Mines Working Paper. – Mining, Minerals and Sustainable Development, 28: C-1–C-20, 3 tab.; London.

Jackson, C. B. & Hach Company (2004): Report on the Validation of Proposed EPA Method 360.3 (Luminescence) for the Measurement of Dissolved Oxygen in Water and Wastewater. – 42 p., 14 fig., 3 tab.; Loveland (Hach Company).

Jacobi, O. & Everling, G. (1981): Praxis der Gebirgsbeherrschung. – 2nd edn., 576 p.; Essen (Glückauf).

Jacobs, M. L. & Denham, R. (2001): Stormwater Management and the Limited Discharge System of the Batu Hijau Mine. – p. 1–10 [CD-ROM], 1 fig., 3 tab., 3 Fot.; Belo Horizonte (Proceedings, IMWA Symposium Belo Horizonte).

Jacobs, M. L. & White, T. (2001): Sediment Control in a Tropical, High Rainfall Environment. – p. 1–11 [CD-ROM], 4 fig., 4 tab., 4 photographs; Belo Horizonte (Proceedings, IMWA Symposium Belo Horizonte).

Jäger, B., Obermann, P. & Wilke, F. L. (1990): Studie zur Eignung von Steinkohlebergwerken im rechtsrheinischen Ruhrkohlebezirk zur Untertageverbringung von Abfall- und Reststoffen. – In: Landesamt für Wasser und Abfall Nordrhein-Westfalen. – 628 p.; Düsseldorf (LWA Studie).

Jakubick, A. T., Jenk, U. & Kahnt, R. (2002): Modelling of mine flooding and consequences in the mine hydrogeological environment: flooding of the Königstein mine, Germany. – Environ. Geol., 42 (2–3): 222–234, 11 fig., 6 tab.; Heidelberg.

Jambor, J. L., Blowes, D. W. & Ritchie, A. I. M. (2003): Environmental Aspects of Mine Wastes. – In: Raeside, R.: Short Course Series Volume 31. – 430 p.; Waterloo, Ontario (Mineralogical Association of Canada).

Janischewskij, E. M. & Konstantinow, W. M. (1962): Der Einfluß tektonischer und lithologischer Faktoren auf die Uranvererzung im Erzgebirge. – Z. Angew. Geol., 8 (3): 124–128, 5 fig.; Berlin.

Janneck, E. & Krüger, H.-G. (1999): Sanierung von Bergbauwässern aus Altablagerungen des Schieferbergbaus – Pilotanlage für die Behandlung saurer Bergbauwässer. – In: Deutsche Bundesstiftung Umwelt FKZ 04600. – 75 p., 27 fig., 12 tab., 5 app.; Freiberg (unpubl. Ergebnisbericht G.E.O.S. Freiberg).

Jarvis, A. & England, A. (2002): Operational and treatment performance of an unique Reducing and Alkalinity Producing System (RAPS) for acidic leachate remediation in Lancashire, UK. – In: Merkel, B. J., Planer-Friedrich, B. & Wolkersdorfer, Ch.: Uranium in the Aquatic Environment. – p. 1019–1026, 4 fig., 1 tab.; Heidelberg (Springer).

Jarvis, A. P. & Younger, P. L. (1999): Design, Construction and Performance of a Full-Scale Compost Wetland for Mine-Spoil Drainage Treatment at Quaking Houses. – Water Environ. Manage., 13 (5): 313–318, 4 fig., 2 tab.; London.

Javeri, V. (1995): Orientierende Analysen zur Naturkonvektion und Dispersion in porösen Medien mit dem Rechenprogramm TOUGH2. – Schriftenreihe Reaktorsicherheit und Strahlenschutz, BMU-1995-416: 121, 50 fig.; Bonn.

Jenk, U., Schreyer, J. & Klinger, C. (2003): Fe(0)/lignitic coal: An efficient and mechanically stable reactive material for purification of water containing heavy metals, radionuclides, and nitroaromatics. – Environ. Sci. Technol., 37 (3): 644–651, 9 fig., 9 tab.; Washington.

Jenk, U., Zimmermann, U. & Ziegenbalg, G. (2005): The use of $BaSO_4$ supersaturated solutions for in-situ immobilization of heavy metals in the abandoned Wismut GmbH uranium mine at Königstein. – In: Merkel, B. J. & Hasche-Berger, A.: Uranium in the Environment. – p. 721–727, 5 fig.; Heidelberg (Springer).

Jensen, E. B. (1983): Mine Water Used to Heat Ventilation Air at Henderson Molybdenum Mine. – Mining Engineering, 35 (1): 17–20, 4 fig., 1 tab.; Littleton.

Jessop, A. M. (1998): Geothermal Energy in Canada. – Geosci. Can., 25 (1): 33–41, 8 fig.; Waterloo.

Jessop, A. M., Macdonald, J. K. & Spence, H. (1995): Clean Energy from Abandoned Mines at Springhill, Nova Scotia. – Energy Sources, 17 (1): 93–106, 5 fig., 2 tab.; Bristol.

Jester, W. A. & Raupach, D. C. (1987): A Survey of applications of non-radioactive but neutron activatable groundwater tracers. – In: IAEA: Isotope techniques in water resources development. – p. 623–633, 4 fig.; Wien (International symposium on the use of isotope techniques in water resources development).

Jiries, A., El-Hasan, T., Al-Hweiti, M. & Seiler, K.-P. (2004): Evaluation of the Effluent Water Quality Produced at Phosphate Mines in Central Jordan. – Mine Water Env., 23 (3): 133–137, 4 fig., 4 tab.; Berlin.

Job, B. (1987): Inrushes at British collieries: 1851 to 1970. – Colliery Guardian, 235 (5, 6): 192–199; 232–235, 8 fig., 2 tab.; London.

Jobst, W., Rentzsch, W., Schubert, W. & Trachbrod, K. (1994): Bergwerke im Freiberger Land. – 2nd edn., 227 p., 167 fig., 38 tab.; Freiberg (Medienzentrum der TU Bergakademie Freiberg).

Johnson, D. B. & Hallberg, K. B. (2002): Pitfalls of passive mine water treatment. – Re/Views in Environmental Science & Bio/Technology, 1: 335–343, 4 fig., 1 tab.; Amsterdam.

Johnson, D. B. (2003): Chemical and microbiological characteristics of mineral spoils and drainage waters at abandoned coal and metal mines. – Water Air Soil Pollut., Focus 3: 47–66, 2 fig., 3 tab.; Dordrecht.

Johnson, K. L. & Younger, P. L. (2000): Abandonment of Frazer's Grove Fluorspar Mine, North Pennines, UK: Prediction and observation of Water level and chemistry changes after closure. – p. 271–279, 4 fig.; Ustron (Proceedings, 7th International Mine Water Association Congress).

Johnson, K. L. & Younger, P. L. (2002): Hydrogeological and geochemical consequences of the abandonment of Frazer's Grove carbonate hosted Pb/Zn fluorspar mine, North Pennines, UK. – Spec. Publ. – Geol. Soc. London, 198: 347–363, 13 fig., 2 tab.; London.

Johnston, D. (2004): A metal mines strategy for Wales. – In: Jarvis, A. P., Dudgeon, B. A. & Younger, P. L.: mine water 2004 – Proceedings International Mine Water Association Symposium 1. – p. 17–23, 1 fig., 2 tab.; Newcastle upon Tyne (University of Newcastle).

Jones, S. G. & Ellis, D. V. (1995): Deep-water STD at the Misima Gold and Silver Mine, Papua, New Guinea. – Mar. Geores. Geotechnol., 13 (1–2): 183–200, 11 fig., 2 tab.; London.

Jordan, G., D'Alessandro, M., Hamor, T., Liedekerke, M. V., Panagos, P., Puura, E., Sommer, S. & Vijdea, A.-M. (2004): Mining, mining waste and related environmental issues: problems and solutions in Central and Eastern European Candidate Countries EUR 20868 EN. – 207 p.; Ispra (Joint Research Centre of the European Commission).

Jordan, H. & Weder, H.-J. (1995): Hydrogeologie – Grundlagen und Methoden. – 2nd edn., 409 p., 427 fig., 178 tab., 3 Kart; Stuttgart (Enke Verlag).

Jovanovic, N. Z., Annandale, J. G., Claassens, A. S., Lorentz, S. A. & Tanner, P. D. (2001): Modeling irrigation with gypsiferous Mine Water – a case study in Botswana. – Mine Water Env., 20 (2): 65–72, 2 fig., 4 tab.; Berlin.

Jung, H. G. (2002): Sekundäre Prozesse in Bergbauhalden und Aufbereitungsabgängen. – 202 p., 148 fig., 25 tab., 18 app.; Hannover (Bundesanstalt für Geowissenschaften und Rohstoffe).

Junker, R., Krause, H., Schumann, K. & Siemroth, J. (1991): Sekundärminerale des Neudorf-Straßberger-Gangzuges im Unterharz. – Der Aufschluss, 42 (2): 95–100, 6 fig.; Heidelberg.

Jurjovec, J., Blowes, D. W. & Ptacek, C. J. (1995): Acid Neutralization in Mill Tailings and the Effect of Natrojarosite Addition. – Proceedings, Sudbury '95 – Mining and the Environment, I: 29–38, 5 fig., 1 tab.; Sudbury, Ontarion, Canada.

Kaden, S. & Schramm, M. (2000): Control model Spree/Schwarze Elster – a tool to optimise rehabilitation of water resources in the Lusatian mining district. – Landsc. Urban Plan., 51 (2–4): 101–108, 3 fig., 2 tab.; Amsterdam.

Kaden, S. (1997): Water Resources Modeling for Decision Support in Open-pit Lignite Mining Areas. – Proceedings, 6th International Mine Water Association Congress, Bled, Slovenia, 2: 385–395, 6 fig., 2 tab.; Ljubljana.

Kahmann, H.-J. & Heinrich, K. (1998): Zur Sauerwasserproblematik bei der Verwahrung der Südharzer Flußspatgrube Straßberg [Vortragsmanuskript]. – Proceedings, Deutsche Rohstoff- und Metalltage 1998: 56, 16 fig.; Clausthal-Zellerfeld.

Kalin, M. (2004): Passive mine water treatment – The correct approach. – Wissenschaftliche Mitteilungen, 25: 107–112; Freiberg.

Kalin, M., Smith, M. P. & Landry, G. (1994): The Application of Ecological Engineering to Acid Coal Seepages in Eastern Canada. – Proceedings,

International Land Reclamation and Mine Drainage Conference, 2: 296–302, 8 fig., 1 tab.; Pittsburgh.

Kalka, H., Märten, H. & Kahnt, R. (2005): Dynamical Models for Uranium Leaching – Production and Remediation Cases. – In: Merkel, B. J. & Hasche-Berger, A.: Uranium in the Environment. – p. 235–245, 7 fig.; Heidelberg (Springer).

Kalka, H., Märten, H. & Münze, R. (2002): Flooding and Post-flooding Scenarios – Dynamics and Geochemistry. – In: Merkel, B. J., Planer-Friedrich, B. & Wolkersdorfer, Ch.: Uranium in the Aquatic Environment. – p. 1037–1044, 7 fig.; Heidelberg (Springer).

Karakas, G., Brookland, I. & Boehrer, B. (2003): Physical characteristics of Acidic Mining Lake 111. – Aquat. Sci., 65 (3): 297–307, 11 fig., 2 tab.; Basel.

Karathanasis, A. D. & Johnson, C. M. (2003): Metal Removal Potential by Three Aquatic Plants in an Acid Mine Drainage Wetland. – Mine Water Env., 22 (1): 22–30, 7 fig., 1 tab.; Berlin.

Karrenberg, H. (1981): Hydrogeologie der nichtverkarstungsfähigen Festgesteine. – 284 p., 83 fig.; Wien u.a. (Springer).

Käss, W. (1964): Die unmittelbare Bestimmung von Uranin-Spuren bei Färbeversuchen. – Steir. Beitr. Hydrogeol., 15/16: 37–65, 7 fig., 6 tab.; Graz.

Käss, W. (1998): Tracing Technique in Geohydrology. – 581 p., 270 fig., 40 tab., 8 Plat.; Rotterdam (Balkema).

Käss, W. (2004): Geohydrologische Markierungstechnik. – In: Matthess, G.: Lehrbuch der Hydrogeologie 9. – 2nd edn., 557 p., 239 fig., 43 tab., 8 plates; Berlin (Borntraeger).

Kecojević, V. J., Mrugala, M. J., Simić, R. D. & Stanković, R. M. (2003): A Computer-Aided System for Design of Drainage Facilities in Surface Mining. – Mine Water Env., 22 (3): 149–154, 10 fig.; Berlin.

Kegel, K. (1950): Bergmännische Wasserwirtschaft einschließlich Grundwasserkunde, Wasserversorgung und Abwasserbeseitigung. – In: Kegel, K., Spackeler, G. & Rammler, E.: Berg- und Aufbereitungstechnik, Band III (Teil 1). – 3rd edn., 374 p., 240 fig., 49 tab.; Halle/Saale (Knapp).

Kelley, D. E., Fernando, H. J. S., Gargett, A. E., Tanny, J. & Özsoy, E. (2003): The Diffusive Regime of Double-Diffusive Convection. – Prog. Oceanogr., 56: 461–481, 7 fig.; Oxford.

Kellner, R., Mermet, J.-M., Otto, M. & Widmer, H. M. (1998): Analytical Chemistry. – 916 p.; Weinheim (Wiley-VCH).

Kelly, D. P. & Wood, A. P. (2000): Reclassification of some species of *Thiobacillus* to the newly designated genera *Acidithiobacillus* gen. nov., *Halothiobacillus* gen. nov. and *Thermithiobacillus* gen. nov. – Int. J. Syst. Evol. Microbiol., 50: 511–516, 1 fig., 1 tab.; Reading.

Kemp, P. & Griffiths, J. (1999): Quaking Houses – Art, science and the community: a collaborative approach to water pollution. – 142 p., 41 fig.; Charlbury (Jon Carpenter).

Kennedy, C. (2002): Alternatives for the Reclamation of Surface Mined Lands. – In: Mudroch, A., Stottmeister, U., Kennedy, C. & Klapper, H.: Remediation of Abandoned Surface Coal Mining Sites. – p. 37–56, 8 tab.; Heidelberg (Springer).

Kepler, D. A. & Mc Cleary, E. C. (1994): Successive Alkalinity-Producing Systems (SAPS) for the Treatment of Acid Mine Drainage. – Proceedings, International Land Reclamation and Mine Drainage Conference, 1: 195–204; Pittsburgh.

Kester, D. R., Byrne, R. H. & Liang, Y.-J. (1975): Redox Reactions and Solution Cornplexes of Iron in Marine Systems. – ACS Symposium Series, 18: 56–79, 6 fig., 9 tab.; Washington.

Kickuth, R. (1977): Degradation and incorporation of nutrients from rural wastewaters by plant hydrosphere under limnic conditions. – Comm. of the Europ. Communitite – Rapports, EUR 5672e: 335–343, 2 fig., 2 tab.; London.

Kiessig, G. & Hermann, E. (2000): Nachfolgelösungen für die konventionelle Behandlung kontaminierter Wässer. – In: Wismut GmbH: Proceedings Wismut 2000 – Bergbausanierung Internationale Konferenz. – 9 p., 10 fig., 2 tab.; Chemnitz (Eigenverlag Wismut GmbH).

Kiessig, G., Kunze, C., Küchler, A., Zellmer, A., Meyer, J. & Kalin, M. (2004): Kostengünstige passive Nachsorgelösung mit einem Constructed Wetland auf der Grundlage von Prognosen der Entwicklung des Grubenwassers der Grube Pöhla. – Wissenschaftliche Mitteilungen, 25: 113–118, 5 fig., 1 tab.; Freiberg.

Kimball, B. A., Bianchi, F., Walton-Day, K., Runkel, R. L., Nannucci, M. & Salvadori, A. (2007): Quantification of Changes in Metal loading from Storm Runoff, Merse River (Tuscany, Italy). – In: Cidu, R. & Frau, F.: Water in Mining Environments. – p. 83–87, 4 fig.; Cagliari (Mako Edizioni).

Kimball, B. A., Runkel, R. L. & Gerner, L. J. (2001): Quantification of mine-drainage inflows to Little Cottonwood Creek, Utah, using a tracer-injection and synoptic-sampling study. – Environ. Geol., 40 (11–12): 1390–1404, 11 fig., 1 tab.; Heidelberg.

Kimball, B. A., Runkel, R. L., Bencala, K. E. & Walton-Day, K. (1999): Use of tracer injections and synoptic sampling to measure metal loading from acid mine drainage. – Water resources investigations report, WRI 99-4018A (1): 31–36, 6 fig.; Reston.

Kimball, B. A., Runkel, R. L., Walton-Day, K. & Bencala, K. E. (2002): Assessment of metal loads in watersheds affected by acid mine drainage by using tracer injection and synoptic sampling: Cement Creek, Colorado, USA. – Appl. Geochem., 17 (9): 1183–1207, 10 fig., 5 tab.; Oxford.

Kindermann, L. (1998): Kontrolle geochemischer Parameter beim Wiedereinbau von Reststoffen in ein stillgelegtes Bergwerk. – Wissenschaftliche Mitteilungen, 7: 196–201, 5 fig.; Freiberg.

Kindsvater, C. E. & Carter, R. W. (1959): Discharge Characteristics of Rectangular Thin-Plate Weirs. – Trans. American Society of Civil Engineers, 124: 772–801, 11 fig., 1 tab.; Reston.

King, C. Y., King, B. S. & Evans, W. C. (1996): Spatial radon anomalies on active faults in California. – Appl. Geochem., 11 (4): 497–510, 15 fig., 1 tab.; Oxford.

Kinzelbach, W. & Rausch, R. (1995): Grundwassermodellierung – Eine Einführung mit Übungen. – 283 p., 223 fig., 15 tab., 2 Disk; Berlin Stuttgart (Borntraeger).

Kipko, E. Y., Spichak, Y. N., Polozov, Y. A., Kipko, A. E. & Hepnar, P. (1993): Grouting of Old Flooded Workings at M. Mayerova Mine in Czechoslovakia. – Mine Water Env., 12 (1–4): 21–26, 1 fig., 2 tab.; Wollongong.

Kirby, C. S. & Cravotta, C. A. I. (2005a): Net alkalinity and net acidity 2: Practical considerations. – Appl. Geochem., 20 (10): 1920–1940, 8 fig., 9 tab.; Amsterdam.

Kirby, C. S. & Cravotta, C. A. I. (2005b): Net alkalinity and net acidity 1: Theoretical considerations. – Appl. Geochem., 20 (10): 1920–1940, 11 fig., 2 tab.; Amsterdam.

Kirschner, F. E. & Williams, R. E. (1993): Hydrogeological evaluation of an in-situ leach cell in abandoned mine and mill waste at the Kellogg, Idaho Superfund site. – p. 77–87, 6 fig.; Jackson Hole (International Biohydrometallurgy Symposium).

Klapper, H. & Geller, W. (2001): Water Quality Management of Mining Lakes – a New Field of Applied Hydrobiology. – Acta Hydrochim. Hydrobiol., 29 (6–7): 363–374, 7 fig., 5 tab.; Weinheim.

Klapper, H. & Schultze, M. (1996): Geogene Versauerung von Bergbaurestseen – Ursachen, Folgen, Möglichkeiten zur Schadensbegrenzung. – In: Arbeitsgruppe des GBL-Gemeinschaftsvorhabens: GBL Gemeinschaftsvorhaben – Vortragsband des 2. GBL-Kolloquiums vom 6.-8. März 1996 in Freiberg – Methoden und Ergebnisse 3. – p. 121–131, 12 fig.; Stuttgart (Schweizerbart).

Klapper, H., Geller, W. & Schultze, M. (1996): Abatement of acidification in mining lakes in Germany. – Lakes and Reservoirs Research and Management, 2 (1–2): 7–16, 6 fig., 3 tab.; Carlton.

Klapperich, H., Meier, G., Sroka, A., Tondera, D. & Busch, W. (2001): 1. Altbergbau-Kolloquium – Tagungsband. – 212 p.; Essen (Glückauf).

Kleinmann, R. L. P. (1990): Acid Mine Water Treatment using Engineered Wetlands. – Int. J. Mine Water, 9 (1–4): 269–276; Castro Verde.

Kleinmann, R. L. P., Tiernan, T. O., Solch, J. G., & Harris, R. L. (1983): A low-cost, low-maintenance treatment system for acid mine drainage using Sphagnum moss and limestone. –In: Carpenter, S. B. & Devore, R. W., Proceedings, 1983 Symposium on Surface Mining, Huydrology, Sedimentology, and Reclamation – p. 241–245, 7 fig., 2 tab., University of Kentucky, Lexington

Klimant, I., Meyer, V. & Kohls, M. (1995): Fibre-optic oxygen microsensors, a new tool in aquatic biology. – Limnol. Oceanogr., 40 (6): 1159–1165, 5 fig., 1 tab.; Waco.

Klinger, C., Hansen, C., Rüterkamp, P. & Heinrich, H. (2000a): In situ tests for interactions between acid mine water and ferrihydrite sludge in the pyrite mine "Elbingerode" (Harz Mts.; Germany). – p. 137–145, 6 fig., 1 tab.; Ustron (Proceedings, 7th International Mine Water Association Congress).

Klinger, C., Jenk, U. & Schreyer, J. (2000b): Investigation of efficacy of reactive materials for reduction of pollutants in acid mine water in the former uranium mine of Königstein (Germany). – Proceedings, 7th International Mine Water Association Congress: 292–298, 1 fig., 2 tab.; Ustron.

Klinger, C., Jenk, U. & Schreyer, J. (2002): Processes in passive mine water remediation with zero-valent iron and lignite as reactive materials. – In: Merkel, B. J., Planer-Friedrich, B. & Wolkersdorfer, Ch.: Uranium in the Aquatic Environment. – p. 569–576, 7 fig., 2 tab.; Heidelberg (Springer).

Knöller, K., Schubert, M., Osenbrück, K. & Weise, S. M. (2005): Natural isotope tracers for assessing the mechanisms of heavy metal mobilization in copper ore tailings. – In: Loredo, J. & Pendás, F.: Mine Water 2005 – Mine Closure. – p. 1–4, 3 fig.; Oviedo (University of Oviedo).

Koetter, G. (2001): Bergbau im Muttental – Geologie und Geschichte des Wittener Bergbauwanderwegs. – 198 p., 108 fig.; Witten (Wöhrle).

Kolitsch, S., Junghans, M., Klemm, W. & Tichomirova, M. (2001): Der Flutungsraum des Grubenfeldes Freiberg – Hydrochemie, Isotopenchemie und Hydraulik. – Wissenschaftliche Mitteilungen, 18: 14–26, 10 fig., 4 tab.; Freiberg.

Kolupaila, S. (1961): Bibliography of hydrometry. – 975 p.; Notre Dame (University of Notre Dame Press).

König, C. & Blömer, C. (1999): Berechnung von Temperatur- und dichteabhängiger Strömung in gefluteten Schächten. – 37 p., 17 fig., 4 tab.; Bochum (GKW-Ingenieurgesellschaft mbH).

Körtzinger, A., Schimanski, J. & Send, U. (2005): High-quality oxygen measurements from profiling floats: A promising new techniques. – J. Atmos. Ocean. Technol., 22 (3): 302–308, 7 fig.; Boston.

Koschorreck, M., Herzsprung, P., Wendt-Potthoff, K., Lorke, A., Geller, W., Luther, G., Elsner, W. & Müller, M. (2002): An In-lake Reactor to Treat an Acidic Lake: the Effect of Substrate Overdosage. – Mine Water Env., 21 (3): 137–149, 10 fig., 3 tab.; Berlin.

Kostenbader, P. D. & Haines, G. F. (1970): High-density sludge treats acid mine water. – Coal Age (September): 90–97, 5 fig., 3 tab.; New York.

Kranawettreiser, J. (1989): Dichteströmungen. – In: Bollrich, G. & Preissler, G.: Technische Hydromechanik 2 (6). – 1st edn., p. 328–350, 9 fig., 1 tab.; Berlin (VEB Verlag für Bauwesen).

Kranjc, A. (1997): Karst Hydrogeologial Investigations in South-Western Slovenia. – Acta Carsologica, 26 (1): 1–388, 188 fig., 83 tab.; Ljubljana.

Kraut, K. H. & Bondareva, O. (2000): Anwendung von Flockungsverfahren bei der Regenwasserbehandlung. – 126 p., 43 fig., 27 tab., 3 app.; Stuttgart (unpubl. Report Universität Stuttgart – FZKA-BWPLUS).

Kreibich, M. & Eisbein, M. (ca. 2003): Georgius Agricola und die Ehrenfriedersdorfer Radpumpe. – 34 p., 24 fig.; Ehrenfriedersdorf (Zweckverband Sächsisches Industriemuseum Zinngrube Ehrenfriedersdorf).

Krentz, O., Witthauer, B., Mittag, R., Novak, E., Wendt, S., Buchholz, P., Korn, M., Funke, S., Forkmann, B., Docekal, A., Donner, F., Heinicke, J. & Temmler, T. (2004): Materialien zur Geologie – Erdbebenbeobachtung im Freistaat Sachsen – Zweijahresbericht 2002–2003. – 54 p., 35 fig.; Dresden (Sächsisches Landesamt für Umwelt und Geologie).

Krischker, A. G. (1990): Die Baryt-Fahlerz-Lagerstätte St. Gertraudi/Brixlegg. – 206 p.; Innsbruck (unpubl. Diploma Thesis Univ. Innsbruck).

Kroker, E. & Farrenkopf, M. (1999): Grubenunglücke im deutschsprachigen Raum: Veröffentlichungen aus dem Deutschen Bergbau-Museum Bochum 79. – 2nd edn., 653 p.; Bochum (Deutsches Bergbau-Museum).

Kroll, A., Amezaga, J. M., Younger, P. L. & Wolkersdorfer, Ch. (2002b): Regulation of Mine Waters in the European Union: Contribution of Scientific Research to Policy Development. – Mine Water Env., 21 (4): 193–200, 1 tab.; Berlin.

Kroll, A., Betlem, G., Brans, E., Getliffe, K., Groen, F., Santamaría, L., Lucas, M. & López, B. (2002a): Environmental Regulation of Mine Waters in the European Union – D 4 European Policies and Mine Waters: ERMITE Report – EU R&D Project EVK1-CT-2000-00078. – 189 p., 5 fig., 8 tab., 7 app.; Sevilla (Institute for Prospective Technological Studies).

Kruse, N. A. S. & Younger, P. L. (2005): Computational methods for acid mine drainage management: decision-making for post-closure decision making. – In: Mine Water 2005 – Mine Closure: Loredo, Jorge; Pendás, Fernando. – p. 427–433, 2 fig., 3 tab.; Oviedo (University of Oviedo).

Kubiak, C. (2007): Characterisation of Low Density Sludge Storage in a flooded Underground Mine. – 24 S., 9 fig., 2 tab., 1 app.; Rennes (unpubl. Master's Thesis Univ. Rennes).

Kugler, J. & Weyer, J. (2002): Der Einfluß des Hochwassers am 12.-14. 08. 2002 auf Anlagen des Montanwesens im Freiberger Revier – Größenordnung potentieller Gefahren und bergmännische Aufgaben nach dem Verbruch des Rothschönberger Stollens. – 37 p., 24 fig.; Freiberg (Eigenverlag).

Kunnert, H. (1974): Bergbauwissenschaft und technische Neuerungen im 18. Jahrhundert – Die "Anleitung zu der Bergbaukunst" von Chr. Tr. Delius (1773). – In: Mitterauer, M.: Österreichisches Montanwesen – Produktion, Verteilung, Sozialformen. – p. 181–198, 1 tab.; München (Oldenburg).

Kunze, C. & Küchler, A. (2003): Passive biological treatment systems of mine waters at WISMUT sites. – In: Nel, P. J. L.: Mine Water and the Environment. – p. 187–199, 10 fig., 1 tab.; Johannesburg (Proceedings, 8th International Mine Water Association Congress).

Kuschka, E. & Franzke, H. J. (1974): Zur Kenntnis der Hydrothermalite des Harzes. – Z. geol. Wiss., 2 (12): 1417–1436, 3 fig.; Berlin.

Kuyumcu, M. & Hartwig, H.-J. (1998): Aufgaben der GVV – Gesellschaft zur Verwahrung und Verwertung von stillgelegten Bergwerksbetrieben mbH. – Bergbau, 49 (1): 18–22, 5 fig., 1 tab.; Gelsenkirchen.

Kwakwa, K. A. (1989): Tracer results. – In: Parker, R. H.: Hot dry rock geothermal energy; Phase 2B final report of the Camborne School of Mines Project (7). – p. 1037–1099, 18 fig., 3 tab.; Oxford (Pergamon Press).

Kwong, Y. T. J. & Van Stempvoort, D. R. (1994): Attenuation of Acid Rock Drainage in a Natural Wetland System. – In: Alpers, C. N. & Blowes, D. W.: Environmental geochemistry of Sulfide Oxidation (25). – p. 382–392, 3 fig., 2 tab.; New York (American Chemical Society).

Lachmar, T. E. (1994): Application of fracture-flow hydrogeology to acid-mine drainage at the Bunker Hill Mine, Kellogg, Idaho. – J. Hydrol., 155 (2): 125–149, 11 fig., 3 tab.; Amsterdam.

Ladwig, K. J., Erickson, P. M., Kleinmann, R. L. P. & Posluszny, E. T. (1984): Stratification in Water Quality in Inundated Anthracite Mines, Eastern Pennsylvania. – Report of Investigations – US Bureau of Mines, 8837: 35, 19 fig., 8 tab.; Washington.

Lai, R. W. M. (1980): Liquid membrane metal extraction process for recovering copper from mine water (Abstract). – Journal of Metals, 32 (12): 49; New York.

Laine, D. M. & Dudeney, A. W. L. (2000): Bullhouse mine-water project. – Trans. Inst. Min. Metall. Sect. A, 109: A224–A227, 3 fig.; London.

Laine, D. M. & Jarvis, A. P. (2003): Engineering design aspects of passive in situ remediation of mining effluents. – Land Contam. Reclam., 11 (2): 113–125, 2 fig., 1 tab.; Richmond.

Lake, P. S., Coleman, D., Mills, B. & Norris, R. (1977): A Reconnaissance of Pollution of the King River in the Comstock-Crotty Area, Western Tasmania. – In: Banks, M. R. & Kirkpatrick, J. B. – p. 157–173, 4 fig., 4 tab.; Hobart (Royal Society of Tasmania).

Landau, L. D. & Lifschitz, E. M. (1991): Hydrodynamik. – In: Ziesche, P.: Lehrbuch der theoretischen Physik VI. – 5th edn., 683 p., 136 fig.; Berlin (Akademie Verlag).

Länderarbeitsgemeinschaft Wasser [LAWA] (2001): Tagebaurestseen – Anforderungen an die Wasserqualität. – 11 p., 1 fig., 2 tab.; Berlin (Kulturbuch-Verlag).

Landesumweltamt Nordrhein-Westfalen (2002): Gewässergütebericht 2001 Nordrhein-Westfalen: Berichtszeitraum 1995–2000. – In: Landesumweltamt Nordrhein-Westfalen. – 274 p., 59 fig., 68 tab.; Essen (Landesumweltamt Nordrhein-Westfalen).

Landriault, D. (2001): Backfill in Underground Mining. – In: Hustrulid, W. A. & Bullock, R. L.: Underground Mining Methods – Engineering Fundamentals and International Case Studies (69). – p. 601–614, 16 fig., 3 tab.; Littleton (Society for Mining, Metallurgy, and Exploration).

Lang, B. (1999): Permanent Sealing of Tunnels to retain Tailings or Acid Rock Drainage. – In: Fernández Rubio, R.: Mine, Water & Environment II. – p. 647–655, 6 fig., 3 tab.; Sevilla (International Mine Water Association).

Lange, G. & Freyhoff, G. (1991): Geologie und Bergbau in der Uranlagerstätte Ronneburg/Thüringen. – Erzmetall, 44: 262–269, 8 fig.; Weinheim.

Lange, G., Mühlstedt, P., Freyhoff, G. & Schröder, B. (1991): Der Uranerzbergbau in Thüringen und Sachsen – ein geologisch-bergmännischer Überblick. – Erzmetall, 44 (3): 162–171; Weinheim.

Langguth, H. R. & Voigt, R. (2004): Hydrogeologische Methoden. – 2nd edn., 1005 p., 304 fig.; Berlin (Springer).

Larew, G. & Skousen, J. (1992) An ounce of prevention is worth a pound of water treatment. – Proceedings, 13th Annual West Virginia Surface Mine Drainage Task Force Symposium, WV Mining and Reclamation Association

LaRosa, P. J., Karnavas, J. A. & Pelczarski, E. A. (1971): Evaluation of a New Acid Mine Drainage Treatment Process. – Water pollution control research series, 14010 DYI 02/71: 155, 60 fig., 23 tab.; Washington.

Lässl, L. (1556): Schwazer Bergbuch. – 378 (396) p., 98 (99) fig.; Schwaz (hand written).

Laubrich, D., Neubert, G., Tomschi, H. & Unland, W. (1996): Sanierung von Altlasten im Bereich der Wismut GmbH – Das Beispiel des ehemaligen Urantagebaus Lichtenberg. – VDF Führungskraft, 1996 (9/10): 33–40, 5 fig., 7 tab.; Essen.

Lausitzer Braunkohle Aktiengesellschaft (2000): Braunkohlenbergbau im Osten Deutschlands. – 64 p.; Senftenberg (Lausitzer Braunkohle Aktiengesellschaft).

Lawson, D. C. & Sonderegger, J. L. (1978): Geothermal data-base study – Mine-water temperatures. – Special Publication – State of Montana Bureau of Mines and Geology, 79: 38, 4 fig., 1 tab., 1 app.; Butte.

Le Carlier de Veslud, C., Royer, J.-J., Gérard, B. & Pagel, M. (1998): Modélisation des transferts hydrologiques et thermiques lors de l'évolution de la paléo-marge ardéchoise (France). – Bull. Soc. géol. France, t. 169 (1): 81–89, 2 fig., 6 tab.; Paris.

Lebecka, J., Chałupnik, S., Michalik, B., Wysocka, M., Skubacz, K. & Mielnikow, A. (1994): Radioactivity of Mine Waters in Upper Silesian Coal Basin and its Influence on Natural Envrionment. – Proceedings, 5th International Mine Water Congress, Nottingham, U.K., 2: 657–662, 1 fig., 1 tab.; Nottingham.

Lebecka, J., Mielnikow, A., Chałupnik, S., Wysocka, M., Skubacz, K. & Michalik, B. (1997): Radium in Mine Waters in Poland – Occurence and Impact on River Waters. – Proceedings, 6th International Mine Water Association Congress, Bled, Slovenia, 2: 415–421, 1 fig.; Ljubljana.

Leblanc, M., Morales, J. A., Borrego, J. & Elbaz-Poulichet, F. (2000): 4,500-Year-Old Mining Pollution in Southwest Spain: Long-Term Implications from Modern Mining Pollution. – Econ. Geol., 95: 655–662, 3 fig., 1 tab.; Lancaster.

Lechner, E. (2003): Heilende Wasser in Tirol – Geschichte und Entwicklung von Heilbäder, Bauernbadln und Kraftquellen. – 153 p.; Innsbruck, Wien (Tyrolia).

Ledin, M. & Pedersen, K. (1996): The environmental impact of mine wastes; roles of microorganisms and their significance in treatment of mine wastes. – Earth-Science Reviews, 41 (1–2): 67–108, 3 fig., 6 tab.; Amsterdam.

Lee, W. L. (1984): Performance assessment of nuclear waste isolation systems: Shaping our energy future. – 59 p.; Knoxville (Proceedings, 11th annual WATTec conference and exposition).

Leitch, R. D. (1935): Sealing of Coal Mines – Will Reduce Acidity of their Effluent Waters. – Coal Age, 40: 323–326, 7 fig.; New York.

Lerner, D. N., Issar, A. S. & Simmers, I. (1990): Groundwater recharge – A guide to understanding and estimating natural recharge: International Contributions to Hydrogeology 8. – 345 p.; Hannover (Heise).

Leupolt, G. & Hocker, M. (ca. 1999): Befahrerhandbuch – Streitschrift zu Arbeitsweisen der praktischen bergbauhistorischen Forschung. – 264 p.; Dresden (Eigenverlag).

Levy, D. B., Custis, K. H., Casey, W. H. & Rock, P. A. (1997): A comparison of metal attenuation in mine residue and overburden material from an abandoned copper mine. – Appl. Geochem., 12 (2): 203–211, 5 fig., 3 tab.; Oxford.

Lewis, R. A. (1990): Borate adsorption by some volcanic tuffs; evaluation of borate as a potential tracer for unsaturated flow at Yucca Mountain: Borate adsorption by some volcanic tuffs; evaluation of borate as a potential tracer for unsaturated flow at Yucca Mountain. – Golden (PhD Thesis Colorado School of Mines).

Lewis, R. L., McGowan, I. R., Hershman, J. I. & Tilk, J. (1997): Numerical simulation of bulkheads to reduce uncontrolled discharge from an underground copper mine. – Mining Engineering, 49 (4): 68–72, 3 fig.; Littleton.

Li, G. & Zhou, W. (1999): Sinkholes in karst mining areas in China and some methods of prevention. – Eng. Geol., 52 (1–2): 45–50, 5 fig.; Amsterdam.

Lieber, D. (2003): Säure- und Basekapazität – Vergleich unterschiedlicher vor-Ort-Bestimmungsmethoden bei unterschiedlichen Bergwerkswässern und Bearbeitern. – 22 fig., 23 tab.; Freiberg (unpubl. Thesis).

Liebisch, K. & Remus, W. (1966): Hydrogeologische Ergebnisse mit Färbversuchen im Kupferschieferbergbau der Mansfelder Mulde. – Z. Angew. Geol., 12 (9): 486–489, 2 fig.; Berlin.

Liedl, R. & Sauter, M. (2000): Charakterisierung von Karstgrundwasserleitern durch Simulation der Aquifergenese und des Wärmetransports. – Grundwasser, 5 (1): 9–16, 8 fig.; Heidelberg.

Liessmann, W. (2002): Die Harzer Erbstollen des 17. und 18. Jahrhunderts – Planung, Bau und wirtschaftliche Bedeutung am Beispiel des St. Andreasberger Reviers. – p. 84–91, 3 fig., 2 tab.; Ridnau/Schneeberg (Proceedings, 5. Internationaler Bergbau Workshop).

Lindgren, G. A. & Destouni, G. (2002): Environmental Regulation of Mine Waters in the European Union – D 3 Institutional research – 3 Sweden: ERMITE Report – EU R&D Project EVK1-CT-2000-00078. – p. 3.1–3.11, 3 fig.; Stockholm (Royal Institute of Technology).

Linsley, R. K., Kohler, M. A. & Paulhus, J. L. H. (1988): Hydrology for engineers – SI Metric edition: McGraw-Hill series in water resources and environmental engineering. – 492 p.; London (McGraw-Hill).

Lippmann, J., Stute, M., Torgersen, T., Moser, D. P., Hall, J. A., Lin, L., Borcsik, M., Bellamy, R. E. S. & Onstott, T. C. (2003): Dating ultra-deep mine waters with noble gases and ^{36}Cl, Witwatersrand Basin, South Africa. – Geochim. Cosmochim. Acta, 67 (23): 4597–4619, 8 fig., 8 tab.; Oxford.

Lloyd, J. W. & Heathcote, J. A. (1985): Natural inorganic chemistry in relation to groundwater – An introduction. – 296 p., 116 fig., 74 tab.; Oxford (Clarendon).

Locher, H. & Keller, R. (1995): Managing impacts of large artificial sediment loads in rivers – a case study. – Proceedings, International Association for Hydraulic Research Congress, 26 (4): 356–361, 4 fig., 1 tab.; London.

Loop, C. M. (2004): Lessons learned from full-scale, non-traditional Placement of Fly Ash. – Proceedings, State Regulation of Coal Combustion By-Product Placement at Mine Sites: A Technical Interactive Forum: 43–46; Harrisburg.

Loop, C. M., Scheetz, B. E. & White, W. B. (2003): Geochemical Evolution of a Surface Mine Lake with Alkaline Ash Addition: Field Observations vs. Laboratory Predictions. – Mine Water Env., 22 (4): 206–213, 5 fig., 3 tab.; Berlin.

Lorax Environmental (2003): Treatment of Sulphate in Mine Effluents. – 129 p., 39 fig., 33 tab.; London (INAP – International Network of Acid Prevention).

Lord, G. J., Pardo-Igúzquiza, E. & Smith, I. M. (2000): A Practical Guide to Wavelets for Metrology. – Reports NPL-CMSC, CMSC 02/00: 39, 26 fig.; Teddington.

Lottermoser, B. (2003): Mine Wastes – Characterization, Treatment and Environmental Impacts. – 277 p., 65 fig., 45 tab.; Heidelberg (Springer).

Luckner, L. & Hummel, J. (1982): Modelling and Prediction of the Quality of Mine Drainage Water used for the Drinking Water Supply in the GDR. – Proceedings, 1st International Mine Water Congress, Budapest, Hungary, D: 25–36; Budapest.

Ludwig, K.-H. & Gruber, F. (1987): Gold- und Silberbergbau im Übergang vom Mittelalter zur Neuzeit – das Salzburger Revier von Gastein und Rauris. – 400 p.; Köln (Böhlen).

Lundgren, D. G. & Silver, M. (1980): Ore leaching by bacteria. – Ann. Rev. Microbiol., 34: 263–283, 1 fig.; Palo Alto.

Lung, T. N. (1986): The history of copper cementation on iron – The world's first hydro metallurgical process from medieval China. – Hydrometallurgy, 17: 113–129, 5 fig., 2 tab.; Amsterdam.

Lusznat, M. (1985): Grube Stahlberg bei Müsen. – Geol. Jb., D 77: 149–154, 2 fig.; Hannover.

Lynch, M. (2002): Mining in world history. – 350 p., 11 fig.; London (Reaktion).

Macadam, J. & Shail, R. (1999): Abandoned Pits and Quarries: a Resource for Research, Education, Leisure and Tourism. – In: Spalding, A.: The conservation value of abandoned pits and quarries in Cornwall (6). – p. 71–80, 8 fig.; Truro (The Historic Environment Service).

MacLeod, R., Bursey, G. G., Aston, T. R. C. & Gale, J. E. (1993): Preliminary modelling of groundwater inflows at Gays River mine, Nova Scotia, Canada. – Trans. Inst. Min. Metall. Sect. A, 102: A153–A164, 15 fig., 3 tab.; London.

Madigan, M. T., Martinko, J. M., Parker, J. & Brock, T. D. (2003): An Overview of Microbial Life: Brock biology of microorganisms (2). – 10th edn., p. 22–37, 24 fig., 1 tab.; Upper Saddle River (Prentice Hall).

Magdziorz, A. & Seweryński, J. (2000): The use of membrane technic in mineralised water treatment for drinking and domestic purposes at "Pokój" coal mine district under liquidation. – Proceedings, 7th International Mine Water Association Congress: 430–442, 4 fig., 4 tab.; Ustron.

Mair, C. (2002): Hydrogeologie, Hydrogeochemie und Isotopie der Grund- und Grubenwässer im Einzugsgebiet des Burgfeyer Stollens bei Mechernich, Eifel. – 270 p., 70 fig., 62 tab.; Aachen (unpubl. Diss. RWTH Aachen).

Mäkinen, T. (1996): Information on Mines, Beneficiation Plants and Disposed Mining Waste in Finland. – (Report of the fifth meeting of HELCOM TC Point, Agenda Items 5.38–5,41 on impact from mining industry).

Małoszewski, P. & Zuber, A. (1984): Interpretation of artificial and environmental tracers in fissured rocks with a porous matrix: Isotope Hydrology 1983 1. – p. 635–654, 6 fig.; Wien (Proceedings Series – International Atomic Energy Agency).

Małoszewski, P. & Zuber, A. (1985): On the theory of tracer experiments in fissured rocks with a porous matrix. – J. Hydrol., 79 (3–4): 333–358, 12 fig.; Amsterdam.

Małoszewski, P. & Zuber, A. (1990): Mathematical Modeling of Tracer Behavior in Short-Term Experiments in Fissured Rocks. – Water Resour. Res., 26 (7): 1517–1528, 11 fig., 2 tab.; Washington.

Małoszewski, P., Harum, T. & Benischke, R. (1992): Mathematical Modelling of Tracer Experiments in the Karst of the Lurbach System. – Steir. Beitr. Hydrogeol., 43: 116–143, 27 fig., 9 tab.; Graz.

Mandelbrot, B. B. (1982): The Fractal Geometry of Nature. – 480 p.; New York (Freeman & Company).

Manyin, T., Williams, F. M. & Stark, L. R. (1997): Effects of iron concentration and flow rate on treatment of coal mine drainage in wetland mesocosms: An experimental approach to sizing of constructed wetlands. – Ecol. Eng., 9: 171–1856, 4 fig., 3 tab.; Amsterdam.

Maree, J. P., Hlabela, P., Nengovhela, R., Geldenhuys, A. J., Mbhele, N., Nevhulaudzi, T. & Waanders, F. B. (2004a): Treatment of Mine Water for Sulphate and Metal Removal Using Barium Sulphide. – Mine Water Env., 23 (4): 196–204, 7 fig., 10 tab.; Berlin.

Maree, J. P., Beer, M. d., Strydom, W. F., Christie, A. D. M. & Waanders, F. B. (2004b): Neutralizing Coal Mine Effluent with Limestone to Decrease Metals and Sulphate Concentrations. – Mine Water Env., 23 (2): 81–86, 3 fig., 6 tab.; Berlin.

Maree, J. P., Günther, P., Strobos, G. & Waanders, F. B. (2004c): Optimizing the Effluent Treatment at a Coal Mine by Process Modelling. – Mine Water Env., 23 (2): 87–90, 6 tab.; Berlin.

Maree, J. P., Strobos, G., Greben, H., Netshidaulu, E., Steyn, E., Christie, A., Günther, P. & Waanders, F. B. (2004d): Treatment of Acid Leachate from Coal Discard using Calcium Carbonate and Biological Sulphate Removal. – Mine Water Env., 23 (3): 144–151, 6 fig., 8 tab.; Berlin.

Mascher, R., Neagoe, A., Voigt, K. D., Fischer, S., Scheiding, W. & Bergmann, H. (2004): Amino alcohols – an alternative tool in phytoremediation of heavy metal polluted soils? – Abstracts, Geomicrobiology in remediation of mine waste (3. Jenaer Sanierungskolloquium): 12:50–13:10; Jena.

Mather, J. D., Gray, D. A. & Jenkins, D. G. (1969): The use of tracers to investigate the relationship between mining subsidence and groundwater occurrence at Aberfan, South Wales. – J. Hydrol., 9 (2): 136–154, 6 fig., 5 tab.; Amsterdam.

Mattsson, K. T. & Stenbacka, S. (2004): The Role of Social Capital in Local Development – The Case of Leksand and Rättvik, Sweden. – 55 p., 3 fig., 1 tab.; Uppsala (EU Project Restructuring in Marginal Rural Areas [RESTRIM]).

Mc Cleary, E. C. & Kepler, D. A. (1994): Ecological Benefits of passive Wetland Treatment Systems designed for Acid Mine Drainage. – Proceedings, International Land Reclamation and Mine Drainage Conference, 3: 111–119; Pittsburgh.

McCourt, J. L. (1999): Environmental Legislation and Water Management Issues during Mine Closure in South Africa. – In: Fernández Rubio, R.: Mine, Water & Environment II. – p. 743–749, 1 tab.; Sevilla (International Mine Water Association).

McKenzie, R. (2005): Software Update to Better Predict Costs of Treating Mine Drainage. – Mine Water Env., 24 (4): 213–215; Berlin.

McKnight, E. T. & Fischer, R. P. (1970): Geology and Ore Deposits of the Picher Field Oklahoma and Kansas. – US Geological Survey Professional Paper, 588: 165, 45 fig., 10 tab., 11 Kart.; Washington, DC.

Meier, G., Sroka, A., Löbel, K.-H., Klapperich, H., Tondera, D., Busch, W., Wagner, H. & Randjbar, B. (2004): 4. Altbergbau-Kolloquium – Tagungsband. – 492 p.; Essen (Glückauf).

Meier, J., Babenzien, H.-D. & Wendt-Potthoff, K. (2004): Microbial cycling of iron and sulfur in sediments of acidic and pH-neutral mining lakes in Lusatia (Brandenburg, Germany). – Biogeochemistry, 67 (2): 135–156, 6 fig., 2 tab.; Dordrecht-Boston-Lancaster.

Meinel, G. (1993): Mögliche Beziehungen zwischen Uranlagerstätten und der Kratonisierung am Beispiel Thüringens. – Geowiss. Mitt. Thüringen, 1: 59–63, 1 fig., 1 tab.; Weimar.

Meinrath, G., Merkel, B. J., Ödegaard-Jensen, A. & Ekberg, C. (2004): Sorption of Iron on Surfaces: Modelling, Data Evaluation, and Measurement Uncertainty. – Acta Hydrochim. Hydrobiol., 32 (2): 154–160, 3 fig., 1 tab.; Weinheim.

Melchiorre, E., Dale, D., Mills Jr., J. & Chapman, B. (2005): Geochemical and Stable Isotope Evidence for Precipitation and Groundwater Sourcing of Recharge at the Green Valley Site, Vigo County, Indiana. – Mine Water Env., 24 (2): 88–100, 13 fig., 5 tab.; Berlin.

MEND [Mine Environment Neutral Drainage Program] (1991): Acid Rock Drainage Prediction Manual – A manual of chemical evaluation procedures for the prediction of acid generation from mine wastes: MEND Project 1.16.lb. – Ottawa (The Mine Environment Neutral Drainage [MEND] Program).

MEND [Mine Environment Neutral Drainage Program] (1993): Review of Canadian and United States Legislation relevant to Decommissioning Acid Mine Drainage Sites. – In: Jacques Whitford Environment Limited: MEND Project 4.2.1. – 52 p.; Ottawa (The Mine Environment Neutral Drainage [MEND] Program).

MEND [Mine Environment Neutral Drainage Program] (1994): Evaluation of alternate dry covers for the inhibition of acid mine drainage from tailings: MEND Project 2.20.1. – Richmond Hill (The Mine Environment Neutral Drainage [MEND] Program).

Mentz, J. W., Warg, J. B. & Skelly and Loy Inc. (1975): Up-dip versus down-dip mining – an evaluation. – Environmental protection technology series, EPA-670/2-75-047: 74; Cincinnati.

Merkel, B. (2002): Flooding of the Königstein uranium mine – Aquifer reactivity versus dilution. – In: Merkel, B. J., Planer-Friedrich, B. & Wolkersdorfer, Ch.: Uranium in the Aquatic Environment. – p. 267–275, 6 fig., 1 tab.; Heidelberg (Springer).

Merkel, B. (2005): Alkalinitätserhöhung in sauren Grubenwässern durch CO_2-Zugabe. – Wissenschaftliche Mitteilungen, 28: 51–55, 3 tab.; Freiberg.

Merkel, B. J. & Helling, C. (1995): Proceedings Uranium-Mining and Hydrogeology 1995, Freiberg, Germany: GeoCongress 1. – 583 p.; Köln (von Loga).

Merkel, B. J. & Planer-Friedrich, B. (2002): Grundwasserchemie – Praxisorientierter Leitfaden zur numerischen Modellierung von Beschaffenheit, Kontamination und Sanierung aquatischer Systeme. – 219 p.; Heidelberg (Springer).

Merkel, B. J. & Sperling, B. (1998): DVWK Schriften – Hydrogeochemische Stoffsysteme Teil 2. – In: Deutscher Verband für Wasserwirtschaft und Kulturbau e.V. (DVWK) 117. – p. 1–415; Bonn (Kommissionsvertrieb Wirtschafts- und Verlagsgesellschaft Gas und Wasser mbH).

Merkel, B. J. (2004): Modelling of Arsenic Removal from Water – Means and Limitations. – Wissenschaftliche Mitteilungen, 25: 119–124, 3 fig., 1 tab.; Freiberg.

Merkel, B. J., Hartsch, J., Pinka, F. & Peters, H. [sic!]. (1997): Chemische Veränderungen nach Auflassen von Uranbergwerken unter besonderer Berücksichtigung von Radionukliden. – 48 p., 32 fig., 9 tab.; Freiberg (unpubl. Manuskript).

Merkel, B. J., Planer-Friedrich, B. & Nordstrom, D. K. (2005b): Groundwater Geochemistry – A Practical Guide to Modeling of Natural and Contaminated Aquatic Systems. – 200 p., 76 fig.; Heidelberg (Springer).

Merkel, B. J., Werner, F. & Wolkersdorfer, Ch. (2005a): Carbon dioxide elimination by using acid mine lakes and calcium oxide suspensions (CDEAL). – Geotechnologien – Science Report, 6: 4–12, 1 fig., 1 tab.; Potsdam.

Merten, D., Kothe, E. & Büchel, G. (2004): Studies on Microbial Heavy Metal Retention from Uranium Mine Drainage Water with Special Emphasis on Rare Earth Elements. – Mine Water Env., 23 (1): 34–43, 5 fig., 5 tab.; Berlin.

Metropolis, N. & Ulam, S. (1949): The Monte Carlo method. – J. Amer. Stat. Assoc., 44 (247): 335–341; Alexandria.

Meyer, J. & Schramm, A. (2004): Grubenflutung Schlema-Alberoda – Stand und Zielsetzung der Flutung der Uranerzgrube Schlema-Alberoda. – 22 overheads, 2 fig.; Bad Schlema (Vortrag Workshop Geothermie).

Meyer, J., Jenk, U., Göhrs, A. S. & Werner (2002): Characterisation of final mine flooding at the Schlema-Alberoda site of WISMUT GmbH with particular emphasis on flooding water quality evolution. – In: Merkel, B. J., Planer-Friedrich, B. & Wolkersdorfer, Ch.: Uranium in the Aquatic Environment. – p. 777–782, 2 fig.; Heidelberg (Springer).

Mielcarek, W. (1988): Application of Deep Diaphragm Walls for Groundwater Control in Open cast Lignite Mining in the German Democratic Republic. – Int. J. Mine Water, 7 (3): 35–47, 9 fig.; Nottingham.

Militzer, H., Schön, J. & Stötzner, U. (1986): Angewandte Geophysik im Ingenieur- und Bergbau. – 2nd edn., 419 p., 341 fig., 53 tab.; Stuttgart (Enke).

Miller, D. J., Semmens, K. J., Viadero Jr., R. C. & Tierney, A. E. (2005): The resource potential of mining discharge water for aquaculture. – World Aquaculture, 35 (4): 57–59 + 70, 2 fig.; Baton Rouge.

Miller, R. W. (1996): Flow measurement engineering handbook. – 3rd edn., 1168 p.; New York (McGraw-Hill).

Miller, S. D., Jeffery, J. J. & Murray, G. S. C. (1990): Identification and Management of Acid Generating Mine Wastes – Procedures and Practices in South-East Asia and the Pacific Regions. – Int. J. Mine Water, 9 (1–4): 57–67, 5 fig.; Castro Verde.

Mills, A. L. (1999): The Role of Bacteria in Environmental Geochemistry. – In: Plumlee, G. S. & Logsdon, M. J.: The Environmental Geochemistry of Mineral Deposits 6A (5). – p. 125–132, 4 fig., 4 tab.; Littleton (Society of Economic Geologists).

Mining Minerals and Sustainable Development [MMSD] Project (2002): Breaking New Ground – Mining, Minerals, and Sustainable Development – The Report of the MMSD Project. – 441 p.; London (Earthscan Publications).

Minister of Public Works and Government Services (2001): Guidance document for flow measurement of metal mining effluents. – Environmental Protection Series, EPS 2/MM/4: 28, 5 fig., 1 tab.; Ottawa.

Modak, J. M., Natarajan, K. A. & Mukhopadhyay, S. (1996): Development of temperature-tolerant strains of *Thiobacillus ferrooxidans* to improve bioleaching kinetics. – Hydrometallurgy, 42 (1): 51–61, 6 fig., 2 tab.; Amsterdam.

Moers, W. (1999): Die 13 ½ Leben des Käpt'n Blaubär. – 703 p.; Frankfurt (Eichborn); English editions available at Vintage, Random House (UK) and The Overlook Press (US); Translation Copyright © 2000 by John Brownjohn.

Mohr, K. (1978): Geologie und Minerallagerstätten des Harzes. – 388 p., 138 fig.; Stuttgart (Schweizerbart).

Morgenstern, P., Friese, K., Wendt-Potthoff, K. & Wennrich, R. (2001): Bulk Chemistry Analysis of Sediments from Acid Mine Lakes by Means of Wavelength Dispersive X-ray Fluorescence. – Mine Water Env., 20 (3): 105–113, 6 fig., 2 tab.; Berlin.

Morton, K. L. & Mekerk, F. A. v. (1993): A Phased Approch to Mine Dewatering. – Mine Water Env., 12 (1–4): 27–33; Wollongong.

Morton, K. L. & Niekerk, F. A. v. (1994): Mine Drainage Control and Environment Protection by using Grouting Technology and the Hydrogeological Approach. – Mine Water Env., 13 (2): 41–43; Wollongong.

Moser, H. & Rauert, W. (1980): Isotopenmethoden in der Hydrologie. – In: Matthess, G.: Lehrbuch der Hydrogeologie 8. – 400 p., 227 fig., 32 tab.; Berlin (Borntraeger).

Motyka, I. & Skibinski, L. (1982): Utilization of Salt Mine Waters in the Means of Environmental Protection. – Proceedings, 1st International Mine Water Congress, Budapest, Hungary, D: 37–47, 2 fig.; Budapest.

Mudroch, A., Stottmeister, U., Kennedy, C. & Klapper, H. (2002): Remediation of Abandoned Surface Coal Mining Sites. – 269 p., 73 fig., 22 tab.; Heidelberg (Springer).

Mueller, D. K., Martin, J. D. & Lopes, T. J. (1997): Quality-Control Design for Surface-Water Sampling in the National Water-Quality Assessment Program. – Open-File Report – US Geological Survey, OF 97-223: 1–17, 1 fig., 1 app.; Reston.

Mühlstedt, P. (1992): Der Uranbergbau in Sachsen und seine Besonderheiten in exogenen Lagerstätten. – Schriftenreihe Gesellschaft Deutscher Metallhütten- und Bergleute, 64: 183–201, 10 fig., 5 tab.; Clausthal-Zellerfeld.

Mulenga, S. C., Fernández-Rubio, R., Leon, A. & Baquero, J. C. (1992): Estimation of Quantitative Water Inflow from Different Sources in Konkola Mine, Zambia. – Mine Water Env., 11 (4): 1–22, 16 fig.; Wollongong.

Müller-Salzburg, L. (1978): Der Felsbau 3. – 945 p., 612 fig., 50 plates; Stuttgart (Enke).

Munro, L. D., Clark, M. W. & McConchie, D. (2004): A Bauxsol™-based Permeable Reactive Barrier for the Treatment of Acid Rock Drainage. – Mine Water Env., 23 (4): 183–194, 6 fig., 4 tab.; Berlin.

Murphy, W. M. (1997): Are pit lakes susceptible to limnic eruptions? – Proceedings, International Conference on Tailings and Mine Waste, 4: 543–548; Rotterdam.

Mutschlechner, G. (1984): Erzbergbau und Bergwesen im Berggericht Rattenberg. – 163 p., 68 fig., 1 tab.; Innsbruck (Eigenverlag Gemeinden Alpbach, Brixlegg, Rattenberg und Reith im Alpbachtal).

Nairn, R. W. & Mercer, M. N. (2000): Alkalinity generation and metals retention in a successive alkalinity producing system. – Mine Water Env., 19 (2): 124–135, 3 fig., 2 tab.; Berlin.

Natural Resources Management Unit Division „Protection of the Environment and Natural Resources“ (1999): Report on the International Round Table on Mining and the Environment („Berlin II Declaration“). – 105 p., 4 tab.; Köln (Carl Duisberg Gesellschaft).

Neal, C., Fox, K. K., Harrow, M. & Neal, M. (1998): Boron in the major UK rivers entering the North Sea. – Sci. Total Environ., 210 (1–6): 41–51; Amsterdam.

Neef, T. (2004): Hydrogeochemischen Verhältnisse in einem natürlichen Feuchtgebiet zur passiven Grubenwasserreinigung. – 141 p.; Freiberg (unpubl. Diploma Thesis TU Bergakademie Freiberg).

Nel, P. J. L. (2000): Cover drilling and cementation of water fissure during shaft sinking and the use of micro fine cement as a grouting medium. – p. 585–594, 4 fig.; Ustron (Proceedings, 7th International Mine Water Association Congress).

Neubert, J. T. (2000): Naturally Degraded Surface Waters Associated with Hydrothermally Altered Terrane in Colorado. – Colorado Geological Survey Open-File Report, 00-16: 1–158, 61 fig., 12 tab.; Denver.

Newton, J. G. (1984): Review of induced sinkhole development. – p. 3–9, 3 fig.; Rotterdam Boston (Balkema).

Nitzsche, O., Meinrath, G. & Merkel, B. (2000): Database uncertainty as a limiting factor in reactive transport prognosis. – J. Contam. Hydrol., 44 (3–4): 223–237, 7 fig., 3 tab.; Amsterdam.

Nixdorf, B. & Deneke, R. (2004): Grundlagen und Maßnahmen zur biogenen Alkalinisierung von sauren Tagebauseen. – 281 p.; Berlin (Weißensee).

Nixdorf, B. (1999): Mining Lakes in East Germany – The Problem of Acidification and Chances for Ecosystem Development. – In: Fernández Rubio, R.: Mine, Water & Environment II. – p. 557–561, 6 fig., 1 tab.; Sevilla (International Mine Water Association).

Niyogi, D. K., McKnight, D. M., Lewis, W. M. J. & Kimball, B. A. (1999): Experimental diversion of acid mine drainage and the effects on a headwater stream. – Water resources investigations report, WRI 99-4018-A: 123–130, 4 fig.; Reston.

Nordstrom, D. K. & Alpers, C. N. (1995): Remedial Investigation, Decisions and Geochemical Consequences at Iron Mountain Mine, California. – Proceedings, Sudbury ‘95 – Mining and the Environment, II: 633–642, 2 fig., 1 tab.; Sudbury, Ontarion, Canada.

Nordstrom, D. K. & Alpers, C. N. (1999): Geochemistry of Acid Mine Waters. – In: Plumlee, G. S. & Logsdon, M. J.: The Environmental Geochemistry of Mineral Deposits 6A (6). – p. 133–160, 7 fig., 14 tab.; Littleton (Society of Economic Geologists).

Nordstrom, D. K. (1977): Hydrogeochemical and microbiological factors affecting the heavy metal chemistry of an acid mine drainage system. – 230 p.; Stanford (Degree: Doctoral).

Nordstrom, D. K. (2000): Advances in the Hydrogeochemistry and Microbiology of Acid Mine Waters. – Int. Geol. Rev., 42: 499–515, 4 fig., 2 tab.; Silver Spring.

Nordstrom, D. K. (2004a): Modeling Low-temperature Geochemical Processes: Treatise on Geochemistry 5 (5.02). – 37–72 p., 10 fig., 3 tab.; Amsterdam (Elsevier).

Nordstrom, D. K. (2004b): From research to remediation: application of hydrogeochemical research for effective mine site remediation. – In: Jarvis, A. P., Dudgeon, B. A. & Younger, P. L.: mine water 2004 – Proceedings International Mine Water Association Symposium 1. – p. 141–148, 3 fig., 2 tab.; Newcastle upon Tyne (University of Newcastle).

Nordstrom, D. K. (2004c): Negative pH, Efflorescent Mineralogy, and the Challence of Environmental Restoration at the Iron Mountain Superfund site or why not plug a mine. – Wissenschaftliche Mitteilungen, 25: 125–131, 3 fig., 1 tab.; Freiberg.

Nordstrom, D. K., Alpers, C. N., Ptacek, C. J. & Blowes, D. W. (2000): Negative pH and Extremely Acidic Mine Waters from Iron Mountain, California. – Environ. Sci. Technol., 34: 254–258, 3 fig., 2 tab.; Washington.

Nordstrom, D. K., Olsson, T., Carlsson, L. & Fritz, P. (1989): Introduction to the hydrogeochemical investigations within the International Stripa Project. – Geochim. Cosmochim. Acta, 53 (8): 1717–1726, fig., 8 tab.; Oxford.

Nordstrom, D. K., Plummer, L. N., Wigley, T. M. L., Wolery, T. J., Ball, J. W., Jenne, E. A., Bassett, R. L., Crerar, D. A., Florence, T. M., Fritz, B., Hoffman, M., Holdren Jr, G. R., Lafon, G. M., Mattigod, S. V., McDuff, R. E., Morel, F., Reddy, M. M., Sposito, G. & Thrailkill, J. (1979): A Comparison of computerized chemical models for equilibrium calculations in aqueous systems. – ACS Symposium Series, 93: 857–892, 12 tab.; Washington.

Norton, D. & Knapp, R. (1977): Transport phenomena in hydrothermal systems: the nature of porosity. – American Journal of Science, 277: 913–936, 10 fig., 3 tab.; New Haven.

Noubactep, C., Meinrath, G., Volke, P., Peter, H.-J., Dietrich, P. & Merkel, B. J. (2002): Mechanism of Uranium Fixation by Zero Valent Iron: The Importance of Coprecipitation. – In: Merkel, B. J., Planer-Friedrich, B. & Wolkersdorfer, Ch.: Uranium in the Aquatic Environment. – p. 581–590, 3 fig., 3 tab.; Heidelberg (Springer).

Novák, J. (2001): Groundwater Remediation in the Stráz Leaching Operation. – Mine Water Env., 20 (4): 158–167, 10 fig., 1 tab.; Berlin.

Novák, J., Marecek, P. & Smetana, R. (1995): System of the Model Solutions of the Physical-Chemical Processes in the Straz Deposit Used for Rock Environment Restoration. – Proceedings, Uranium-Mining and Hydrogeology, Freiberg, Germany; GeoCongress, 1: 423–432, 8 fig.; Freiberg.

Novak, P. (1994): Improvement of water quality in rivers by aeration at hydraulic structures. – In: Hino, M.: Water Quality and its Control. – p. 147–168, 7 fig., 3 tab.; Rotterdam (Balkema).

Nunes, J. F., Richards, D. G. & Gama, H. A. (1990): Mine Water and Environmental Protection – The Somincor Case. – Int. J. Mine Water, 9 (1–4): 237–253, 5 fig., 2 tab.; Castro Verde.

Nuttall, C. A. & Younger, P. L. (2004): Hydrochemical stratification in flooded underground mines: an overlooked pitfall. – J. Contam. Hydrol., 69 (1–2): 101–114, 6 fig., 3 tab.; Amsterdam.

Nuttall, C. A., Adams, R. & Younger, P. L. (2002): Integrated hydraulic-hydrogeochemical assessment of flooded deep mine voids by test pumping at the Deerplay (Lancashire) and Frances (Fife) Colliery. – Spec. Publ. – Geol. Soc. London, 198: 315–326, 11 fig., 5 tab.; London.

O'Sullivan, A. D., Murray, D. A. & Otte, M. L. (2004): Removal of Sulfate, Zinc, and Lead from Alkaline Mine Wastewater Using Pilot-scale Surface-Flow Wetlands at Tara Mines, Ireland. – Mine Water Env., 23 (2): 58–65, 3 fig., 1 tab.; Berlin.

Oberste-Brink, K. (1940): Die Frage der Hebungen bei Bodenbewegungen infolge Bergbaues [Surface uplift due to ground movements as a result of mining]. – Glückauf, 76 (18): 249–256, 14 fig., 2 tab.; Essen.

Oelsner, O. W. (1958): Die erzgebirgischen Granite, ihre Vererzung und die Stellung der Bi-Co-Ni-Formation innerhalb der Vererzung. – Geologie, 7 (3–6): 682–697, 5 fig., 6 photographs; Berlin.

Oertel, H. (2001): Introduction to fluid mechanics. – 1st edn., 167 p.; Braunschweig (Viehweg).

Office of Surface Mining (2002): Office of Surface Mining – 2002 Annual Report (CD Version). – 93 p.; Washington (Office of Surface Mining).

Office of Surface Mining Reclamation and Enforcement (2004): Abandoned Mine Land Inventory System. – Internet Database (http://ismhdqa02.osmre.gov/OSM.HTM).

Ogasawara, H., Fujimori, K., Koizumi, N., Hirano, N., Fujiwara, S., Otsuka, S., Nakao, S., Nishigami, K., Taniguchi, K., Iio, Y., Nishida, R., Oike, K. & Tanaka, Y. (2002): Microseismicity induced by heavy rainfall around flooded vertical ore veins. – Pure Appl. Geophys., 159 (1–3): 91–109, 11 fig., 2 tab.; Basel.

Ohimain, E. I. (2003): Available options for the bioremediation and restoration of abandoned pyritic dredge spoils causing the death of fringing mangroves in the Niger Delta. – 9 p., 2 fig., 1 tab.; Athen (Proceedings International Biohydrometallurgy Symposium).

Ohly, M. (2005): Schutzwürdigkeit und Schutzfähigkeit des Wassereinzugsgebietes des Teichtalstollens in Königshütte/Ldkr. Wernigerode. – 69 p., 15 fig., 2 tab., 2 app.; Freiberg (unpubl. Dipl.-Arbeit TU Bergakademie Freiberg).

Ossandón, G. C., Fréraut, R. C., Gustafson, L. B., Lindsay, D. D. & Zentilli, M. (2001): Geology of the Chuqicamata Mine – A Progress Report. – Econ. Geol. Bull. Soc. Econ. Geol., 96 (2): 249–270, 8 fig., 1 tab.; Lancaster.

Ott, A. N. (1986): Estimating iron and aluminum content of acid mine discharge from a north-central Pennsylvania coal field by use of acidity titration curves. – Water resources investigations report, WRI 84-4335: 25, 9 fig., 9 tab.; Washington.

Ott, A. N. (1988): Dual-acidity titration curves – fingerprint, indicator of redox state, and estimator of iron and aluminum content of acid mine drainage and related waters. – US Geological Survey Water-Supply Paper, W 2330: 19–33, 11 fig., 5 tab.; Reston.

Otto, M. (2000): Analytische Chemie. – 2nd edn., 671 p., 326 fig., 104 tab.; Weinheim (Wiley-VCH).

Pack, D. (1992) Hydrologic Characteristics of Abandoned Coal Mines as Sources of Public Water in McDowell County, West Wirginia. – US Geological Survey Water Resource Investigation, WRI 92–4073.

Palme, R. & Ingenhaeff-Berenkamp, W. (1993): Stollen, Schächte, fahle Erze – Zur Geschichte des Schwazer Bergbaus. – 3rd edn., 111 p., 96 fig.; Schwaz (Berenkamp).

Palme, R., Gstrein, P. & Ingenhaeff, W. (2002): Glück Auf! – Faszination Schwazer Silberbergwerk. – 111 p., 56 fig.; Innsbruck (Berenkamp).

Parish, R. V. (1977): The metallic elements. – 254 p., 124 fig., 89 tab.; New York (Longman).

Parkhurst, D. L. (1987): Chemical analyses of water samples from the Picher mining area, Northeast Oklahoma and Southeast Kansas. – Open-File Report – US Geological Survey, OF 87-0453: 43, 3 fig., 6 tab.; Denver.

Parkhurst, D. L. (1988): Mine-water discharge, metal loading, and chemical reactions. – Open-File Report – US Geological Survey, OF 86-0481: D5–D9, 4 fig., 1 tab.; Reston.

Parliament of the Commonwealth of Australia (1970): Water Pollution in Australia – Report from the Senate Select Committee on Water Pollution. – 214 p., 13 fig., 6 tab., 14 plates; Canberra (Commonwealth Printing Office).

Parry, W. T., Forster, C. B., Solomon, D. K. & James, L. P. (2000): Ownership of Mine-Tunnel Discharge. – Ground Water, 38 (4): 487–496; Westerville.

Parsons, A. S. & Hunter, M. D. (1972): Investigation into the movement of groundwater from Bryn Pit above Ebbw Vale, Monmouthshire. – Water Pollut. Control, 71: 568–572, 2 fig., 1 tab.; London.

Pascaly, P. (1997): Forum – Leserbriefe. – Grundwasser, 2 (4): 185; Heidelberg.

Paschoa, A. S. (1998): Potential Environmental and Regulatory Implications of Naturally Occurring Radioactive Materials (NORM). – Appl. Radiat. Isot., 49 (3): 189–196, 3 fig., 3 tab.; Oxford.

Paul, M., Gengnagel, M., Vogel, D. & Kuhn, W. (2002): Four years of flooding WISMUT's Ronneburg uranium mine – a status report. – In: Merkel, B. J.,

Planer-Friedrich, B. & Wolkersdorfer, Ch.: Uranium in the Aquatic Environment. – p. 783–792, 4 fig., 4 tab.; Heidelberg (Springer).

Pauley, E., Shumaker, T. & Cole, B. (2003): Non-Injury Mine Inundation Accident (Entrapment) July 24, 2002 – Black Wolf Coal Company, Inc. Quecreek No. 1 Mine – CMAP # 56981301 – Quecreek, Somerset County, Pennsylvania. – Report of Investigations – Pennsylvanian Department of Environmental Protection Bureau of Deep Mine Safety: 52, 26 fig., 23 app.; Pittsburgh.

Pearce, P. F. & Ries, E. R. (1982): Protection of environmental resources by effective mine water management. – Proceedings, 1st International Mine Water Congress, Budapest, Hungary, C: 80–95, 6 fig., 3 tab.; Budapest.

Pelzel, J., Hemmann, M. & Seifert, G. (1972): Ursachen und Entwicklung der Senkungserscheinungen und Erdfälle bei der Flutung der Grube Friedenshall bei Bernburg. – Ber. Dtsch. Ges. Geol. Wiss., Reihe A, Geol. Paläontol., 17 (2): 191–219, 17 fig., 4 tab.; Berlin.

Penrose, R. G. J. (1974): Limestone selection for permeable plug mine seals. – Papers Presented Before the Symposium on Coal Mine Drainage Research, 5: 133–145, 6 fig., 1 tab.; Washington.

Perrin, C. J., Wilkes, B. & Richardson, J. S. (1992): Stream periphyton and benthic insect responses to additions of treated and mine drainage in a continuous-flow on-site mesocosm. – Environ. Toxicol. Chem., 11 (11): 1513–1525, 3 fig., 5 tab.; Pensacola.

Perry, E. F. & Evans, R. S. (1999): Application of geochemical modeling and hydrologic techniques to interpret sources, mixing and evolution of mine drainage: Mining and Reclamation for the Next Millennium 2. – p. 444–452, 1 fig., 4 tab.; Boulder (Proceedings, 16th Annual National Meeting – American Society for Surface Mining and Reclamation).

Perry, E. F. (2000): Mine Drainage Prediction, a Summary of Concepts and Processes. – In: Kleinmann, R. L. P.: Prediction of Water Quality at Surface Coal Mines (7). – p. 168–194, 5 fig., 6 tab.; Morgantown (National Mine Land Reclamation Center).

Perry, E. F., Gardner, M. D. & Evans, R. S. (1997): Effects of acid material handling and disposal on coal mine drainage quality 3. – p. 1007–1025, 8 fig., 8 tab.; Vancouver (Proceedings, 4th International Conference on Acid Rock Drainage).

Perry, R. S., Dudeney, A. W. L. & Leak, D. J. (2001): Microbiology and 16S rDNA molecular ecology of a polluting neutral mine drainage sediment. – In: Ciminelli, V. S. T. & Garcia Jr, O.: Biohydrometallurgy – Fundamentals, Technology and Sustainable Development 11A. – p. 335–344; Amsterdam (Elsevier).

Peter, G. (1983): Untersuchungen zu Dynamik, Chemismus und Temperaturregime von Wässern ersoffener Grubenbaue. – Freiberg (unpubl. Thesis, Bergakademie Freiberg).

Peukert, D., Reisner, H. & Kunze, W. (1982): Application of a continuously operating Model for Forecasting a Regional Groundwater Flow Situation Caused by Open-

Cast Mine Dewatering Measures. – Proceedings, 1st International Mine Water Congress, Budapest, Hungary, A: 209–225, 6 fig.; Budapest.

Pezdič, J. (1998): Simulation of Solubility and Equilibration Rate of relevant Coal Seam: a stable Isotope and chemical Study. – In: Nel, P. J. L.: Mine Water and Environmental Impacts 1. – p. 43–49, 3 fig.; Johannesburg (Proceedings International Mine Water Association Symposium).

Pfingsten, W. & Mull, R. (1990): Transportprozesse in Kluftgrundwasserleitern. – Deutsche Gewässerkundl. Mitt., 34 (4): 116–123, 9 fig., 1 tab.; Koblenz.

Pietsch, W. & Schötz, A. (2004): Constructed Wetlands – ein Beitrag zur Behandlung extrem saurer Tagebaugewässer. – In: Nixdorf, B. & Deneke, R.: Grundlagen und Maßnahmen zur biogenen Alkalinisierung von sauren Tagebauseen (11). – p. 175–198, 7 fig., 5 tab.; Berlin (Weißensee).

Piper, A. M. (1944): A graphic procedure in the geochemical interpretation of water analyses. – Trans. Amer. Geophys. Union, 25: 914–923, 6 fig., 2 Tab; Washington.

Piper, A. M. (1953): A graphic procedure in the geochemical interpretation of water analyses. – US Geological Survey Ground Water Notes – Geochemistry, 12: 1–14, 6 fig., 2 Tab; Washington.

PIRAMID Consortium (2003): Engineering Guidelines for the Passive Remediation of Acidic and/or Metalliferous Mine Drainage and similar Wastewaters. – 166 p., 10 fig., 27 tab.; Newcastle Upon Tyne (University of Newcastle Upon Tyne).

Pirkl, H. (1961): Geologie des Trias-Streifens und des Schwazer Dolomits südlich des Inn zwischen Schwaz und Wörgl (Tirol). – Jb. Geol. B.-A., 104 (1): 1–150, 9 fig., 7 plates; Wien.

Playton, S. J. & Davis, R. E. (1978): Quality of water in abandoned zinc mines in northeastern Oklahoma and southeastern Kansas. – US Geological Survey Professional Paper, 1100: 111, 4 fig., 4 tab.; Washington.

Playton, S. J., Davis, R. E. & McClaflin, R. G. (1978): Water quality in abandoned zinc mines in the Picher Field, Tri-State mining district. – Abstracts with Programs – Geological Society of America, 10 (1): 24; Boulder.

Playton, S. J., Davis, R. E. & McClaflin, R. G. (1980): Chemical Quality of Water in abandoned Zinc Mines in Northeastern Oklahoma and Southeastern Kansas. – Oklahoma Geological Survey Circular, 82: 1–49, 13 fig., 19 tab.; Norman.

Plinke, M. & Wildhagen, E. (2003): Mining and water in Germany – Legal framework and technical aspects. – In: Nel, P. J. L.: Mine Water and the Environment. – p. 279–286; Johannesburg (Proceedings, 8th International Mine Water Association Congress).

Plotnikov, I. I. (1988): Method of protecting mineral deposits in karst rocks from inundation. – Soviet Eng. Geol., 1: 51–56, 1 fig.; New York.

Plotnikov, N. I., Roginets, I. I. & Viswanathan, S. (1989): Hydrogeology of Ore Deposits: Russian translations series 72. – 290 p., 76 fig., 14 tab.; Rotterdam (Balkema).

Pluta, I. (2000): Barium discharged in waters from the closed Zory Coal Mine in the Upper Silesian Coal Basin, Poland. – Proceedings, 7th International Mine Water Association Congress: 595–603, 6 fig., 2 tab.; Ustron.

Pohl, W. L. (2005): W. u. W. E. Petrascheck's Lagerstättenlehre – Mineralische und Energie-Rohstoffe – Eine Einführung zur Entstehung und nachhaltigen Nutzung von Lagerstätten. – 5th edn., 527 p., 189 fig.; Stuttgart (Schweizerbart).

Pokryszka, Z., Tauzičde, C., Lagny, C., Guise, Y., Gobillot, R., Planchenault, J.-M. & Lagarde, R. (2005): Gas migration from closed coal mines to the surface – Risk assessment methodology and prevention means. – p. 1–15 [CD-ROM], 8 fig.; Nancy (Proceedings, Post-Mining 2005).

Poland, J. F. (1984): Guidebook to studies of land subsidence due to ground-water withdrawal: Studies and reports in hydrology 40. – 305 p.; Paris (UNESCO).

Pöttgens, J. J. E. (1985): Bodenhebung durch ansteigendes Grubenwasser [Uplift as a result of rising mine waters]: The Development Science and Art of Minerals Surveying 2. – p. 928–938, 7 fig.; Harrogate (Proceedings, 6th International Congress International Society for Mine Surveying).

Potvin, Y., Thomas, E. G. & Fourie, A. B. (2005): Handbook on mine fill. – 179 p.; Nedlands (Australian Centre for Geomechanics).

Prediction Workgroup of the Acid Drainage Technology Initiative (2000): Prediction of Water Quality at Surface Coal Mines. – In: Kleinmann, R. L. P. – 241 p.; Morgantown (National Mine Land Reclamation Center).

Prestor, J., Štrucl, S. & Pungartnik, M. (2003): Mežica lead and zinc mine closure impact on hydrogeological conditions in upper Meža valley. – RMZ – Materials and Geoenvironment, 50 (1): 313–316, 1 fig., 1 tab.; Ljubljana.

Pring, G. W. (1999): International law and mineral resources: Mining, Environment and Development. – 55 p.; Genf (UNCTAD – United Nations Conference on Trade and Development).

Puura, E., Marmo, L. & D'Alessandro, M. (2003): Workshop on Mine and Quarry Waste – The Burden from the Past. – In: European Commisson Joint Research Centre EUR 20661 EN. – 79 p.; Luxembourg (Office for Official Publications of the European Communities).

Qiang, W., Yujie, J., Baolin, T., Dean, L. & Ynoghua, X. (1992): An application of artificial tracers in determining vertical groundwater recharge locations in a coal mine. – In: Hötzl, H. & Werner, A.: Tracer Hydrology – 6th International Symposium on Water Tracing. – p. 365–367, 1 fig., 1 tab.; Rotterdam (Balkema).

Quinn, J. J., Negri, M. C., Hinchman, R. R., Moos, L. P., Wozniak, J. B. & Gatliff, E. G. (2001): Predicting the Effect of Deep-Rooted Hybrid Poplars on the Groundwater Flow System at a Large-Scale Phytoremediation Site. – Int. J. Phytoremediation, 3 (1): 41–60, 8 fig., 2 tab.; Boca Raton.

Rackley, R. I. (1976): Origin of Western-States Type Uranium Mineralization. – In: Wolf, K. H.: Handbook of strata-bound and stratiform ore deposits 9 (3). – p. 89–156, 22 fig.; Amsterdam (Elsevier).

Rammlmair, D. & Grissemann, C. (2000): Natural attenuation in slag heaps versus remediation. – In: Rammlmair, D., Mederer, J., Oberthür, T., Heimann, R. B. & Pentinghaus, H.: Applied Mineralogy in Research, Economy, Technology, Ecology and Culture 2. – p. 645–648, 2 fig.; Göttingen (Proceedings, 6th International Congress on Applied Mineralogy).

Randall, M. M., Lanyon, G. W., Nicholls, J. & Willis-Richards, J. (1990): Evaluation of jointing in the carnmenellis granite. – In: Baria, R.: Hot dry rock: Geothermal Energy. – p. 108–123, 7 fig., 2 tab.; Camborne (Robertson Scientific Publications).

Ranville, J. F. & Schmiermund, R. L. (1999): General Aspects of Aquatic Colloids in Environmental Geochemistry. – In: Plumlee, G. S. & Logsdon, M. J.: The Environmental Geochemistry of Mineral Deposits 6A (8). – p. 183–199, 8 fig., 4 tab.; Littleton (Society of Economic Geologists).

Rätsep, A. & Liblik, V. (1998): Changes in physico-chemical composition of mine water in closed Kiviőli oil shale mine. – Oil Shale, 15 (4): 341–352, 1 fig., 3 tab.; Tallin.

Rauche, H., Jockel, A. & Rauche, H. (2002): Monitoring von Altkalischächten – Instrument zur Beurteilung der Gefährdungspotentiale. – In: Busch, W., Maas, K., Klapperich, H., Tondera, D., Sroka, A. & Meier, G.: 2. Altbergbau-Kolloquium – Tagungsband. – p. 232–246, 4 fig.; Clausthal-Zellerfeld (Papierflieger).

Rawlings, D. E. (2002): Heavy Metal Mining using Microbes. – Ann. Rev. Microbiol., 56: 56–91, 5 fig.; Palo Alto.

Rawlings, D. E. (2005): Characteristics and adaptability of iron- and sulfur-oxidizing microorganisms used for the recovery of metals from minerals and their concentrates. – Microbial Cell Factories, 4 (13): [1–15], 3 fig.; London.

Razowska, L. (2001): Changes of groundwater chemistry caused by the flooding of iron mines (Czestochowa Region, Southern Poland). – J. Hydrol., 244: 17–32, 8 fig., 4 tab.; Amsterdam.

Rees, B. & Connelly, R. (2003): Review of design and performance of the Pelenna wetland systems. – Land Contam. Reclam., 11 (2): 293–300, 5 fig., 3 tab.; Richmond.

Rees, S. B. & Bowell, R. J. (1999): Stable Isotopic Modelling of the Longevity of Treatment Processes Operating in a Constructed Wetland for the Amelioration of Acid Mine Drainage. – In: Fernández Rubio, R.: Mine, Water & Environment II. – p. 585–589, 4 fig., 3 tab.; Sevilla (International Mine Water Association).

Rees, S. B., Bowell, R. J. & Wiseman I. (2002): Influence of mine hydrogeology on minewater discharge chemistry. – Spec. Publ. – Geol. Soc. London, 198: 379–390, 7 fig., 3 tab.; London.

Regenspurg, S., Meima, J. A., Kassahun, A. & Rammlmair, D. (2005): Krustenbildung in Bergbauhalden. – Wissenschaftliche Mitteilungen, 28: 63–67, 3 fig., 2 tab.; Freiberg.

Regionaler Planungsverband Westsachsen (1998): Braunkohlenplan Tagebau Vereinigtes Schleenhain. – 94 p., 2 fig., 2 app., 7 Kart.; Bautzen (Lausitzer Druck- und Verlagshaus).

Reimus, P. W. & Haga, M. J. (1999): Analysis of Tracer Responses in the BULLION Forced-Gradient Experiment at Pahute Mesa, Nevada. – Los Alamos Report, LA-13615-MS: 32, 16 fig., 3 tab., 2 app.; Los Alamos.

Ren, G.-L., Izadi, B., King, B. & Dowding, E. (1996): Preferential Transport of Bromide in undisturbed Cores under different Irrigation Methods. – Soil Sci., 161 (4): 214–225, 5 fig., 6 tab.; Baltimore.

Reznik, Y. M. (1990): Determination of groundwater paths using method of streaming potentials. – In: Graves, D. H.: National symposium on mining: a new beginning. – p. 217–221, 2 fig.; Knoxville, TN (Proceedings National Symposium on Mining).

Riedel, L. (1993): Zur Geschichte des Bergbaues um Chemnitz. – Veröff. Mus. Naturk. Chemnitz, 16: 1–88, 28 fig., 1 plates; Chemnitz.

Riemann, C. (1913): Die deutschen Salzlagerstätten – ihr Vorkommen, ihre Entstehung und die Verwertung ihrer Produkte in Industrie und Landwirtschaft. – 97 p., 29 fig.; Leipzig (Teubner).

Robbins, E. I., Cravotta, C. A., Savela, C. E. & Nord, G. L. (1999): Hydrobiogeochemical interactions in 'anoxic' limestone drains for neutralization of acidic mine drainage. – Fuel, 78 (2): 259–270, 4 fig., 2 tab.; Oxford.

Robbins, E. I., Rodgers, T. M., Alpers, C. N. & Nordstrom, D. K. (2000): Ecogeochemistry of the subsurface food web at pH 0–2.5 in Iron Mountain, California, USA. – Hydrobiologia, 433 (1–3): 15–23, 3 fig., 2 tab.; Dordrecht.

Robertson, A. M. & Shaw, S. C. (1999): The Concept of Custodial Transfer of Mined Land. – In: Fernández Rubio, R.: Mine, Water & Environment II. – p. 765–771, 3 fig., 1 tab.; Sevilla (International Mine Water Association).

Robertson, A. M., Devenny, D. & Shaw, S. C. (1998): Post Mining Sustainable use Plans vs Closure Plans: Mine Reclamation and Remediation. – p. 95–110, 4 fig., 1 tab.; Penticton (22nd Annual British Columbia Mine Reclamation Symposium).

Robillard, L., Nguyen, T. H. & Vasseur, P. (1986): Free Convection in a Two-Dimensional Porous Loop. – J. Heat Transfer, 108 (2): 277–283, 4 fig.; New York.

Roehl, K. E. (2004): Passive in situ Remediation of Contaminated Groundwater: Permeable Reactive Barriers – PRBS: Conference Papers 35. – p. 57–70, 4 fig.; Wien (Umweltbundesamt).

Rogoz, M. (1994): Computer Simulation of the Process of Flooding up a Group of Mines. – Proceedings, 5th International Mine Water Congress, Nottingham, U.K., 1: 369–377, 3 fig., 1 tab.; Nottingham.

Rohwerder, T., Gehrke, T., Kinzler, K. & Sand, W. (2003): Bioleaching review part A – Progress in bioleaching – fundamentals and mechanisms of bacterial metal sulfide oxidation. – Appl. Microbiol. Biotechnol., 63: 239–248, 4 fig.; Berlin.

Rolland, W., Wagner, H., Chmielewski, R. & Grünewald, U. (2001): Evaluation of the long term Groundwater Pollution by the open cast Lignite mine Jänschwalde (Germany). – J. Geochem. Explor., 73 (2): 97–111, 12 fig., 2 tab.; Amsterdam.

Rommel, K. (2004): Oxygen Measurement: Optically or Electrochemically? – Asian Environ. Technol., 8 (4): 32–33, 2 fig.; Hertfordshire.

Roschlau, H. (1994): Bohr- und Sprengarbeiten im Uranerzbergbau der AG/SDAG Wismut. – Erzmetall, 47 (6/7, 8, 10): 411–416, 480–488, 627–631, 28 fig., 4 tab.; Weinheim.

Rose, A. W. & Dietz, J. M. (2002): Case Studies of Passive Treatment Systems – Vertical Flow Systems. – p. 776–797, 7 fig., 4 tab.; Lexington (Proceedings, Annual National Meeting – American Society for Surface Mining and Reclamation).

Rose, A. W. (2004): Vertical Flow Systems – Effects of Time and Acidity Relations. – p. 1595–1616, 6 fig., 5 tab.; Lexington (Proceedings, Annual National Meeting – American Society for Surface Mining and Reclamation).

Roseneck, R. (2001): Der Rammelsberg als UNESCO-Weltkulturerbe. – In: Roseneck, R.: Der Rammelsberg – Tausend Jahre Mensch-Natur-Technik 1. – p. 16–39, 39 fig.; Goslar (Goslarsche Zeitung).

Rossum, J. R. (1975): Checking the accuracy of water analyses through the use of conductivity. – J. Am. Water Works Assoc., 67 (4): 204–205, 2 tab.; New York.

Rothenhöfer, P., Sahin, H. & Peiffer, S. (2000): Verringerung der Schwermetall- und Sulfatbelastung in sauren bergbaubelasteten Gewässern durch Aluminiumpräzipitate? – Acta Hydrochim. Hydrobiol., 28 (3): 136–144, 8 fig., 2 tab.; Weinheim.

Roy, S. & Worral, F. (1999): Pyrite Oxidation in Coal-Bearing Strata – The Use of P-Blocking Techniques. – In: Fernández Rubio, R.: Mine, Water & Environment II. – p. 547–550, 1 fig., 1 tab.; Sevilla (International Mine Water Association).

Rüde, T. (2004): Natural acid drainage laboratory: lessons from the Rötlbach (Zillertal Alps/Austria). – In: Jarvis, A. P., Dudgeon, B. A. & Younger, P. L.: mine water 2004 – Proceedings International Mine Water Association Symposium 1. – p. 219–224, 3 fig., 1 tab.; Newcastle upon Tyne (University of Newcastle).

Rudisell, M. T., Stuart, B. J., Novak, G., Payne, H. & Togni, C. S. (2001): Use of flue gas desulfurization by-product for mine sealing and abatement of acid mine drainage. – Fuel, 80 (6): 837–843, 6 fig., 1 tab.; Oxford.

Rudolph, C. (1997): Die untertägige Sanierung der abgebauten Uranerzlagerstätte Niederschlema-Alberoda als komplexer Sanierungsprozeß. – Glückauf, 133 (5): 250–254, 2 fig., 3 tab.; Essen.

Rushton, K. R. (2003): Groundwater hydrology – Conceptual and computational models. – 416 p.; Chichester (Wiley & Sons).

Rüterkamp, P. & Messer, J. (2000): Untersuchungen zur hydraulischen und hydrochemischen Situation in den drei Teilrevieren der gefluteten Flussspatgrube

Straßberg. – For: BST Mansfeld GmbH & Co KG Niederröblingen. – 46 p., 14 app., 3 app.; Essen (Deutsche Montan Technologie GmbH).

Sabir, I. H., Torgersen, J., Haldorsen, S. & Alestrom, P. (1999): DNA tracers with information capacity and high detection sensitivity tested in groundwater studies. – Hydrogeol. J., 7 (3): 264–272, 6 fig., 1 tab.; Berlin.

Sächsisches Landesamt für Umwelt und Geologie (1997): Merkblätter zur Grundwasserbeobachtung, Grundwasserprobenahme. – Materialien zur Wasserwirtschaft, L II-1/6 (4): 1–20, 7 app.; Radebeul.

Salkield, L. U. (1987): A technical history of the Rio Tinto mines – some notes on exploitation from pre-Phoenician times to the 1950s. – 114 p., 22 fig., 12 app.; London (Institution of Mining and Metallurgy).

Salmon, S. U. & Destouni, G. (2001): Environmental Regulation of Mine Waters in the European Union – D 1 National Case Studies – Sweden: ERMITE Final Report – EU R&D Project EVK1-CT-2000-00078. – 21 p., 6 fig., 5 tab.; Stockholm (Royal Institute of Technology).

Salmon, S. U. & Malmström, M. E. (2002): MiMi – Steady state, geochemical box model of a tailings impoundment – Application to Impoundment 1. Kristineberg, Sweden, and prediction of effect of remediation MiMi 2002:2. – 88 p., 32 fig., 23 tab.; Luleå (MiMi Print).

Sammarco, O. (1986): Spontaneous Inrushes of Water in Underground Mines. – Int. J. Mine Water, 5 (2): 29–41, 6 fig.; Granada.

Sammarco, O. (1995): Storage of Water in the Voids of Abandoned Mines and Fractured Rocks. – In: Hotchkiss, W. R., Downey, J. S., Gutentag, E. D. & Moore, J. E.: Water resources at risk. – p. IMWA.102–114, 4 fig., 1 tab.; Minneapolis (American Institute of Hydrology).

Sams, J. I. I. & Veloski, G. A. (2003): Evaluation of Airborne Thermal Infrared Imagery for Locating Mine Drainage Sites in the Lower Kettle Creek and Cooks Run Basins, Pennsylvania, USA. – Mine Water Env., 22 (2): 85–93, 7 fig., 2 tab.; Berlin.

Sams, J. I. I., Veloski, G. A. & Ackman, T. E. (2003): Evaluation of Airborne Thermal Infrared Imagery for Locating Mine Drainage Sites in the Lower Youghiogheny River Basin, Pennsylvania, USA. – Mine Water Env., 22 (2): 94–103, 10 fig., 2 tab.; Berlin.

Sand, W. & Gehrke, T. (2006): Extracellular polymeric substances mediate bioleaching/biocorrosion via interfacial processes involving iron(III) ions and acidophilic bacteria. – Research in Microbiology, 157: 49–56, 4 fig., 1 tab.; Amsterdam.

Sand, W., Gehrke, T., Hallmann, R. & Schippers, A. (1995): Sulfur chemistry, biofilm, and the (in)direct attack mechanism – critical evaluation of bacterial leaching. – Appl. Microbiol. Biotechnol., 43: 961–966; Berlin.

Sandén, P., Karlsson, S., Duker, A., Ledin, A. & Lundman, L. (1997): Variations in hydrochemistry, trace metal concentration and transport during a rain storm event in a small catchment. – J. Geochem. Explor., 58 (2–3): 145–155, 4 fig., 2 tab.; Amsterdam-New York.

Sander, W. & Herbert, H.-J. (1985): NaCl crystallization at the $MgCl_2$ / NaCl solution boundary; a possible natural barrier to the transport of radionuclides. – Mineral. Mag., 49 (2): 265–270, 11 fig.; London.

Sares, M. A., Hauff, P. L., Peters, D. C., Coulter, D. W., Bird, D. A., III, F. B. H. & Prosh, E. C. (2004): Characterizing Sources of Acid Rock Drainage and Resulting Water Quality Impacts Using Hyperspectral Remote Sensing – Examples from the Upper Arkansas River Basin, Colorado. – Proceedings, 2004 Advanced Integration of Geospatial Technologies in mining and Reclamation: 1–20, 10 fig., 1 tab.; Atlanta.

Sarpkaya, T., Kolupaila, S., Powell, R. W., Paull, J. W., Oki, I., Carstens, M. R., Kindsvater, C. E. & Carter, R. W. (1959): Discharge Characteristics of Rectangular Thin-Plate Weirs – Discussion. – Trans. American Society of Civil Engineers, 124: 802–822, 5 fig., 1 tab.; Reston.

Saul, H. (1948): Mine Water. – Transactions of the Institution of Mining Engineers, 107: 294–310, 7 fig., 6 tab.; London.

Savic, D. A. & Walters, G. A. (1999): Hydroinformatics, data mining and maintenance of UK water networks. – Anti-Corros. Methods Mater., 46 (6): 415–425, 3 fig., 1 tab.; Bradford.

Sawada, A., Uchida, M., Shimo, M., Yamamoto, H., Takahara, H. & Doe, T. W. (2000): Non-sorbing tracer migration experiments in fractured rock at the Kamaishi Mine, Northeast Japan. – Eng. Geol., 56 (1–2): 75–96, 16 fig., 2 tab.; Amsterdam.

Schalich, J., Schneider, F. K. & Stadler, G. (1986): Die Bleierzlagerstätte Mechernich – Grundlage des Wohlstandes, Belastung für den Boden. – Fortschr. Geol. Rheinld. u. Westf., 34: 11–91, 25 fig., 5 tab., 2 plates; Krefeld.

Schiffner, C. & Weidig, M. (1909): Radioaktive Wässer in Sachsen II. – p. 59–144, 35 fig., 36 tab.; Freiberg (Craz & Gerlach).

Schiffner, C. (1908): Radioaktive Wässer in Sachsen I. – p. 1–57, 16 fig., 15 tab.; Freiberg (Craz & Gerlach).

Schiffner, C., Weidig, M. & Friedrich, R. (1911): Radioaktive Wässer in Sachsen III. – p. 145–216, 51 fig., 52 tab.; Freiberg (Craz & Gerlach).

Schindler, I., Kutschera, U. & Wiesenberger, H. (2004): Medienübergreifende Umweltkontrolle in ausgewählten Gebieten – Brixlegg: Monographien des Umweltbundesamtes Wien 168 (2). – p. 81–118, 14 fig., 15 tab.; Wien (Umweltbundesamt).

Schindler, S. (2003): Hydrogeochemische Untersuchung des Grubenwassers im Hagenbachtal bei Gernrode/Harz. – 146 p., 77 fig., 11 tab., 29 app.; Freiberg (unpubl. Diploma Thesis TU Bergakademie Freiberg).

Schindler, W., Gerlach, F. & Kaden, H. (1995): pH-Sensor für Präzisionsmessungen in Tiefenwässern. – Sensor Magazin, 1995 (3): 6–9, 6 fig.; Bad Nenndorf.

Skousen, J. G., Simmons, J., McDonald, L. M. & Ziemkiewicz, P. (2002): Acid-Base Accounting to Predict Post-Mining Drainage Quality on Surface Mines. – J. Environ. Qual., 31 (6): 2034–2044, 6 fig., 7 tab.; Madison.

Skowronek, E. & Zmij, M. (1977): Okreslenie pochodzenia wody wyplywajacej zza obmurza szybu znakowaniem rodamina B [Determination of the Origin of Water Flowing Out from Behind a Shaft Lining Traced with Rhodamine B]. – Przegl. Gorn., 33 (1): 13–19, 4 fig., 1 tab.; Gliwice, Poland.

Skowronek, J., Lipowczan, A. & Dubiński, J. (2004): NORM IV Conference – Naturally Occurring Radioactive Materials (Abstracts). – Quarterly Research Reports, 1: 15–153; Katowice.

Smetana, R., Muzák, J. & Novák, J. (2002): Environmental Impact of Uranium ISL in Northern Bohemia. – In: Merkel, B. J., Planer-Friedrich, B. & Wolkersdorfer, Ch.: Uranium in the Aquatic Environment. – p. 703–712, 7 fig.; Heidelberg (Springer).

Smith, K. S. & Huyck, H. L. O. (1999): An Overview of the Abundance, Relative Mobility, Bioavailability, and Human Toxicity of Metals. – In: Plumlee, G. S. & Logsdon, M. J.: The Environmental Geochemistry of Mineral Deposits 6A (2). – p. 29–70, 17 fig., 16 tab.; Littleton (Society of Economic Geologists).

Smith, K. S., Filipek, L. H., Updegraff, D. M. & Papp, C. S. E. (1988): Distribution of microorganisms and selected metals in mine drainage, stream water, and sediment. – Open-File Report – US Geological Survey, OF 86-0481: D11–D15, 3 fig.; Reston.

Smith, M. W. & Brady, K. B. C. (1998): Alkaline Addition. – In: The Pennsylvanian Department of Environmental Protection: Coal mine drainage prediction and pollution prevention in Pennsylvania (13). – p. 13.1–13.13, 4 fig., 1 tab.; Harrisburg (Pennsylvanian Bureau of Mining and Reclamation).

Smith, M. W. (1988): Establishing baseline pollution load from preexisting pollutional discharges for remining in Pennsylvania. – Bureau of Mines Information Circular, IC-9183: 311–318, 8 fig., 1 tab.; Washington.

Sobek, A. A., Schuller, W. A., Freeman, J. R. & Smith, R. M. (1978): Field and laboritory methods applicable to overburden and minesoils 600/2-78-054. – 203 p.; Washington (U.S. Environmental Protection Agency).

Sohrabi, M. (1998): The State-of-the-art on Worldwide Studies in some Environments with Elevated Naturally Occurring Radioactive Materials (NORM). – Appl. Radiat. Isot., 49 (3): 169–188, 5 fig., 4 tab.; Oxford.

Somerfield, P. J., Gee, J. M. & Warwick, R. M. (1994): Benthic Community Structure in Relation to an Instantaneous Discharge of Waste Water from a Tin Mine. – Mar. Pollut. Bull., 28 (6): 363–369, 6 fig., 2 tab.; Oxford-New York.

Spence, C. & McPhie, M. (1997): Streamflow measurement using salt dilution in tundra streams, Northwest Territories, Canada. – J. Am. Water Resour. Assoc., 33 (2): 285–292, 6 fig., 1 tab.; Herndon.

Spiessl, S. M., Prommer, H., Sauter, M. & Zheng, C. (2002): Numerical simulation of uranium transport in flooded underground mines. – In: Merkel, B. J., Planer-

Friedrich, B. & Wolkersdorfer, Ch.: Uranium in the Aquatic Environment. – p. 277–286, 4 fig., 1 tab.; Heidelberg (Springer).

Spruill, T. B. (1983): Effects of abandoned lead and zinc mines on water quality in Kansas. – US Geological Survey Professional Paper, 1375: 1–226; Washington, DC.

Spruill, T. B. (1987): Assessment of Water Resources in Lead-Zinc Mined Areas in Cherokee County, Kansas, and Adjacent Areas. – US Geological Survey Water-Supply Paper, W 2268: 1–68, 23 fig., 26 tab.; Reston, VA.

Sroka, A., Löbel, K.-H., Klapperich, H., Tondera, D., Meier, G. & Busch, W. (2003): 3. Altbergbau Kolloquium – Tagungsband. – 443 p.; Essen (Glückauf).

Stark, L. R., Wenerick, W. R., Williams, F. M., Stevens Jr, S. E. & Wuest, P. J. (1994): Restoring the capacity of spent mushroom compost to treat coal mine drainage by reducing the inflow rate: a microcosm experiment. – Water Air Soil Pollut., 75 (3–4): 405–420, 9 fig., 1 tab.; Dortrecht.

Stark, L. R., Williams, F. M., Wenerick, W. R., Wuest, P. J. & Urban, C. (1996): The effects of substrate type, surface water depth, and flow rate on manganese retention in mesocosm wetlands. – J. Environ. Qual., 25 (1): 97–106, 10 fig., 3 tab.; Madison.

Stefan, H. G. & Gu, R. (1992): Efficiency of Jet Mixing of Temperature-Stratified Water. – J. Environm. Engineering, 118 (3): 363–379, 17 fig., 1 tab.; New York.

Stemprok, M. (1992): Geochemical development of the Krusné hory/Erzgebirge granite pluton exemplified on its Czechoslovak part. – Geophys. u. Geol., IV (4): 51–63, 10 fig.; Berlin.

Stengel-Rutkowski, W. (1993): Trinkwasserversorgung aus Grubengebäuden des ehemaligen Roteisensteinbergbaus im Lahn-Dill-Kreis (Rheinisches Schiefergebirge). – Geol. Jb. Hessen, 121: 125–140, 1 fig., 2 tab.; Wiesbaden.

Stengel-Rutkowski, W. (2002): Trinkwasserversorgung aus Grubengebäuden des ehemaligen Bergbaus im Rheingau-Taunus-Kreis (Rheinisches Schiefergebirge). – Jb. nass. Ver. Naturk., 123: 125–138, 3 fig., 3 tab.; Wiesbaden.

Stevens, C. L. & Lawrence, G. A. (1998): Stability and meromixis in a water-filled mine pit. – Limnol. Oceanogr., 43 (5): 946–954, 10 fig.; Waco.

Stober, I. (1995): Die Wasserführung des kristallinen Grundgebirges. – 191 p., 81 fig., 16 tab.; Stuttgart (Enke).

Stoddart, P. R., Cadusch, P. J., Pearce, J. B., Vukovic, D., Nagarajah, C. R. & Booth, D. J. (2005): Fibre optic distributed temperature sensor with an integrated background correction function. – Meas. Sci. Technol., 16: 1299–1304, 6 fig., 1 tab.; Bristol.

Stoertz, M. W., Hughes, M. L., Wanner, N. S. & Farley, M. E. (2001): Long-term water quality trends at a sealed, partially flooded underground mine. – Environ. Eng. Geosci., 7 (1): 51–65, 12 fig.; Texas.

Stoica, L., Iuhas, T., Bardan, N., Razvan, A., Filip, D. & Oproiu, G. (1998): Removal of ^{226}Ra(II) from the Uranium Mine and Pond Waters. – Proceedings, Uranium-

Mining and Hydrogeology II, Freiberg, Germany; GeoCongress, 5: 331–338, 4 fig., 4 tab.; Köln.

Stoll, R. & Bauer, D. (1991): Anwendung geophysikalischer Verfahren zur Kontrolle und Überwachung technologischer Teilprozesse im untertägigen Erzbergbau. – Neue Bergbautechnik, 21 (12): 427–431, 11 fig., 1 tab.; Leipzig.

Stoltz, E. & Greger, M. (2002): Cottongrass Effects on Trace Elements in Submersed Mine Tailings. – J. Environ. Qual., 31 (5): 1477–1483, 5 fig., 2 tab.; Madison.

Straskraba, V. & Moran, R. E. (1990): Environmental Occurence and Impacts of Arsenic at Gold Mining Sites in the Western United States. – Int. J. Mine Water, 9 (1–4): 181–191, 2 fig., 1 tab.; Castro Verde.

Strongman, J. (2000): Mine Closure – An Overview of the Issues. – p. 1–29; Jakarta (Presentation Indonesia Mine Closure Workshop).

Strunz, H. & Nickel, E. H. (2001): Strunz Mineralogical Tables – Chemical-Structural Mineral Classification System. – 9th edn., 870 p.; Stuttgart (Schweizerbart).

Stuart, W. T. & Simpson, T. A. (1961): Variations of pH with depth in anthracite mine-water pools in Pennsylvania; Article 37. – US Geological Survey Professional Paper, 0424-B: B82–B84, 1 tab.; Reston.

Stumm, W. & Morgan, J. J. (1996): Aquatic chemistry – Chemical Equilibria and Rates in Natural Waters. – 3rd edn., 1022 p.; New York (Wiley & Sons).

Suman, A. L., Freeman, S., Erdner, B., Sharp, M., Meier, L., Datin, D., Dalgarn, J., Garvin, M., Wood, F., Keheley, E., Rainbolt, J. & Crawford, J. (2000): Mine Shaft Subcommittee Final Report to Governor Frank Keating's Tar Creek Superfund Task Force. – 64 p., 9 fig., 2 tab., 3 app.; Oklahoma City (Tar Creek Superfund Task Force Mine Shaft Subcommittee).

Sumioka, S. S. (1991): Quality of water in an inactive uranium mine and its effects on the quality of water in Blue Creek, Stevens County, Washington, 1984–85. – Water resources investigations report, WRI 89-4110: 1–70, 8 fig., 26 tab.; Reston.

Sundborg, Å. (1956): The River Klarälven: a study of fluvial processes. – Geografiska Annaler, 38: 127–316; Stockholm.

Sustainable Development Advisory Council & Copper Mines of Tasmania (1995): Copper Mines of Tasmania Project – Final assessment report. – 93 p.; Hobart (The Council).

Šušteršič, F. (1996): Poljes and Caves of Notranjska. – Acta Carsologica, 15: 252–289, 6 fig., 3 tab.; Ljubljana.

Systat Software Inc. (2002): TableCurve 2D 5.01 for Windows User's Manual – Automated curve fitting and equation discovery. – p. 1.1–13.14; Richmond (SYSTAT Software Inc. Marketing Department).

Szurowski, H. (1968): Petrographische, mineralogische und geochemische Untersuchungen in den Graptolithenschiefern im Raum südlich von Ronneburg. – 168 p., 34 fig., 19 tab.; Freiberg (unpubl. Diploma Thesis TU Bergakademie Freiberg).

Takai, K., Moser, D. P., Onstott, T. C., Spoelstra, N., Pfiffner, S. M., Dohnalkova, A. & Fredrickson, J. K. (2001): *Alkaliphilus transvaalensis* gen. nov., sp nov., an extremely alkaliphilic bacterium isolated from a deep South African gold mine. – Int. J. Syst. Evol. Microbiol., 51: 1245–1256, 4 fig., 3 tab.; Reading.

Tamme, S. (2002): Naturnahe Reinigung kontaminierter Bergwerkswässer – Bemessung und Entwurf eines konstruierten Feuchtgebietes bei Gernrode/Harz. – 132 p., 44 fig., 13 tab., 18 app.; Freiberg (unpubl. Diploma Thesis TU Bergakademie Freiberg).

Tang, D. H., Frind, E. O. & Sudicky, E. A. (1981): Contaminant Transport in Fractured Porous Media: Analytical Solution for a Single Fracture. – Water Resour. Res., 17 (3): 555–564, 10 fig.; Washington.

Tar Creek Superfund Task Force Water Quality Subcommittee (2000a): Final Report Governor Frank Keating's Tar Creek Superfund Task Force Water Quality Subcommittee Surface & Ground Water Monitoring. – 21 p., 17 fig., 42 tab., 4 app.; Oklahoma City (Tar Creek Superfund Task Force Water Quality Subcommittee).

Tar Creek Superfund Task Force Water Quality Subcommittee (2000b): Final Report Governor Frank Keating's Tar Creek Superfund Task Force Water Quality Subcommittee Water Quality Improvement Alternatives. – 35 p., 2 fig., 4 tab.; Oklahoma City (Tar Creek Superfund Task Force Water Quality Subcommittee).

Taylor, B. J. (1978): British Regional Geology – Northern England. – 4th edn., 125 p., 29 fig., 4 tab., 13 plates; London (Her Majesty's Stationery Office).

Taylor, J. & Waring, C. (2001): The Passive Prevention of ARD in Underground Mines by Displacement of Air with a Reducing Gas Mixture: GaRDS. – Mine Water Env., 20 (1): 2–7, 1 fig.; Berlin.

Taylor, J. R., Weaver, T. R., McPhail, D. C. & Murphy, N. C. (1996): Characterisation and impact assessment of mine tailings in the King River system and delta, Western Tasmania. – In: Supervising Scientist 105. – 135 p., 32 fig., 14 tab., 8 app.; Canberra (Supervising Scientist Report).

Teaf, C., Merkel, B., Mulisch, H.-M., Kuperberg, M. & Wcislo, E. (2006): Industry, mining and military sites. – In: World Health Organisation: Protecting groundwater for health: managing the quality of drinking-water sources (11, 23). – 640 p.; London (IWA Publishing).

Terzaghi, K. (1925): Erdbaumechanik auf bodenphysikalischer Grundlage. – 399 p., 65 fig., 63 tab.; Leipzig (Deuticke).

The Pennsylvanian Department of Environmental Protection (1998): Coal mine drainage prediction and pollution prevention in Pennsylvania. – 398 p., 68 fig., 193 tab.; Harrisburg (Pennsylvanian Bureau of Mining and Reclamation).

Theis, C. V. (1935): The relation between the lowering of the piezometric surface and the rate and duration of discharge of a well using groundwater storage. – Trans. Amer. Geophys. Union, 16: 519–524, 3 fig.; Washington.

Theis, C. V. (1940): The source of water derived from wells – Essential factors controlling the response of an aquifer to development. – Civil Engineering, 10 (5): 277–280, 5 fig.; New York.

Thieme, G. (2005): Wismut muss neuen Stollen auffahren – Flutung der Grube Gittersee seit über zwei Jahren gestoppt – Grund: unkontrollierter Wasseraustritt. – Freie Presse, 20.12.2005: 2; Chemnitz.

Thies, M. (2003): Vereinbarkeit von Fledermausschutz und berghistorischer Forschung. – In: Reger, K.: 10 Jahre – Besucherbergwerk Grube Wohlfahrt. – p. 178–180, 1 fig., 1 tab.; Rescheid (Tagungsband – 6. Internationaler Bergbau-Workshop).

Thomas, C. W. (1959): Errors in Measurement of Irrigation Water. – Trans. American Society of Civil Engineers, 124: 319–332, 6 fig., 4 tab.; Reston.

Thomas, R. C. & Romanek, C. S. (2002): Passive Treatment of low-pH, Ferric Iron-Dominated Acid Rock Drainage in a Vertical Flow Wetland I: Acidity Neutralization and Alkalinity Generation. – p. 723–748, 3 fig., 3 tab.; Lexington (Proceedings, Annual National Meeting – American Society for Surface Mining and Reclamation).

Thomas, R. C., Romanek, C. S. & Paddock, L. S. (2004): Trace Metal Retention in LBOS-Amended, Vertical-Flow Constructed Wetlands Treating Low pH, Ferric Iron Dominated Acid Rock Drainage. – p. 1862–1893, 12 fig., 2 tab.; Lexington (Proceedings, Annual National Meeting – American Society for Surface Mining and Reclamation).

Tietze, K. (1978): Geophysikalische Untersuchung des Kivusees und seiner ungewöhnlichen Methangaslagerstätte – Schichtung, Dynamik und Gasgehalt des Seewassers. – 149 p.; Kiel (Naturwiss. Diss. Univ. Kiel).

Timmermann, J. G. & Langaas, S. (2004): Environmental Information in European Transboundary Water Management. – 258 p.; London (IWA Publishing).

Tiwary, R. K. (2001): Environmental Impact of coal Mining on Water Regime and its Management. – Water Air Soil Pollut., 132 (1–2): 185–199, 1 fig., 3 tab.; Dordrecht.

Torrence, C. & Compo, G. P. (1998): A Practical Guide to Wavelet Analysis. – Bulletin of the American Meteorological Society, 79 (1): 61–78, 11 fig., 2 tab.; Boston.

Tröger, K., Knöller, K. & Strauch, G. (2005): Application of stable isotopes to assess reducing processes in lignite mining dumps. – In: Loredo, J. & Pendás, F.: Mine Water 2005 – Mine Closure. – p. 261–164, 3 fig.; Oviedo (University of Oviedo).

Tsang, Y. W. & Tsang, C. F. (1987): Channel Model of Flow Through Fractured Media. – Water Resour. Res., 23 (3): 467–479, 15 fig.; Washington.

Tuovinen, O. H. & Bhatti, T. M. (1999): Microbiological leaching of uranium ores. – Minerals Metallurg. Process., 16 (4): 51–60, 4 fig., 1 tab.; Littleton.

Turek, M. & Gonet, M. (1996): Nanofiltration in the utilization of coal-mine brines. – Desalination, 108: 171–177, 1 tab.; Amsterdam.

Turek, M. (2004): Electrodialytic desalination and concentration of coal-mine brine. – Desalination, 162 (1–3): 355–359, 1 fig., 4 tab.; Amsterdam.

Turek, M., Mrowiec-Bialon, J. & Gnot, W. (1995): Utilization of coal mine brines in the chlorine production process. – Desalination, 101: 57–67, 2 fig., 8 tab.; Amsterdam.

Tutu, H., Cukrowska, E., McCarthy, T. S., Mphephu, N. F. & Hart, R. (2003): Determination and modelling of geochemical speciation of uranium in gold mine polluted land in South Africa. – In: Nel, P. J. L.: Mine Water and the Environment. – p. 137–155, 10 fig.; Johannesburg (Proceedings, 8th International Mine Water Association Congress).

Twining, J. R. & Cameron, R. F. (1997): Decision-making processes in ecological risk assessment using copper pollution of Macquarie Harbour from Mt. Lyell, Tasmania, as a case study. – Hydrobiologia, 5: 207–218, 5 fig., 2 tab.; Dordrecht.

U.S. Army Corps of Engineers (2004): Tar Creek and Lower Spring River Watershed Management Plan – Reconnaissance Phase Draft. – 21 p., 10 fig., 9 app. (Unpublished Report).

U.S. Departmentof the Interior – Bureau of Reclamation (2001): Water Measurement Manual. – 3 revisedth edn., 272 p., 155 fig., 25 tab.; Washington (U.S. Government Printing Office).

U.S. Environmental Protection Agency & Office of Water (1988): REMINE – Coal Remining – Best Professional Judgement Analyses. – 80 p.; Washington (U.S. Environmental Protection Agency).

U.S. Environmental Protection Agency (1994): Guidance for the Data Quality Objectives Process EPA-600/R-96/055. – 69 p., 14 fig., 4 tab.; Washington (U.S. Environmental Protection Agency).

U.S. Environmental Protection Agency (2000): Abandoned Mine Site Characterization and Cleanup Handbook EPA 910-B-00-001. – 408 p., 5 fig., 18 tab.; Denver, San Fancisco, Seattle (U.S. Environmental Protection Agency).

U.S. Environmental Protection Agency (2003): EPA and Hardrock Mining – A Source Book for Industry in the Northwest and Alaska. – 57 p., 3 fig., 11 tab.; Seattle (U.S. Environmental Protection Agency).

U.S. Environmental Protection Agency, Office of Water & Office of Science and Technology Engineering and Analysis Division (2001): Coal Remining best Management Practices Guidance Manual EPA-821-B-01-010. – 618 p., 71 fig., 76 tab.; Washington (U.S. Environmental Protection Agency).

U.S. Geological Survey (2000): Interim report on the scientific investigations in the Animas River watershed, Colorado to facilitate remediation decisions by the U.S. Bureau of Land Management and the U.S. Forest Service. – Open-File Report – US Geological Survey, OF 00-245: 1–34, 14 fig., 4 tab.; Reston.

Uerpmann, E.-P. (1980): Hydrogeologische Fragen bei der Endlagerung radioaktiver Abfälle. – 128 p., 21 fig., 5 tab.; Clausthal (unpubl. Diss. TU Clausthal).

Uhlmann, W., Büttcher, H., Totsche, O. & Steinberg, C. E. W. (2004): Buffering of Acidic Mine Lakes: The Relevance of Surface Exchange and Solid-Bound Sulphate. – Mine Water Env., 23 (1): 20–27, 7 fig., 4 tab.; Berlin.
Unger, K. (2002): Hydrodynamische Verhältnisse im gefluteten Unterbau des Bergkwerks Großkogel/Tirol – Numerische Modellierung mit ANSYS-FLOTRAN. – 117 p., 52 fig., 37 tab., 2 app.; Freiberg (unpubl. Diploma Thesis TU Bergakademie Freiberg).
United Nations Conference on Environment and Development [UNCED] (1992): Agenda 21. – In: United Nations A/CONF. 151/26. – 600 p.; New York (United Nations).
United Nations Department of Economic and Social Affairs [UNDESA] & United Nations Environment Programme Industry and Environment [UNEP] (2002): Berlin II Guidelines for Mining and Sustainable Development. – 52 p., 1 tab., 11 app. (United Nations).
United Nations Department of Technical Co-operation for Development & Development Policy Forum of the German Foundation for International Development (1992): Mining and the Environment – The Berlin Guidelines – A study based on an international Round Table in June 1991. – 180 p.; London (Mining Journal Books).
United Nations Environment Programme [UNEP] (1994): Environmental management of mine sites – Training Manual. – Technical report, 30; Kenia.
United States Department of Agriculture (1979): Field manual for research in agricultural hydrology. – In: Brakensiek, D. L., Osborn, H. B. & Rawls, W. J.: US Department of Agriculture Handbook 224. – 547 p., 355 fig., 40 tab.; Washington (US Department of Agriculture).
Unland, W., Eckart, M., Paul, M., Kuhn, W. & Ostermann, R. (2002): Groundwater rebound compatible with the aquatic environment – technical solutions at WISMUT's Ronneburg mine. – In: Merkel, B. J., Planer-Friedrich, B. & Wolkersdorfer, Ch.: Uranium in the Aquatic Environment. – p. 667–676, 5 fig., 1 tab.; Heidelberg (Springer).
Valenzuela, F., Basualto, C., Tapia, C. & Sapag, J. (1999): Application of hollow-fiber supported liquid membranes technique to the selective recovery of a low content of copper from a Chilean mine water. – J. Membr. Sci., 155 (1): 163–168, 3 fig.; Amsterdam.
Valenzuela, F., Vega, M. A., Yanez, M. F. & Basualto, C. (2002): Application of a mathematical model for copper permeation from a Chilean mine water through a hollow fiber-type supported liquid membrane. – J. Membr. Sci., 204 (1–2): 385–400, 9 fig., 7 tab.; Amsterdam.
van Zyl, H. C., Maree, J. P., van Niekerk, A. M., van Tonder, G. J. & Naidoo, C. (2001): Collection, Treatment and re-use of Mine Water in the Olifants River Catchment. – J. S. Afr. Inst. Min. Metall., 101 (1): 41–46, 2 fig., 7 tab.; Johannesburg.

Varnell, C. J., Van Brahana, J. & Steele, K. (2004): The influence of coal quality variation on utilization of water from abandoned coal mines as a municipal water source. – Mine Water Env., 23 (4): 204–208, 3 fig.; Berlin.

Vasan, S. S., Modak, J. M. & Natarajan, K. A. (2001): Some recent advances in the bioprocessing of bauxite. – Int. J. Miner. Process., 62 (1–4): 173–186, 3 fig., 4 tab.; Amsterdam.

Vega, E., Lesikar, B. & Pillai, S. D. (2003): Transport and Survival of Bacterial and Viral Tracers Through Submerged-flow Constructed Wetland and Sand-filter System. – Bioresour. Technol., 89 (1): 49–56.

Veil, J. A., Kupar, J. M. & Puder, M. G. (2003): Use of Mine Pool Water for Power Plant Cooling. – For: U.S. Department of Energy. – 52 p., 7 fig., 10 tab.; Argonne (Argonne National Laboratory).

Veit, U., Petzold, B. & Piehl, H.-D. (1987): Klimatologische Normalwerte 1951/80: Klimadaten der Deutschen Demokratischen Republik – Ein Handbuch für die Praxis – Reihe B 14. – 111 p., 47 tab.; Potsdam (Meteorologischer Dienst der Deutschen Demokratischen Republik).

Veselič, M. & Norton, P. J. (1997): Successes and Failures in Prediction of Groundwater Rebound in Abandoned Mines – Some Theory and Case Studies. – Proceedings, 6th International Mine Water Association Congress, Bled, Slovenia, 1: 127–147, 4 fig.; Ljubljana.

Veselič, M., Pregl, M., Vižintin, G. & Supovec, I. (2001): Environmental Regulation of Mine Waters in the European Union – D 1 National Case Studies – Slovenia: ERMITE Final Report – EU R&D Project EVK1-CT-2000-00078. – 25 p., 7 fig., 4 tab.; Ljubljana (IRGO Slovenia).

Vieira, F. M. C. C., Diering, D. H. & Durrheim, R. J. (2001): Methods to Mine the Ultra-Deep Tabular Gold-Bearing Reefs of the Witwatersrand Basin, South Africa. – In: Hustrulid, W. A. & Bullock, R. L.: Underground Mining Methods – Engineering Fundamentals and International Case Studies (76). – p. 691–704, 21 fig., 3 tab.; Littleton (Society for Mining, Metallurgy, and Exploration).

Villas Bôas, R. C. & Barreto, M. L. (2000): Mine Closure in Iberoamerica. – 455 p.; Rio de Janeiro (CYTED – IMACC – UNIDO).

Vinci, B. J. & Schmidt, T. W. (2001): Passive, Periodic Flushing Technology for Mine Drainage Treatment Systems. – p. 611–625, 6 fig.; Lexington (Proceedings, 18th Annual National Meeting – American Society for Surface Mining and Reclamation).

Vogt, W. (2001): Alte Vorarlberger Heilbäder – Eine Reise durch die Vorarlberger Bäderlandschaft. – 1st edn., 187 p., 193 fig.; Feldkirch (Benvenuti).

von Rohden, C. & Ilmberger, J. (2001): Tracer experiment with sulfur hexafluoride to quantify the vertical transport in a meromictic pit lake. – Aquatic Sciences, 63 (4): 417–431, 9 fig.; Basel.

Voznjuk, G. G. & Gorshkov, V. A. (1983): Mine Water Utilization in the USSR National Economy. – Int. J. Mine Water, 2 (1): 23–30, 1 fig.; Granada.

Vutukuri, V. S. & Singh, R. N. (1993): Recent Developments in Pumping Systems in Underground Metalliferous Mining. – Mine Water Env., 12 (1–4): 71–94, 18 fig.; Wollongong.

Vutukuri, V. S. & Singh, R. N. (1995): Mine Inundation – Case Histories. – Mine Water Env., 14 (1–4): 107–130, 12 fig., 2 tab.; Wollongong.

Wackwitz, T. (2002): Multitracerversuch im gefluteten Blindschacht des Bergwerks Großkogel/Tirol: Vorbereitung, Durchführung und GIS unterstützte Auswertung. – 104 p., 37 fig., 17 tab., 48 app.; Freiberg (unpubl. Diploma Thesis TU Bergakademie Freiberg).

Waddel, R. K., Jr., Parizek, R. R., & Buss, D. R. (1986) Acidic drainage abatement through surficial application of limestone quarry waste and limeplant flue dust – Proceedings, 7th Annual West Virginia Surface Mine Drainage Task Force Symposium, WV Mining and Reclamation Association

Walder, I. F., Nilssen, S., Räisänen, M. L., Heikkinen, P., Pulkkinen, K., Korkka-Niemi, K., Salonen, V.-P., Destouni, G., Hasche, A., Wolkersdorfer, Ch., Witkowski, A. J., Blachére, A., Morel, S., Lefort, D., Midžić, S., Silajdžić, I., Coulton, R. H., Williams, K. P., Rees, B., Hallberg, K. B. & Johnson, D. B. (2005): Contemporary Reviews of Mine Water Studies in Europe, Part 2. – Mine Water Env., 24 (1): 2–37, 10 fig., 13 tab.; Berlin.

Walter, R., Giese, P., Walther, H. W. & Dill, H. G. (1992): Geologie von Mitteleuropa. – 5th edn., 561 p.; Stuttgart (Schweizerbart).

Walton-Day, K. (1999): Geochemistry of the Processes that Attenuate Acid Mine Drainage in Wetlands. – In: Plumlee, G. S. & Logsdon, M. J.: The Environmental Geochemistry of Mineral Deposits 6A (10). – p. 215–228, 4 fig.; Littleton (Society of Economic Geologists).

Waring, C. L. & Taylor, J. R. (1999): A new technique for building in-situ sub-surface hydrologic barriers: NBT. – In: Fernández Rubio, R.: Mine, Water & Environment II. – p. 663–665, 3 fig., 1 tab.; Sevilla (International Mine Water Association).

Waterhouse, J., Murie, A. & Thomson, D. (2003): Groundwater inflows and management at Emily Ann and Maggie Hays underground nickel mines, Western Australia. – In: Nel, P. J. L.: Mine Water and the Environment. – p. 73–81, 5 fig.; Johannesburg (Proceedings, 8th International Mine Water Association Congress).

Watzlaf, G. R., Kairies, C. L., Schroeder, K. T., Danehy, T. & Beam, R. (2002): Quantitative Results from the Flushing of four Reducing and Alkalinity-Producing Systems. – Proceedings, West Virginia Surface Mine Drainage Task Force Symposium, 23: 1–11, 5 fig., 2 tab.; Morgantown.

Watzlaf, G. R., Schroeder, K. T. & Kairies, C. L. (2000): Long-term performance of anoxic limestone drains. – Mine Water Env., 19 (2): 98–110, 3 fig., 5 tab.; Berlin.

Watzlaf, G. R., Schroeder, K. T., Kleinmann, R. L., Karies, C. L. & Nairn, R. W. (2004) The Passive Treatment of Coal Mine Drainange, U.S. Department of Energy/NETL-2004/1202; http://www.netl.doe.gov/technologies/coalpower/ewr/water/pdfs/Passive%20Treatment.pdf

Waugh, W. J., Smith, G. M., Bergman-Tabbert, D. & Metzler, D. R. (2001): Evolution of Cover Systems for the Uranium Mill Tailings Remedial Action Project, USA. – Mine Water Env., 20 (4): 190–197, 3 fig.; Berlin.

Weber, L. (2004): Die Nassereither Bergwasserexplosion als Folge einer unzureichenden Schließung eines Altbergbaus. – In: Meier, G., Sroka, A., Löbel, K.-H., Klapperich, H., Tondera, D., Busch, W., Wagner, H. & Randjbar, B.: 4. Altbergbau-Kolloquium – Tagungsband. – p. 270–280, 9 fig.; Essen (Glückauf).

Webster, J. G., Swedlund, P. J. & Webster, K. S. (1998): Trace metal adsorption onto acid mine drainage iron oxide. – In: Arehart, G. B. & Hulston, J. R.: Water-Rock Interaction 9. – p. 951–954, 2 fig.; Rotterdam u.a. (Balkema).

Weidig, M. (1912): Radioaktive Wässer in Sachsen IV. – p. 217–359, 73 fig., 81 tab.; Freiberg (Craz & Gerlach).

Weiersbye, I. M. & Witkowski, E. T. F. (2003): Acid rock drainage (ARD) from gold tailings dams on the Witwatersrand basin impacts on tree seed fate, viability, inorganic content and seedling morphology. – In: Nel, P. J. L.: Mine Water and the Environment. – p. 311–328, 4 fig., 3 tab.; Johannesburg (Proceedings, 8th International Mine Water Association Congress).

Weissenborn, B. & Benthin, A. (2003): Idee, Planung und bergmännische Arbeiten zur Nutzung untertägiger Wetter für die Klimatisierung des Kreiskrankenhauses Freiberg. – In: Sroka, A., Löbel, K.-H., Klapperich, H., Tondera, D., Meier, G. & Busch, W.: 3. Altbergbau Kolloquium. – p. 345–358; Essen (Glückauf).

Wendel, G. (1998): Radioactivity in mines and mine water – sources and mechanisms. – J. S. Afr. Inst. Min. Metall., 98 (2): 87–92, 4 fig., 2 tab.; Johannesburg.

Wendland, E. & Himmelsbach, T. (2002): Underground Waste Repositories in Exhausted Coal Mines – Recent Developments. – Mine Water Env., 21 (4): 160–167, 11 fig.; Berlin.

Werner, F., Bilek, F. & Luckner, L. (2001): Implications of Predicted Hydrologic Changes on Lake Senftenberg as Calculated Using Water and Reactive Mass Budgets. – Mine Water Env., 20 (3): 129–139, 8 fig., 4 tab.; Berlin.

Werner, F., Bremstahler, F. & Luckner, L. (1999): Predictive Modeling of Water Quality in Post Mining Environments. – In: Fernández Rubio, R.: Mine, Water & Environment II. – p. 445–448, 6 fig., 1 tab.; Sevilla (International Mine Water Association).

West Virginia Division of Environmental Protection – Office of Mining and Reclamation (2001): West Virginia Acid Mine Drainage Study 2000 – A Quantitative Inventory and Interpretive Review of Water Quality on Active Mining Operations in West Virginia. – 15 p., 10 fig., 11 tab.; Morgantown (West Virginia Division of Environmental Protection).

Weyer, J. (2003): Collapse of the adit "Rothschoenberger Stollen" as a result of the flood on 12/13 August 2002. – In: Nel, P. J. L.: Mine Water and the Environment. – p. 477–492, 24 fig., 1 tab.; Johannesburg (Proceedings, 8th International Mine Water Association Congress).

Whitworth, K. R. (2002): The monitoring and modelling of mine water recovery in UK coalfields. – Spec. Publ. – Geol. Soc. London, 198: 61–73, 15 fig., 2 tab.; London.

Wildeman, T., Brodie, G. & Gusek, J. (1993): Wetland Design for Mining Operations. – 408 p., 67 fig., 40 tab.; Richmond, BC (BiTech).

Williams, D. J., Bigham, J. M., Cravotta, C. A., Traina, S. J., Anderson, J. E. & Lyon, J. G. (2002): Assessing Mine Drainage pH from the Color and Spectral Reflectance of Chemical Precipitates. – Appl. Geochem., 17 (10): 1273–1286; Oxford.

Williams, M. (2001): Arsenic in mine waters: an international study. – Environ. Geol., 40 (3): 267–278, 8 fig., 2 tab.; New York, Berlin.

Williams, R. E. & Kirschner, F. E. J. (1992): Hydrogeological Evaluation of a Prototype In-situ Leach Cell in Unsaturated Lead-Zinc Mine and Mill Wastes. – Mine Water Env., 11 (4): 23–33, 7 fig.; Wollongong.

Wilson, D. & Brown, D. (1997): Fast-track Provision of St. Aidan's Reverse-Osmosis Water-Treatment Plant. – Water Environ. Manage., 11: 271–276, 2 fig., 4 tab.; London.

Wilson, G. V. & Flett, J. S. (1921): The Lead, Zinc, Copper and Nickel Ores of Scotland – Islay. – Memoirs of the Geological Survey, Scotland, XVIII: 65–73, 1 fig., 1 tab.; Edinburgh.

Winkelmann, H. (1956): Schwazer Bergbuch – Codex Vindobonensis 10.852. – 284 p.; Lünen (Gewerkschaft Eisenhütte Westfalia).

Winkler, C. (2001): Einsatz passiver Reinigunsmethoden zur Behandlung von Bergbauwässern am Beispiel der Grube Strassberg/Harz. – Wissenschaftliche Mitteilungen, 16: 179–192, 4 fig., 3 tab.; Freiberg.

Wireman, M., Davies, G. J. & Stover, B. K. (2001): Characterizing Ground-Water Flow Paths in high-altitude fractured Rock Settings impacted by Mining Activities. – In: Kueper, B. H., Novakowski, B. H. & Reynolds, D. A. – p. 1–11 [CD-ROM], 9 fig.; Toronto (Proceedings, International Conference on Fractured rock).

Wirsing, G. (1995): Hydrogeologische Untersuchungen im Grubengebäude des Schauinslandbergwerks – Beurteilung des Grundwasservorkommens und Abgrenzung eines Wasserschutzgebietes für die Quelle 14 der Freiburger Energie- und Wasserversorgungs-AG. – For: Umweltschutzamt Stadt Freiburg i. Br. – 45 p., 7 tab., 16 app.; Freiburg (Geologisches Landesamt Baden-Württemberg).

Wismut GmbH (1994): Ergebnisse der Umweltüberwachung und Sanierungstätigkeit 1993 – Sanierungsbetrieb Aue. – 68 p., 16 app.; Chemnitz (Eigenverlag Wismut GmbH).

Wisotzky, F. (2001): Prevention of Acidic Groundwater in Lignite Overburden Dumps by the Addition of Alkaline Substances: Pilot-scale Field Experiments. – Mine Water Env., 20 (3): 122–128, 6 fig., 2 tab.; Berlin.

Witkowski, A. J., Motyka, J. & Wróbel, J. (2003): Assessment of potential risk of groundwater contamination in areas subjected to the intensive mining drainage

– Case study from Poland – Olkusz Zn-Pb ore mining region. – In: Nel, P. J. L.: Mine Water and the Environment. – p. 221–234, 9 fig., 4 tab.; Johannesburg (Proceedings, 8th International Mine Water Association Congress).

Wittrup, M. B., Kyser, T. K. & Danyluk, T. (1986): The use of stable isotopes to determine the source of brine in Saskatchewan potash mines. – Special Publication – Saskatchewan Geological Society, 8: 159–165, 5 fig.; Regina.

Wöbking, H. (1982): Untertägige Eigenpotentialmessungen in den Bergbauen Großkogel bei Brixlegg und Falkenstein bei Schwaz. – Berg- u. Hüttenm. Mh, 127 (12): 476–483, 8 fig.; Wien.

Wolkersdorfer, Ch. (1991): Geschichte des Bergbaues im westlichen Mieminger Gebirge/Tirol. – Der Aufschluss, 42 (6): 359–379, 18 fig.; Heidelberg.

Wolkersdorfer, Ch. (1995): 2. Bericht zum Kooperationsvertrag zwischen der WISMUT GmbH und der Abteilung Ingenieurgeologie. – 107 p., 45 fig., 16 tab.; Clausthal (unpubl. Report).

Wolkersdorfer, Ch. (1996): Hydrogeochemische Verhältnisse im Flutungswasser eines Uranbergwerks – Die Lagerstätte Niederschlema/Alberoda. – Clausthaler Geowissenschaftliche Dissertationen, 50: 1–216, 131 fig., 61 tab.; Clausthal.

Wolkersdorfer, Ch. (2001): Tracer Tests in Flooded Underground Mines. – In: Seiler, K.-P. & Wohnlich, S.: New Approches Characterizing Groundwater Flow 1. – p. 229–233, 4 fig., 3 tab.; Rotterdam (Balkema).

Wolkersdorfer, Ch. (2002a): Rare Earth Elements (REEs) as Natural Tracers in Mine Waters. – In: Merkel, B. J., Planer-Friedrich, B. & Wolkersdorfer, Ch.: Uranium in the Aquatic Environment. – p. 967–974, 2 fig., 1 tab.; Heidelberg (Springer).

Wolkersdorfer, Ch. (2002b): Mine water tracing. – Spec. Publ. – Geol. Soc. London, 198: 47–61, 5 fig., 6 tab.; London.

Wolkersdorfer, Ch. (2004): Mine Water Literature in ISI's Science Citation Index Expanded™. – Mine Water Env., 23 (2): 96–99, 3 fig., 3 tab.; Berlin.

Wolkersdorfer, Ch. (2006): Acid Mine Drainage Tracer Tests: ICARD 2006 7. – p. 2490–2501 [CD-ROM]; St. Louis (Proceedings, International Conference of Acid Rock Drainage [ICARD]).

Wolkersdorfer, Ch., Feldtner, N. & Trebušak, I. (1997b): LydiA – A new Method for tracing Mine Water. – Proceedings, 6th International Mine Water Association Congress, Bled, Slovenia, 1: 43–55, 3 fig., 2 tab.; Ljubljana.

Wolkersdorfer, Ch. & Göbel, J. (2001a, b): Environmental Regulation of Mine Waters in the European Union – D 2 Overview of the EU and Eastern Europe – 22 Malta | 31 Turkey: ERMITE Final Report – EU R&D Project EVK1-CT-2000-00078. – 4 p. | 4 p.; Freiberg (TU Bergakademie Freiberg).

Wolkersdorfer, Ch. & Göbel, J. (2004): Hydrogeologie der Troianischen Landschaft – Eine Bestandsaufnahme. – Studia Troica, 14: 157–167, 9 fig., 1 tab.; Mainz.

Wolkersdorfer, Ch. & Hasche, A. (2001a): Tracer Test in the abandoned Fluorspar Mine Straßberg/Harz Mountains, Germany. – Wissenschaftliche Mitteilungen, 16: 57–67, 5 fig., 5 tab.; Freiberg.

Wolkersdorfer, Ch. & Hasche, A. (2001b): Tracer Test in the abandoned Fluorspar Mine Straßberg/Harz Mountains, Germany. – p. 1–11 [CD-ROM], 5 fig., 5 tab.; Belo Horizonte (Proceedings, IMWA Symposium Belo Horizonte).

Wolkersdorfer, Ch. & Hasche, A. (2001c): Environmental Regulation of Mine Waters in the European Union – D 1 National Case Studies – Germany: ERMITE Final Report – EU R&D Project EVK1-CT-2000-00078. – 33 p., 5 fig., 5 tab.; Freiberg (TU Bergakademie Freiberg).

Wolkersdorfer, Ch. & Hasche, A. (2003): Multitracerversuch Königshütte/Harz. – For: GEOMONTAN Gesellschaft für Geologie und Bergbau mbH & Co. KG Sachsen. – 13 p., 5 tab., 7 app.; Freiberg (TU Bergakademie Freiberg, unpublished).

Wolkersdorfer, Ch. & Hasche, A. (2004): Tracer Investigations in flooded mines – The Straßberg/Harz Multitracer Test: Conference Papers 35. – p. 45–56, 5 fig., 5 tab.; Wien (Umweltbundesamt).

Wolkersdorfer, Ch., Hasche, A., Unger, K. & Wackwitz, T. (2002): Tracer Techniken im Bergbau – Georgi-Unterbau bei Brixlegg/Tirol. – Wissenschaftliche Mitteilungen, 19: 37–43, 4 fig., 1 tab.; Freiberg/Sachsen.

Wolkersdorfer, Ch., Neumann, C. & Hasche-Berger, A. (2007): Tracer Tests in the flooded Himmelfarth Fundgrube Underground Mine (Freiberg/Saxony). – Wissenschaftliche Mitteilungen, 35: 157–162, 3 fig., 1 tab.; Freiberg.

Wolkersdorfer, Ch., Tamme, S. & Hasche, A. (2003): Natural attenuation of iron rich mine water by a surface brook. – In: Nel, P. J. L.: Mine Water and the Environment. – p. 433–439, 7 fig.; Johannesburg (Proceedings, 8th International Mine Water Association Congress).

Wolkersdorfer, Ch. & Thiem, G. (1999): Ground Water Withdrawal and Land Subsidence in Northeastern Saxony (Germany). – Mine Water Env., 18 (1): 81–92, 3 fig., 4 tab.; Lakewood, CO.

Wolkersdorfer, Ch., Trebušak, I. & Feldtner, N. (1997a): Development of a Tracer Test in a flooded Uranium Mine using *Lycopodium clavatum*. – In: Kranjc, A.: Tracer Hydrology 97 7. – p. 377–385, 4 fig., 2 tab.; Rotterdam (Balkema).

Wolkersdorfer, Ch. & Younger, P. L. (2002): Passive Grubenwassereinigung als Alternative zu aktiven Systemen. – Grundwasser, 7 (2): 67–77, 7 fig., 2 tab.; Heidelberg.

Wolkersdorfer, Ch., Younger, P. L. & Bowell, R. (2004): PADRE – Partnership for Acid Drainage Remediation in Europe. – Mine Water Env., 23 (4): 181–182; Berlin.

Wong, M. H. (2003): Ecological restoration of mine degraded soils, with emphasis on metal contaminated soils. – Chemosphere, 50 (6): 775–780, 4 tab.; Oxford.

Wood, S. C., Younger, P. L. & Robins, N. S. (1999): Long-term changes in the quality of polluted minewater discharges from abandoned underground coal workings in Scotland. – Q. J. Eng. Geol., 32: 69–79, 9 fig., 2 tab.; London.

Worrall, F. (1999): Chemical Fingerprinting of acid Minewaters: the Use of Rare Earth Elements (REE) Patterns. – In: Fernández Rubio, R.: Mine, Water &

Environment II. – p. 509–511, 3 fig.; Sevilla (International Mine Water Association).

Worthington, S. R. H. (1999): A comprehensive strategy for understanding flow in carbonate aquifer. – In: Palmer, A. N., Palmer, M. V. & Sasowsky, I. D.: Karst Modeling. – p. 30–37, 4 fig., 5 tab.; Charles Town (Karst Waters Institute).

Wright, J. B., Nilsen, D. N., Hundley, G. & Galvan, G. J. (1995): Field-Test of liquid Emulsion Membrane Technique for Copper Recovery from Mine Solutions. – Miner. Eng., 8 (4–5): 549–556, 2 fig., 2 tab.; Oxford.

Wrobel, M., Rüde, T. R. & Wohnlich, S. (2000): Small scale mining in the area of the Choapa river, Central Chile: an environmental risk? – Proceedings, 7th International Mine Water Association Congress: 195–203, 5 fig.; Ustron.

Yager, D. B., Lund, K., Luedke, R. G., Bove, D. J., O'Neill, M. J. & Sole, T. C. (1998): Digital geologic compilations of the upper Animas River and Boulder River watersheds; geographic information systems technology used as a scientific interpretation tool. – Open-File Report – US Geological Survey, OF 98-0297: 27; Reston.

Yang, X. J., Gu, Z. M. & Fane, A. G. (1999): Extraction and group separation of rare earths by a combined extraction/electrostatic pseudo liquid membrane from simulated rare earths mine water. – Materials Science Forum, 315-3: 275–281, 2 fig., 4 tab.; Aedermannsdorf.

Yefimochkina, E. (2004): Hydrochemische Untersuchungen an Sicker- und Grubenwässern des Freiberger Bergbaureviers und ihre ökologische Bewertung. – 126 p., 95 fig., 7 tab.; Freiberg (unpubl. Diploma Thesis TU Bergakademie Freiberg).

Yelpatyevsky, P. V. (1995): Factors controlling metal content of mining waters. – In: Kharaka, Y. K. & Chudaev, O. V.: Water-Rock Interaction 8. – p. 901–904, 2 fig.; Rotterdam u.a. (Balkema).

Younger, P. L. & Adams, R. (1999): Predicting Mine Water Rebound. – 109 p., 32 fig., 88 tab.; Bristol (Environment Agency).

Younger, P. L. & Elliott, A. (2003): The Shilbottle Spoil Heap Permeable Reactive Barrier. – London (Proceedings of the CL:AIRE Annual Project Conference 2003, University College).

Younger, P. L. & Harbourne, K. J. (1995): "To Pump or Not to Pump" – Cost-Benefit Analysis of Future Environmental Management Options for the Abandoned Durham Coalfield. – Water Environ. Manage., 9 (4): 405–415, 5 fig., 7 tab.; London.

Younger, P. L. & LaPierre, A. B. (2000): "Uisge Mèinne" – Mine Water hydrogeology in the Celtic lands, from *Kernow* (Cornwall, UK) to *Ceap Breattain* (Cape Breton, Canada). – Spec. Publ. – Geol. Soc. London, 182: 35–52, 7 fig., 2 tab.; London.

Younger, P. L. & Robins, N. S. (2002): Challenges in the characterisation and prediction of the hydrogeology and geochemistry of mined ground. – Spec. Publ. – Geol. Soc. London, 198: 1–16, 1 fig., 2 tab.; London.

Younger, P. L. & Sapsford, D. J. (2006): Acid Drainage Prevention Guidelines for Scottish Opencast Coal Mining: the Primacy of the Conceptual Model: ICARD 2006. – p. 1–20 [CD-ROM], 1 fig., 1 tab.; St. Louis (Proceedings, International Conference of Acid Rock Drainage ICARD).

Younger, P. L. (1993): Possible Environmental Impact of the Closure of two Collieries in County Durham. – Water Environ. Manage., 7 (5): 521–531, 2 fig., 3 tab.; London.

Younger, P. L. (1995): Hydrogeochemistry of minewaters flowing from abandoned coal workings in the Durham coal field. – Q. J. Eng. Geol., 28 (4): S101–S113, 4 fig., 4 tab.; London.

Younger, P. L. (1997): The longevity of minewater pollution – a basis for decision-making. – Sci. Total Environ., 194–195: 457–466, 5 fig.; Amsterdam.

Younger, P. L. (1998): Adit Hydrology in the Long-Term: Observations from the Pb-Zn Mines of Nothern England. – In: Nel, P. J. L.: Mine Water and Environmental Impacts 2. – p. 347–356, 4 fig., 2 tab.; Johannesburg (Proceedings International Mine Water Association Symposium).

Younger, P. L. (1998): Design, construction and initial operation of full-scale compost-based passive systems for treatment of coal mine drainage and spoil leachate in the UK. – In: Mine Water and Environmental Impacts 2. – p. 413–424; Johannesburg (Proceedings International Mine Water Association Symposium).

Younger, P. L. (2000a): Holistic remedial strategies for short- and long-term water pollution from abandoned mines. – Trans. Inst. Min. Metall. Sect. A, 109: A210–A218, 5 fig., 1 tab.; London.

Younger, P. L. (2000b): Nature and practical implications of heterogeneities in the geochemistry of zinc-rich, alkaline mine waters in an underground F-Pb mine in the UK. – Appl. Geochem., 15 (9): 1383–1397, 5 fig., 4 tab.; Amsterdam.

Younger, P. L. (2000c): The adoption and adaptation of passive treatment technologies for mine waters in the United Kingdom. – Mine Water Env., 19 (2): 84–97, 5 fig., 1 tab.; Berlin.

Younger, P. L. (2000d): Predicting temporal changes in total iron concentrations in groundwaters flowing from abandoned deep mines: a first approximation. – J. Contam. Hydrol., 44: 47–69, 5 fig., 3 tab.; Amsterdam.

Younger, P. L. (2002a): Shilbottle Colliery remediation scheme: the UK's first large-scale permeable reactive barrier for mine spoil leachate remediation. – CL:AIRE View, Autumn 2002: 1–3, 3 fig.; London.

Younger, P. L. (2002b): Coalfield closure and the water environment in Europe. – Trans. Inst. Min. Metall. Sect. A, 111: A201–A209, 1 fig., 5 tab.; London.

Younger, P. L. (2002c): A reducing and alkalinity-producing system (RAPS) for the passive treatment of acidic, aluminium-rich leachates emanating from revegetated colliery spoil materials at Bowden Close, County Durham. – 21 p.; London (Proceedings of the CL:AIRE Annual Project Conference, Imperial College).

Younger, P. L. (2002d): Deep mine hydrogeology after closure – insights from the UK. – In: Merkel, B. J., Planer-Friedrich, B. & Wolkersdorfer, Ch.: Uranium in the Aquatic Environment. – p. 25–40, 1 fig., 2 tab.; Heidelberg (Springer).

Younger, P. L. (2002e): Mine Water Pollution from Kernow to Kwazulu-Natal: Geochemical Remedial Options and their Selection in Practice. – Proceedings, Ussher Society, 10: 255–266, 8 fig., 3 tab.; Callington.

Younger, P. L. (2003): Frazer's Grove: the life and death of the last great Weardale fluorspar mine. – In: Fairbairn, R. A.: Fluorspar in the North Pennines. – p. 47–72, 6 fig., 3 tab.; Durham (Friends of Killhope).

Younger, P. L. (2004a): Environmental impacts of coal mining and associated wastes – a geochemical perspective. – Spec. Publ. – Geol. Soc. London, 236: 169–209, 6 fig., 7 tab.; London.

Younger, P. L. (2004b): "Making water": the hydrogeological adventures of Britain's early mining engineers. – Spec. Publ. – Geol. Soc. London, 225: 121–157, 4 fig., 2 tab.; London.

Younger, P. L., Banwart, S. A. & Hedin, R. S. (2002): Mine Water – Hydrology, Pollution, Remediation. – 464 p.; Dordrecht (Kluwer).

Younger, P. L., Coulton, R. H. & Froggatt, E. C. (2005): The Contribution of Science to Risk-Based Decision-Making: Lessons from the Development of Full-Scale Treatment Measures for Acidic Mine Waters at Wheal Jane, UK. – Sci. Total Environ., 338 (1–2): 137–154, 3 fig., 3 tab.; Amsterdam.

Younger, P. L., Jayaweera, A. & Wood, R. (2004): A full-scale reducing and alkalinity-producing system (RAPS) for the passive treatment of acidic, aluminium-rich mine site drainage at Bowden Close, County Durham. – p. 1–13, 5 fig., 3 tab.; London (Proceedings of the CL:AIRE Annual Project Conference, Imperial College).

Younger, P. L., Jayaweera, A., Elliot, A., Wood, R., Amos, P., Daugherty, A. J., Martin, A., Bowden, L., Aplin, A. C. & Johnson, D. B. (2003): Passive treatment of acidic mine waters in subsurface-flow systems – exploring RAPS and permeable reactive barriers. – Land Contam. Reclam., 11 (2): 127–135, 3 fig., 2 tab.; Richmond.

Younger, P. L., Jenkins, D. A., Rees, S. B., Robinson, J., Jarvis, A. P., Ralph, J., Johnston, D. N. & Coulton, R. H. (2004): Mine waters in Wales – pollution, risk management and remediation. – Geological Series – National Museum & Art Galleries of Wales, 23: 138–154, 9 fig., 4 tab.; Cardiff.

Zamora, A. (1999): Small-scale mining. – In: Natural Resources Management Unit Division „Protection of the Environment and Natural Resources": Report on the International Round Table on Mining and the Environment ("Berlin II Declaration"). – p. 47–49; Köln (Carl Duisberg Gesellschaft).

Zeissler, K.-O., Nindel, K. & Hertwig, T. (2005): Hydrochemical Aspects of the Flooding of the Mine Königstein – A Water Mixing Model for Recognizing the Influence of Groundwater by Contaminated Water. – In: Uranium in the Environment. – p. 713–719, 4 fig.; Heidelberg (Springer).

Zentralinstitut für Physik der Erde (1989): Silicic Magmatism and Metallogenesis of the Erzgebirge. – Veröffentlichungen des Zentralinstituts für Physik der Erde, 107: 1–316; Berlin.

Zhou, J., Sun, J. & Zhang, F. (1993): Origin of Mine Water by Using Environmental Isotope Techniques in Water-Logged Coal Mines, Pingdingshan, China. – Mine Water Env., 12 (1–4): 63–69, 2 fig., 6 tab.; Wollongong.

Zhou, W. F. (1997): The formation of sinkholes in karst mining areas in China and some methods of prevention. – Environ. Geol., 31 (1–2): 50–58, 7 fig., 1 tab.; New York, Berlin.

Zhu, Y., Merkel, B. J., Stober, I. & Bucher, K. (2003): The Hydrogeochemistry of Arsenic in the Clara Mine, Germany. – Mine Water Env., 22 (3): 110–117, 6 fig.; Berlin.

Ziegenbalg, G. (1999): In-situ remediation of heavy metal contaminated soil or rock formations and sealing of water inflows by directed and controlled crystallization of natural occurring minerals. – In: Fernández Rubio, R.: Mine, Water & Environment II. – p. 667–672, 11 fig.; Sevilla (International Mine Water Association).

Ziegenbalg, G. (2005): A novel technology for sealing and immobilization – the use of precipitation processes from supersaturated solutions. – In: Merkel, B. J. & Hasche-Berger, A.: Uranium in the Environment. – p. 449–458, 8 fig.; Heidelberg (Springer).

Ziehr, H. (1983): Zur Geschichte des Steinkohlenbergbaues in Stockheim/Nordostbayern. – Der Aufschluss, 34 (12): 541–547, 3 fig.; Heidelberg.

Ziemkiewicz, P. F., Skousen, J. G. & Simmons, J. (2003): Long-term Performance of Passive Acid Mine Drainage Treatment Systems. – Mine Water Env., 22 (3): 118–129, 1 fig., 7 tab.; Berlin.

Zittnan, P., Ravinger, R. & Kováč, L. (1990): Radioaktive Tracer-Meßgeräte und ihre Anwendung unter extremen Bedingungen. – Freiberger Forsch.-H., C 442: 97–104, 5 fig.; Freiberg.

Zosel, J. (1997): Funktion und Anwendung elektrochemischer Sensoren in stömenden Medien unter Einbeziehung hydrodynamischer Messungen mittels Laser-Doppler-Anemometrie. – 111 p., 54 fig., 7 tab., 1 app.; Freiberg (unpubl. Diss. TU Bergakademie Freiberg).

Zötl, J. G. (1995): Badgastein spa – Austrian central Alps. – Environ. Geol., 26 (4): 240–245, 6 fig., 2 tab.; New York-Berlin.

Zötl, J. G. (2001): Badgastein Spa of Austria. – In: LaMoreaux, P. E. & Tanner, J. T.: Springs and Bottled Waters of the World – Ancient History, Source, Occurence, Quality and Use (8.25). – p. 230–233, 6 fig., 1 tab.; Berlin (Springer).

Zoumis, T. & Calmano, W. (2000): Development of fly ash granules for heavy metal removal from mine waters. – Proceedings, 7th International Mine Water Association Congress: 492–504, 13 fig., 3 tab.; Ustron.

Zoumis, T. (2003): Entwicklung aktiver Barrieren für die Entfernung von Schwermetallen aus Grubenwässern am Beispiel der Freiberger Grube, Sachsen. – 159 p., 112 fig., 69 tab.; Düsseldorf (VDI-Verl.).

Zoumis, T., Calmano, W. & Förstner, U. (1999): Naturnahe Demobilisierung von Schwermetallen aus Grubenwässern. – In: Fachgruppe Wasserchemie in der Gesellschaft Deutscher Chemiker: Kurzreferate und Teilnehmerverzeichnis – Jahrestagung 1999 (V27). – p. 138–142, 2 fig., 1 tab.; Regensburg (Eigenverlag Gesellschaft Deutscher Chemiker).

Zoumis, T., Calmano, W. & Förstner, U. (2000): Demobilization of heavy metals from mine waters. – Acta Hydrochim. Hydrobiol., 28 (4): 212–218, 7 fig., 4 tab.; Weinheim.

Zschocke, K., Preuschen, E., Pittioni, R., Firbas, F., Kisser, J. & Netolitzky, F. (1932): Das urzeiliche Bergbaugebiet von Mühlbach-Bischofshofen. – Materialien zur Urgeschichte Österreichs, 6: 1–287, 19 fig., 28 plates, 6 maps; Wien.

Zuber, A., Grabczak, J. & Garlicki, A. (2000): Catastrophic and dangerous inflows to salt mines in Poland as related to the origin of water determined by isotope methods. – Environ. Geol., 39 (3–4): 299–311, 10 fig., 5 tab.; Heidelberg.

Zurl, R., Eckart, D., Vogel, D. & Kuhl, J. (1992): Sanierung des Tagebau [sic!] Lichtenberg im Thüringer Bergbaugebiet. – 9 p., 1 fig., 1 tab.; Chemnitz (unpubl. presentation Workshop Wismut).

14 Appendices

Appendix I: Electrical Conductivity Calculations

The principles of comparing chemical analyses by the use of the electrical conductivity and the limiting ionic molar conductivities are described in section 9.3.2. There is a simple method to do this, based on the theoretical investigations of Debye and Hückel for calculating the electrical conductivity of a solution with weak and strong electrolytes. Appendix I extracts the most important parts of the publications of Rossum (1975) as well as the compiled table of limiting ionic molar conductivities in Coury (1999). Table 31 lists only those ions that might be important for mine water analyses. Other molar conductivities can be found in the literature cited by Rossum (1975) and Coury (1999) – (according to a pers. comm. by Peter T. Kissinger, Chairman and CEO Bioanalytical Systems Inc., the second part of Coury's article was never published).

$$\kappa = \kappa_+^0 + \kappa_-^0 - \left[\frac{\Lambda^0 \times Z_+ \times Z_-}{115.2 \times (Z_+ + Z_-)} \times \left(\frac{2 \times Q}{1 + \sqrt{Q}} \right) + 0.668 \right] \times \left[(Z_+ + Z_-) \times C \right]^{1.5} \tag{85}$$

$$Q = \frac{\Lambda_0 \times Z_+ \times Z_-}{(Z_+ + Z_-) \times (Z_+ \times \lambda_+ + Z_- \times \lambda_-)} \tag{86}$$

$$Z = \frac{\sum (c \times z^2)}{\sum (c \times z)} \tag{87}$$

$$\lambda = \frac{\kappa^0}{\sum c} \tag{88}$$

$$\kappa^0 = \sum (c \times \lambda^0) \tag{89}$$

$$\Lambda^0 = \lambda_+ + \lambda_- \tag{90}$$

$$C = \frac{\sum c_+ + \sum c_-}{2} \tag{91}$$

The use of those equations shall be exemplified by a simple calculation for a given water analyses from the abandoned Schwaz/Austria silver mine (analyses SWZ-2511-MB09).

Table 30. Results of the SWZ-2511-MB09 mine water analyses and the necessary calculations for the electrical conductivity. Differences are due to rounding.

Parameter / element	c_m, mg L^{-1}	c, meq L^{-1}	$c \cdot \lambda^0$	$c \cdot z^2$	$c \cdot z$
pH	7.75				
elec. cond.	1086 µS cm^{-1}				
Ca^{2+}	66	3.29	195.97	13.174	6.587
Mg^{2+}	80.1	6.59	349.99	26.365	13.182
Na^+	20.2	0.88	44.02	0.879	0.879
K^+	8.61	0.22	16.19	0.220	0.220
H^+		$10^{-7.75}$	0.01	0.000	0.000
	sum	10.98	606.17	40.63	20.87
HCO_3^-	174	2.85	126.90	2.852	2.852
SO_4^{2-}	432	8.99	717.72	35.976	17.988
Cl^-	2.77	0.08	5.96	0.078	0.078
NO_3^-	2.84	0.05	3.27	0.046	0.046
OH^-		$10^{-6.25}$	0.11	0.001	0.001
	sum	11.97	853.96	38.96	20.97

Calculation based on the tabulated values in table 30 und table 31:

$$Z_+ = \frac{40.63}{20.87} = 1.947; \quad Z_- = \frac{38.96}{20.97} = 1.858$$

$$\lambda_+ = \frac{606.17}{10.98} = 55.207; \quad \lambda_- = \frac{853.96}{11.97} = 71.342$$

$$\Lambda^0 = 55.207 + 71.759 = 126.548$$

$$C = \frac{10.98 + 11.97}{2} = 11.475$$

$$Q = \frac{126.549 \times 1.947 \times 1.858}{(1.947 + 1.858) \times (1.947 \times 55.207 + 1.858 \times 71.342)} = 0.498$$

$$\kappa = 606.17 + 853.96 - \left[\frac{126.548 \times 1.947 \times 1.858}{115.2 \times (1.947 + 1.858)} \times \left(\frac{2 \times 0.498}{1 + \sqrt{0.498}}\right) + 0.668\right] \times$$

$$\left[(1.947 + 1.858) \times 11.475\right]^{1.5} = 1091.45$$

Table 31. Limiting ionic molar conductivities for common mine water constituents at 25 °C. λ^0 has to be divided by the ionic charge when used in equation 89.

Cation	λ^0_+, S × cm² × mol⁻¹	Anion	λ^0_-, S × cm² × mol⁻¹
H^+	349.6	OH^-	199.1
Li^+	38.7	F^-	55.4
Na^+	50.10	Cl^-	76.35
K^+	73.50	Br^-	78.1
Rb^+	77.8	I^-	76.8
Cs^+	77.2	NO_2^-	71.8
Ag^+	61.9	NO_3^-	71.46
NH_4^+	73.5	HCO_3^-	44.5
Be^{2+}	90	$H_2PO_4^-$	57
Mg^{2+}	106.0	HSO_4^-	50
Ca^{2+}	119.0	CO_3^{2-}	138.6
Sr^{2+}	118.9	HPO_4^{2-}	66
Ba^{2+}	127.2	SO_4^{2-}	160.0
Fe^{2+}	108.0	$C_2O_4^{2-}$	148.2
Cu^{2+}	107.2	PO_4^{3-}	207
Zn^{2+}	105.6		
Pb^{2+}	42.0		
UO_2^{2+}	64		
Al^{3+}	183		
Fe^{3+}	204		
La^{3+}	209.1		
Ce^{3+}	209.4		

As can be seen the result from the calculation of the electrical conductivity, 1091 µS cm^{-1}, is very close to the measured electrical conductivity of 1086 µS cm^{-1}, the total error being only 0.5%.

Appendix II: Weir Equations

Introductory Remarks

This appendix shall not add another solution to the several hundreds of solutions already published but shall give mine water researchers a quick tool for measuring flows over thin-plated weirs without going too deeply into discharge measurement. Section 9.2.1.4 gives an overview of several prerequisites and limitations of weir constructions that are commonly used in mine water research. Usually, these simple techniques are satisfactory but if more accuracy is needed, the following set of equations should be used instead of the simplified equations. All of the equations are based on the published literature (e.g. Ackers et al. 1978; Herschy 1995) but the tables and graphs for getting the relevant coefficients have been updated, compared with other published data, and brought into equations that can be used in spreadsheets or routines for flow calculation. For coefficient fitting, the software package TableCurve has been used and all of the equations have been selected to minimise error, usually to less than 0.5% for the reported data. These equations can only be used in cases where the limitations of the original researchers are met, and therefore these constraints are listed here as well. They can only be used for "normal" mine water under "normal" temperature conditions (usually +4 to +29 °C). Sludge flows or mine waters with extremely high solid contents cannot be measured without calibrating the weir to the special conditions on site.

Besides those limitations, the crest thickness, the roughness of the weir plate, the rounding of the crest, the fluids viscosity, and – to a lesser extent – the velocity distribution of the flow approaching the weir crest, influence the discharge (Ackers et al. 1978). Furthermore, weir plates must be installed vertically, they have to be regularly cleaned (biofilms and precipitates must be removed), and the overflowing water must be ventilated, as the flow depends on the pressure conditions around the weir crest. Finally, the surface tension and the viscosity must be taken into account, for which analytical solutions for thin-plated rectangular weirs have been discussed as well (e.g. equation 47 in Sarpkaya et al. 1959). Finally, keep in mind that even if thin-plate weirs "look" simple, measuring flows with such devices is not!

Symbols and Notation for Appendix II

A	Cross-sectional area of stream bed (Cipolletti weir only), m^2
B	Bed width, m
b	Crest breath, m
b_e	Effective crest breath, m
C_d	Coefficient of discharge, –
C_v	Coefficient of velocity of approach
f	Factor
g	Gravitational acceleration ($\approx$ 9.807–9.815 m s^{-2}), m s^{-2}
h	Gauged head related to weir crest, m
h_e	Effective head related to weir crest, m
h_{max}	Maximum expected head related to weir crest, m
k_b	Width correction factor, m
k_v	Head correction factor, m
L_h	Distance from weir to position of head measurement, m
m	Coefficients 1–5 (Thomson weir)
n	Coefficients 1–5 (Thomson weir)
o	Coefficients 1–4 (Thomson weir, simplified discharge equation)
P	Crest height of weir above mean bed level, m
Q	Discharge, m^3 s^{-1}
r	Coefficients 1–2 for calculating the coefficient of discharge
s	Coefficients 1–4 (contracted Poncelet weir)
t	Coefficients 1–9 (contracted Poncelet weir)
u	Coefficients 1–7 (contracted Poncelet weir)
v	Coefficients 1–3 (Cipolletti weir)
α	Angle of V-notch weir (full opening), ° (degrees)

V-notch (Thomson) Weirs (90°, ½ 90°, ¼ 90°)

Nowadays, most researches use the Kindsvater-Shen equation for calculating the flow over a V-notch weir. It has been simplified by many authors to a form as given in equation 92. Compared to a 90° V-notch weir, the ½ 90° and ¼ 90° V-notch weirs have a flow which is half and a quarter of the flow over a 90° V-notch weir at he same gauged head, respectively. All the calculations for C_d and k_v given here, are based on the originally published graphical relations by Shen (1981) and were used for the curve fitting (Table 32). They are only relevant if the following limitations are met (compare inset in Fig. 126):

$h \geq 0.05$ m
$P > 0.45$ m
$B > 0.9$ m
$20° \leq \alpha \leq 100°$
$4\ °C \leq \theta \leq 29\ °C$

$$Q = \frac{8}{15} \times \sqrt{2 \times g} \times C_d \times \tan\frac{\alpha}{2} \times h_e^{2.5} \tag{92}$$

$$h_e = h + k_v \tag{93}$$

$$C_d = m_1 + m_2 \times h + m_3 \times h^2 + m_4 \times h^3 + m_5 \times h^4 \tag{94}$$

$$k_v = n_1 + n_2 \times \alpha^{1.5} + n_3 \times \alpha^2 + n_4 \times \alpha^{-1.5} \tag{95}$$

There is also a calculation for the simplified equations given in section 9.2.1.4. If the simplified equations have the general form of

$$Q = f(\alpha) \times h^{2.5} \tag{96}$$

where α is given in degrees, then $f(\alpha)$ can be written as

$$f(\alpha) = \frac{o_1 + o_2 \times \alpha}{1 + o_3 \times \alpha + o_4 \times \alpha^2} \tag{97}$$

with $20° \leq \alpha \leq 100°$.

Table 32. Coefficients for equations 94, 95, and 97.

Coefficient	*m*	*n*	*o*
1	$6.057 \cdot 10^{-1}$	$1.900 \cdot 10^{-3}$	$6.886 \cdot 10^{-3}$
2	$-7.524 \cdot 10^{-4}$	$-4.757 \cdot 10^{-6}$	$1.211 \cdot 10^{-2}$
3	$2.809 \cdot 10^{-6}$	$3.575 \cdot 10^{-7}$	$9.483 \cdot 10^{-4}$
4	$4.082 \cdot 10^{-8}$	$1.198 \cdot 10^{-1}$	$-3.506 \cdot 10^{-5}$
5	$-1.831 \cdot 10^{-10}$	–	–

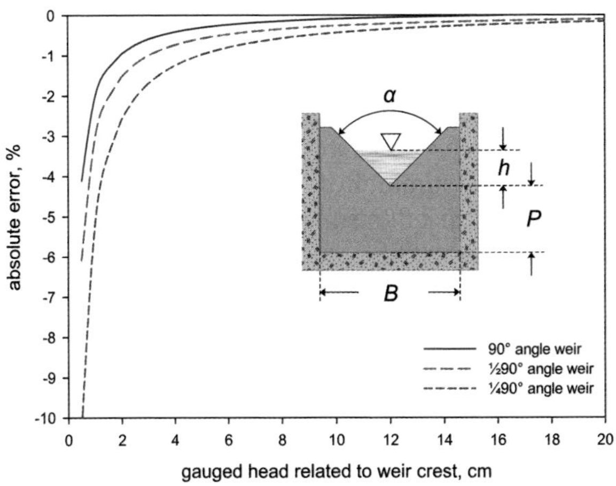

Fig. 126. Absolute errors of discharge measurements by using the simplified equations 54–56 for V-notch weirs. Inset: typical layout of a V-notch weir.

Rectangular, full Width Notch (Poncelet) Weirs

Though several hundreds of publications about weirs are available, it is still not possible to give *one* equation that can be recommended for all possible boundary conditions. Therefore, Ackers et al. (1978), based on investigations conducted by the Hydraulics Research Station Wallingford England, suggested using equation 98 for full-width weirs with the following limits of application:

$h \geq 0.02$ m
$b \geq 0.20$ m
$P \geq 0.15$ m

$$\frac{h}{P} \leq 2.2$$

$$2.5 \leq \frac{L_h}{P} \leq 3$$

$$Q = 0.564 \times \sqrt{g} \times \left(1 + 0.150 \times \frac{h}{P}\right) \times b \times h_e^{1.5} \qquad (98)$$

$$h_e = h + 0.001 \qquad (99)$$

This equation is less accurate for weirs with $\frac{h}{P} > 1.8$ and in cases where h is smaller than 0.075 m.

Rectangular, Contracted Notch (Poncelet) Weirs

As in the case of full-width weirs, different researchers have developed different experimental solutions for the general discharge equation. The first investigations for flows over rectangular weirs were conducted in Munich, Dresden, Leipzig, and the Walchensee/Germany hydraulic laboratory by Kirchhoff, Kötter, Ott, Zschiedrich, Kirschmer, and Esterer (Kolupaila in Sarpkaya et al. 1959). Yet, frequently, the Kindsvater equation is used and it will therefore be given here, too. Ackers et al. (1978) showed that the flow over a contracted rectangular weir is a function of many variables defining the geometry of such a flow measurement installation. Therefore, the number of coefficients and functions to calculate the flow through a contracted rectangular weir is far greater than for the other cases. All efforts to reduce the number of parameters that resulted from the curve fitting code were unsuccessful; therefore, 20 coefficients are necessary to calculate the flow through such a type of weir (Table 33). Sarpkaya (in Sarpkaya et al. 1959) derived an equation for calculating C_d that includes the Reynolds and Weber numbers, Re and We, for taking into account the surface tension and viscosity of the fluid. They are not considered here because viscosity and surface tension only become important if the head above the weir crest and the weir width are comparatively small. Such situations are rare in most mine water measurement cases, and so are not considered here.

$h \geq 0.03$ m
$b \geq 0.15$ m
$P > 0.10$ m
$\frac{h}{P} \leq 2$
$\frac{B-b}{2} \geq 0.1$ m
$4 \leq \frac{L_h}{h_{max}} \leq 5$

$$Q = \frac{2}{3} \times \sqrt{2 \times g} \times C_d \times b_e \times h_e^{1.5} \quad (100)$$

$$C_d = r_1 + r_2 \times \frac{h}{P} \quad (101)$$

$$b_e = b + k_b \quad (102)$$

$$h_e = h + 0.001 \quad (103)$$

$$r_1 = \frac{s_1 + s_2 \times f}{1 + s_3 \times f + s_4 \times f^2} \quad (104)$$

$$r_2 = \frac{t_1 + t_2 \times f^2 + t_3 \times f^4 + t_4 \times f^6 + t_5 \times f^8}{1 + t_6 \times f^2 + t_7 \times f^4 + t_8 \times f^6 + t_9 \times f^8} \quad (105)$$

$$k_b = \frac{u_1 + u_2 \times f^2 + u_3 \times f^4 + u_4 \times f^6}{1 + u_5 \times f^2 + u_6 \times f^4 + u_7 \times f^6} \quad (106)$$

$$f = \frac{b}{B} \quad (107)$$

Table 33. Coefficients for equations 104, 105, and 106.

Coefficient	*s*	*t*	*u*
1	$5.877 \cdot 10^{-1}$	$-3.260 \cdot 10^{-3}$	$2.390 \cdot 10^{-3}$
2	$-5.279 \cdot 10^{-1}$	$-2.131 \cdot 10^{-1}$	$-8.640 \cdot 10^{-3}$
3	$-9.108 \cdot 10^{-1}$	2.545	$1.766 \cdot 10^{-2}$
4	$1.005 \cdot 10^{-2}$	$1.227 \cdot 10^{1}$	$-1.148 \cdot 10^{-2}$
5	–	-5.907	-3.893
6	–	$20.15 \cdot 10^{1}$	6.560
7	–	$4.783 \cdot 10^{2}$	-3.596
8	–	$-7.995 \cdot 10^{2}$	–
9	–	$4.165 \cdot 10^{2}$	–

Trapezoidal (Cipolletti) Weirs

Both side slopes of a Cipolletti weir have to be at an angle of 1:4, with the intent of reducing the side contraction effects (Herschy 1995). Though Ackers et al. (1978) concludes that Cipolletti weirs do not have an advantage in practice, the simpler empirical weir equation compared to Poncelet weirs seems to prove something else (Table 34). Yet, it is a matter of habit if one uses a Cipolletti weir or a rectangular weir.

$0.06\ \text{m} \leq h \leq 0.6\ \text{m}$
$P < 2\ h_{\max}$
$P \geq 0.30\ \text{m}$
$\frac{h}{b} \leq 0.5$
$4 \leq \frac{L_h}{h_{\max}} \leq 5$

$$Q = \frac{2}{3} \times \sqrt{2 \times g} \times C_v \times C_d \times b \times h^{1.5} \tag{108}$$

$$C_d = 0.63 \tag{109}$$

$$C_v = v_1 + v_2 \times e^{\left(\frac{-f}{v_3}\right)} \tag{110}$$

$$f = \frac{C_d \times b \times h}{A} \tag{111}$$

Coefficient	v
1	$9.846 \cdot 10^{-1}$
2	$1.261 \cdot 10^{-2}$
3	$-2.743 \cdot 10^{-1}$

Table 34. Coefficients for equation 110.

Appendix III: The Berlin I and II Guidelines

Introduction

In 1991, the first International Round Table on Mining and the Environment (Berlin I) was organised by the UN Department of Economic and Social Development (DESD) and the Development Policy Forum of the German Foundation for International Development. Mainly meant for developing countries, the Berlin I Guidelines were also adopted by mining companies in developed countries and some of the changes in the mining industry can clearly be related to those guidelines. Because those guidelines are important for mine planning and advance mine closure planning, their central aspects shall be cited here (United Nations Department of Technical Co-operation for Development and Development Policy Forum of the German Foundation for International Development 1992). Meanwhile, a second round table was organised in 1999 and their recommendations were also published (Natural Resources Management Unit Division „Protection of the Environment and Natural Resources"1999; United Nations Department of Economic and Social Affairs [UNDESA] and United Nations Environment Programme Industry and Environment [UNEP] 2002). During the Berlin II meeting, the delegates realised that:

> Since the publication of the first edition of the Guidelines, continuing changes have occurred within the mining sector – particularly in the evolution of legal, fiscal and regulatory policies, accompanying the trend of liberalization and privatization of the industry. Among other things, environmental conditions attached to credit and insurance have raised the profile of environmental planning and management of sustainable development. There have also been major advances in voluntary actions by industry and the standardization of environmental management systems.
>
> During this period, significant advances have been made in defining and promoting best practices in the sustainable management and operation of mining projects. These advances have often been stimulated by the attention of pressure groups, the media, and public opinion. Many of the major mining companies have now started to work in partnership with regulators and the local communities to maximise corporate, national and community benefits, while minimising the social and physical impacts.

Yet, the delegates of the Berlin II round table realised that the life of a mine also has to deal with closure, and therefore put forward guidelines for mine closure and abandoned mines (Balkau 1999).

Environmental Management Guidelines for Mining

World-wide long-term economic development can best be achieved through the pursuit of sustainable development policies comprising a balance of economic, socio-cultural, and environmental protection measures. While taking into account global environmental concern, each country should apply this concept to meet the needs of their environmental and economic circumstances. Sustainable mining activities require good environmental stewardship in all activities, from exploration and processing to decommissioning and reclamation. It acknowledges the importance of integrating environmental and economic considerations in the decision-making process and the fact that the mineral deposits are unique in their occurrence. It recognises the importance of mining to the social, economic, and material needs of society, in particular for developing countries, and that minerals, notably metals, offer great potential for the use of future generations through increased recycling programs. Sustainable mining under appropriate environmental guidelines is based on interaction between industry, governments, non-governmental organizations and the public, directed towards optimising economic development while minimising environmental degradation. The need for such guidelines is recognised by industry, governments, and international agencies. It is also recognised that the political will of governments, together with the commitment of industry management and of the community, are the essential conditions needed to enforce environmental legislation and more importantly, to ensure compliance with all applicable laws for the protection of the environment, employees and the public.

Environmental Guidelines for Action Addressed to the Mineral Sector (Berlin I Guidelines, revised in 2000; United Nations Department of Economic and Social Affairs [UNDESA] and United Nations Environment Programme Industry and Environment [UNEP] 2002)

Governments, mining companies and the minerals industries should as a minimum:

1. Recognise environmental management as a high priority, notably during the licencing process and through the development and implementation of environmental management systems. These should include early and comprehensive environmental impact assessments, pollution control and other preventive and mitigative measures, monitoring and auditing activities, and emergency response procedures.

2. Recognise the importance of socio-economic impact assessments and social planning in mining operations. Social-economic impacts should be taken into account at the earliest stages of project development. Gender issues should also be considered at a policy and project level.
3. Establish environmental accountability in industry and government at the highest management and policy-making levels.
4. Encourage employees at all levels to recognise their responsibility for environmental management and ensure that adequate resources, staff and requisite training are available to implement environmental plans.
5. Ensure the participation of and dialogue with the affected community and other directly interested parties on the environmental and social aspects of all phases of mining activities and include the full participation of women and other marginalised groups.
6. Adopt best practices to minimise environmental degradation, notably in the absence of specific environmental regulations.
7. Adopt environmentally sound technologies in all phases of mining activities and increase the emphasis on the transfer of appropriate technologies that mitigate environmental impacts, including those from small-scale mining operations.
8. Seek to provide additional funds and innovative financial arrangements to improve environmental performance of existing mining operations.
9. Adopt risk analysis and risk management in the development of regulation and in the design, operation, and decommissioning of mining activities, including the handling and disposal of hazardous mining and other wastes.
10. Reinforce the infrastructure, information systems service, training and skills in environmental management in relation to mining activities.
11. Avoid the use of such environmental regulations that act as unnecessary barriers to trade and investment.
12. Recognise the linkages between ecology, socio-cultural conditions and human health and safety, the local community and the natural environment.
13. Evaluate and adopt, wherever appropriate, economic and administrative instruments such as tax incentive policies to encourage the reduction of pollutant emissions and the introduction of innovative technology.
14. Explore the feasibility of reciprocal agreements to reduce transboundary pollution.
15. Encourage long-term mining investment by having clear environmental standards with stable and predictable environmental criteria and procedures.

15 Index

D

E

G

H

I

M

N

O

S

Y

Z

Printing: Krips bv, Meppel, The Netherlands
Binding: Stürtz, Würzburg, Germany